Semantics of Probabilistic Processes

Yuxin Deng

Semantics of Probabilistic Processes

An Operational Approach

上海交通大學出版社
SHANGHAI JIAO TONG UNIVERSITY PRESS

Springer

Yuxin Deng
Shanghai Jiao Tong University
Shanghai
China

ISBN 978-3-662-51598-3 ISBN 978-3-662-45198-4 (eBook)
DOI 10.1007/978-3-662-45198-4

Jointly published with Shanghai Jiao Tong University Press
ISBN: 978-7-313-12083-0 Shanghai Jiao Tong University Press

Springer Heidelberg New York Dordrecht London

Softcover reprint of the hardcover 1st edition 2014

Printed on acid-free paper

Springer is part of Springer Science + Business Media (www.springer.com)

Preface

Probabilistic concurrency theory aims to specify and analyse quantitative behaviour of concurrent systems, which necessarily builds on solid semantic foundations of probabilistic processes. This book adopts an operational approach to describing the behaviour of nondeterministic and probabilistic processes, and that the semantic comparison of different systems are based on appropriate behavioural relations such as bisimulation equivalence and testing preorders.

It mainly consists of two parts. The first part provides an elementary account of bisimulation semantics for probabilistic processes from metric, logical and algorithmic perspectives. The second part sets up a general testing framework and specialises it to probabilistic processes with nondeterministic behaviour. The resulting testing semantics is treated in depth. A few variants of it are shown to coincide, and they can be characterised in terms of modal logics and coinductively defined simulation relations. Although in the traditional (nonprobabilistic) setting, simulation semantics is in general finer than testing semantics because it distinguishes more processes for a large class of probabilistic processes, the gap between simulation and testing semantics disappears. Therefore, in this case, we have a semantics where both negative and positive results can be easily proved: to show that two processes are not related in the semantics, we just give a witness test, and to prove that two processes are related, we only need to establish a simulation relation.

While most of the results have been announced before, they are spread over several papers in the period from 2007 to 2014, and sometimes with different terminology and notation. This prevents us from having a comprehensive understanding of the bisimulation and testing semantics of probabilistic processes. In order to improve the situation, the current work brings all the related concepts and proof techniques to form a coherent and self-contained text.

Besides presenting the recent research advances in probabilistic concurrency theory, the book exemplifies the use of many mathematical techniques to solve problems in Computer Science, which is intended to be accessible to postgraduate students in Computer Science and Mathematics. It can also be used by researchers and practitioners either for advanced study or for technical reference. The reader is assumed to have some basic knowledge in discrete mathematics. Familiarity with real analysis is not a prerequisite, but would be helpful.

Most of the work reported in this book was carried out during the last few years with a number of colleagues. The testing semantics for probabilistic processes was developed in conjunction with Rob van Glabbeek, Matthew Hennessy, Carroll Morgan and Chenyi Zhang. The various characterisations of probabilistic bisimulation in Chap. 3 are based on joint work with Wenjie Du.

The BASICS laboratory at Shanghai Jiao Tong University has offered a creative and pleasant working atmosphere. Therefore, I would like to express my gratitude to Yuxi Fu and all other members of the laboratory. Thanks go also to Barry Jay, Matthew Hennessy and Carroll Morgan for having read parts of the first draft and provided useful feedback. My research on probabilistic concurrency theory has been sponsored by the National Natural Science Foundation of China under grants 61173033 and 61261130589, as well as ANR 12IS02001 "PACE".

Finally, my special gratitude goes to my family, for their unfailing support.

Shanghai Yuxin Deng
September, 2014

Contents

List of Symbols

Chapter 1
Introduction

Abstract This introduction briefly reviews the history of probabilistic concurrency theory and three approaches to the semantics of concurrent systems: denotational, axiomatic and operational. This book focuses on the last one and more specifically on (bi)simulation semantics and testing semantics. The second section surveys the contents and main results for other chapters of the book.

Keywords Probabilistic concurrency theory · Semantics · Bisimulation · Testing

1.1 Background

Computer science aims to explain in a rigorous way how computational systems should behave, and then to design them so that they do behave as expected. Nowadays the notion of computational systems includes not only *sequential systems* but also *concurrent systems*. The attention of computer scientists goes beyond single programs in free-standing computers. For example, computer networks, particles in physics and even proteins in biology can all be considered as concurrent systems. Some classical mathematical models (e.g. the λ-calculus [1]) are successful for describing sequential systems, but they turn out to be insufficient for reasoning about concurrent systems, because what is more important now is how different components of a system interact with each other rather than their input–output behaviour.

In the 1980s *process calculi* (sometimes also called *process algebras*), notably calculus of communicating systems (CCS) [2], communicating sequential processes (CSP) [3] and algebra of communicating processes (ACP) [4, 5], were proposed for describing and analysing concurrent systems. All of them were designed around the central idea of *interaction* between processes. In those formalisms, complex systems are built from simple subcomponents, using a small set of primitive operators such as *prefix, nondeterministic choice, restriction, parallel composition* and *recursion*. Those traditional process calculi were designed to specify and verify *qualitative behaviour* of concurrent systems.

Since the 1990s, there has been a trend to study the *quantitative behaviour* of concurrent systems. Many probabilistic algorithms have been developed in order to

DOI 10.1007/978-3-662-45198-4_1

gain efficiency or to solve problems that are otherwise impossible to solve by deterministic algorithms. For instance, probabilities are introduced to break symmetry in distributed coordination problems (e.g. the dining philosophers' problem, leader election and consensus problems). Probabilistic modelling has helped to analyse and reason about the correctness of probabilistic algorithms, to predict system behaviour based on the calculation of performance characteristics and to represent and quantify other forms of uncertainty. The study of probabilistic model checking techniques has been a rich research area.

A great many probabilistic vari ants of the classical process calculi have also appeared in the literature. The typical approach is to add probabilities to existing models and techniques that have already proved successful in the nonprobabilistic settings. The distinguishing feature of probabilistic process calculi is the presence of a *probabilistic-choice* operator, as in the probabilistic extensions of CCS [6, 7], the probabilistic CSP [8], the probabilistic ACP [9] and the probabilistic asynchronous π-calculus [10].

In order to study a programming language or a process calculus, one needs to assign a consistent meaning to each program or process under consideration. This meaning is the *semantics* of the language or calculus. Semantics is essential to verify or prove that programs behave as intended. Generally speaking, there are three major approaches for giving semantics to a programming language. The *denotational* approach [11] seeks a valuation function that maps a program to its mathematical meaning. This approach has been very successful in modelling many sequential languages; programs are interpreted as functions from the domain of input values to the domain of output values. However, the nature of interaction is much more complex than a mapping from inputs to outputs, and so far the denotational interpretation of concurrent programs has not been as satisfactory as the denotational treatment of sequential programs.

The *axiomatic* approach [12, 13] aims at understanding a language through a few axioms and inference rules that help to reason about the properties of programs. It offers an elegant way of gaining insight into the nature of the operators and the equivalences involved. For example, the difference between two notions of program equivalence may be characterised by a few axioms, particularly if adding these axioms to a complete system for one equivalence gives a complete system for the other equivalence. However, it is often difficult and even impossible to achieve a fully complete axiomatic semantics if the language in question is beyond a certain expressiveness.

The *operational* approach has been shown to be very useful for giving semantics of concurrent systems. The behaviour of a process is specified by its *structural operational semantics* [14], described via a set of labelled transition rules inductively defined on the structure of a term. In this way each process corresponds to a labelled *transition graph*. The shortcoming of operational semantics is that it is too concrete, because a transition graph may contain many states that intuitively should be identified. Thus, a great number of equivalences have been proposed, and different transition graphs are compared modulo some equivalence relations. Usually there is no agreement on which is the best equivalence relation; in formal verification different equivalences might be suitable for different applications. Sometimes an equivalence is induced by a preorder relation, by taking the intersection of the preorder with its inverse relation, instead of being directly defined.

Among the various equivalences, *bisimilarity* [2, 15] is one of the most important ones as it admits beautiful characterisations in terms of fixed points, modal logics, coalgebras, pseudometrics, games, decision algorithms, etc. In this book we will characterise bisimilarity for probabilistic processes from metric, logical and algorithmic perspectives.

Preorders can be used to formalise a "better than" relation between programs or processes, one that has its origins in the original work assigning meanings to programs and associating a logic with those meanings [12, 13]. Usually that relation is expressed in two different ways: either to provide a witness for the relation or to provide a testing context to make obvious that one program is actually not better than another.

Two important kinds of preorders are *testing preorders* [16, 17] and *simulation preorders*. They give rise to *testing semantics* and *simulation semantics*, respectively. In a testing semantics, two processes can be compared by experimenting with a class of tests. Process P is deemed "better" than process Q if the former passes every test that the latter can pass. In contrast, to show that P is not "better" than Q it suffices to find a test that Q can pass but P cannot. In a simulation semantics, process P can simulate Q if Q performs an action and evolves into Q' then P is able to exhibit the same action and evolve into P' such that P' can simulate Q' in the next round of the simulation game. Simulation is coinductively defined and comes along with a proof principle called *coinduction*: to show that two processes are related it suffices to exhibit a simulation relation containing a pair consisting of the two processes. In the nonprobabilistic setting, simulation semantics is in general finer than testing semantics in that it can distinguish more processes. However, in this book we will see that for a large class of probabilistic processes, the gap between simulation and testing semantics disappears. Therefore, in this case we have a semantics where both negative and positive results can easily be proved: to show that two processes are not related in the semantics we just give a witness test, while to show that two processes are related we construct a relation and argue that it is a simulation relation.

1.2 Synopsis

The remainder of the book is organised as follows. Chapter 2 collects some fundamental concepts and theorems in a few mathematical subjects such as lattice theory, topology and linear programming. They are briefly reviewed and meant to be used as references for later chapters. Most of the theorems are classic results, and thus are stated without proofs as they can be easily found in many standard textbooks in mathematics. It is not necessary to go through the whole chapter; readers can refer to relevant parts of this chapter when it is mentioned elsewhere in the book.

Chapter 3 introduces an operational model of probabilistic systems called *probabilistic labelled transition systems*. In this model, a state might make a non-deterministic choice among a set of available actions. Once an action is taken, the state evolves into a distribution over successor states. Then in order to compare the behaviour of two states, we need to know how to compare two distributions. There is a nice lifting operation that turns a relation between states into a relation between distributions. This operation is closely related to *the Kantorovich metric* in mathematics and the *network flow* problem in optimisation theory. We give an elementary account of the lifting operation because it entails a neat notion of probabilistic bisimulation that can be characterised by behavioural pseudometrics and decided by polynomial algorithms over finitary systems. We also provide modal characterisations of the probabilistic bisimulation in terms of probabilistic extensions of *the Hennessy–Milner logic* and the *modal mu-calculus*.

Starting from Chap. 4 we investigate the testing semantics of probabilistic processes. We first set up a general testing framework that can be instantiated into a vector-based testing or scalar testing approach, depending on the number of actions used to indicate success states. A fundamental theorem is that for finitary systems the two approaches are equally powerful. In order to prove this result we make use of a notion of reward testing as a stepping stone. The *separation hyperplane theorem* from discrete geometry plays an important role in the proof.

Chapter 5 investigates the connection between testing and simulation semantics. For *finite processes*, i.e. processes that correspond to probabilistic labelled transition systems with finite tree structures, testing semantics is not only sound but also complete for simulation semantics. More specifically, may testing preorder coincides with simulation preorder and must testing preorder coincides with failure simulation preorder. Therefore, unlike the traditional (nonprobabilistic) setting, here there is no gap between testing and simulation semantics. To prove this result we make use of logical characterisations of testing preorders. For example, each state s has a characteristic formula ϕ_s in the sense that another state t can simulate s if and only if t satisfies ϕ_s. We can then turn this formula ϕ_s into a characteristic test T_s so that if t is not related to s via the may testing preorder then T_s is a witness test that distinguishes t from s. Similarly for the case of failure simulation and must testing. We also give a complete axiom system for the testing preorders in the finite fragment of a probabilistic CSP. This chapter paves the way for the next chapter.

In Chap. 6 we extend the results in the last chapter from finite processes to *finitary processes*, i.e. processes that correspond to probabilistic labelled transition systems that are *finite-state* and *finitely branching* possibly with loops. The soundness and completeness proofs inherit the general schemata from the last chapter. However, the technicalities are much more subtle and more interesting. For example, we make a significant use of subdistributions. A key topological property is that from any given subdistribution, the set of subdistributions reachable from it by weak transitions can be finitely generated. The proof is highly nontrivial and involves techniques from *Markov decision processes* such as rewards and static policies. This result enables us to approximate coinductively defined relations by stratified inductive relations. As a consequence, if two processes behave differently we can tell them apart by a finite test.

We also introduce a notion of real-reward testing that allows for negative rewards. It turns out that real-reward may preorder is the inverse of real-reward must preorder, and vice versa. More interestingly, for finitary convergent processes, real-reward must testing preorder coincides with nonnegative-reward testing preorder.

In Chap. 7 we introduce a notion of weak probabilistic bisimulation simply by taking the symmetric form of the simulation preorder given in Chap. 6. It provides a sound and complete proof methodology for an extensional behavioural equivalence, a probabilistic variant of the traditional *reduction barbed congruence* well known in concurrency theory.

References

1. Barendregt, H.: The Lambda Calculus: Its Syntax and Semantics. North-Holland, Amsterdam (1984)
2. Milner, R.: Communication and Concurrency. Prentice Hall, Englewood Cliffs (1989)
3. Hoare, C.A.R.: Communicating Sequential Processes. Prentice Hall, Englewood Cliffs (1985)
4. Bergstra, J.A., Klop, J.W.: Process algebra for synchronous communication. Inf. Comput. **60**, 109–137 (1984)
5. Baeten, J.C.M., Weijland, W.P.: Process Algebra, Cambridge Tracts in Theoretical Computer Science, vol. 18. Cambridge University Press, Cambridge (1990)
6. Giacalone, A., Jou, C.C., Smolka, S.A.: Algebraic reasoning for probabilistic concurrent systems. Proceedings of IFIP TC2 Working Conference on Programming Concepts and Methods (1990)
7. Hansson, H., Jonsson, B.: A calculus for communicating systems with time and probabilities. Proceedings of IEEE Real-Time Systems Symposium. IEEE Computer Society Press, 278–287 (1990)
8. Lowe, G.: Probabilities and priorities in timed CSP. Ph.D. Thesis, Oxford (1991)
9. Andova, S.: Process algebra with probabilistic choice. Tech. Rep. CSR 99-12, Eindhoven University of Technology (1999)
10. Herescu, O.M., Palamidessi, C.: Probabilistic asynchronous pi-calculus. Tech. Rep., INRIA Futurs and LIX (2004)
11. Scott, D., Strachey, C.: Toward a mathematical semantics for computer languages. Technical Monograph PRG-6, Oxford University Computing Laboratory (1971)
12. Floyd, R.W.: Assigning meanings to programs. Proc. Am. Math. Soc. Symp. Appl. Math. **19**, 19–31 (1967)
13. Hoare, C.A.R.: An axiomatic basis for computer programming. Commun. ACM **12**(10), 576–580 (1969)
14. Plotkin, G.: The origins of structural operational semantics. J Log Algebraic Program 60–61, 3–15 (2004)
15. Park, D.: Concurrency and automata on infinite sequences. Proceedings of the 5th GI-Conference on Theoretical Computer Science, Lecture Notes in Computer Science, vol. 104, Springer, 167–183 (1981)
16. De Nicola, R., Hennessy, M.: Testing equivalences for processes. Theor. Comput. Sci. **34**, 83–133 (1984)
17. Hennessy, M.: Algebraic Theory of Processes. The MIT Press, Cambridge (1988)

Chapter 2
Mathematical Preliminaries

Abstract We briefly introduce some mathematical concepts and associated important theorems that will be used in the subsequent chapters to study the semantics of probabilistic processes. The main topics covered in this chapter include the Knaster–Tarski fixed-point theorem, continuous functions over complete lattices, induction and coinduction proof principles, compact sets in topological spaces, the separation theorem, the Banach fixed-point theorem, the π-λ theorem and the duality theorem in linear programming. Most of the theorems are stated without proofs because they can be found in many textbooks.

Keywords Lattice · Induction · Coinduction · Topology · Metric · Probability · Linear programming

2.1 Lattice Theory

A very useful tool to study the semantics of formal languages is the Knaster–Tarski fixed-point theorem, which will be used many times in subsequent chapters. We begin with some basic notions from lattice theory.

Definition 2.1 A set X with a binary relation $\sqsubseteq$ is called a *partially ordered set*, if the following holds for all $x, y, z \in X$:

1. $x \sqsubseteq x$ (reflexivity);
2. If $x \sqsubseteq y$ and $y \sqsubseteq x$, then $x = y$ (antisymmetry);
3. If $x \sqsubseteq y$ and $y \sqsubseteq z$, then $x \sqsubseteq z$ (transitivity).

An element $x \in X$ is called an *upper bound* for a subset $Y \subseteq X$, if $y \sqsubseteq x$ for all $y \in Y$. Dually, x is a *lower bound* for Y if $x \sqsubseteq y$ for all $y \in Y$. Note that a subset might not have any upper bound or lower bound.

Example 2.1 Consider the set $X = \{0, 1, 2\}$ with the binary relation $\sqsubseteq$ defined by $\sqsubseteq = Id_X \cup \{(0, 1), (0, 2)\}$, where Id_X is the identity relation $\{(0, 0), (1, 1), (2, 2)\}$. It constitutes a partially ordered set. The subset $\{1, 2\}$ has no upper bound, as Fig. 2.1a shows.

Often we can use *Hasse diagrams* to describe finite partially ordered sets. For a partially ordered set $(X, \sqsubseteq)$, we represent each element of X as a vertex and draw

DOI 10.1007/978-3-662-45198-4_2

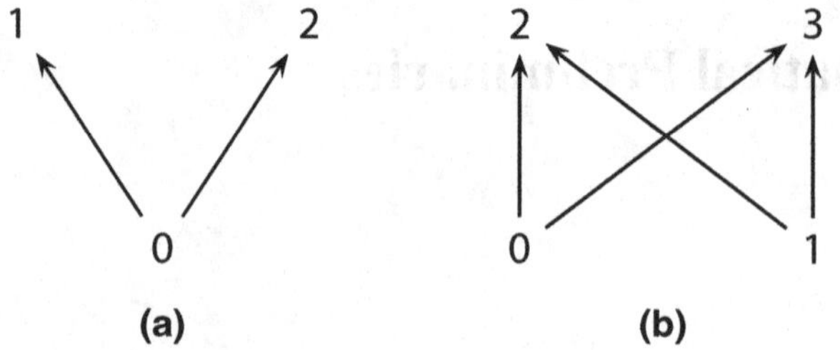

Fig. 2.1 Partially ordered sets

an arrow that goes upward from x to y if y is an immediate successor of x, i.e. $x \sqsubset y$ and there is no z with $x \sqsubset z \sqsubset y$, where $\sqsubset$ is obtained from $\sqsubseteq$ by removing elements (x, x) for all x. For instance, the partially ordered set given in the above example can be depicted by the diagram in Fig. 2.1a.

If Y has an upper bound that is also an element of Y, then it is said to be the *greatest element* in Y. We can dually define the *least element*. In the presence of a least element we speak of a *pointed* partially ordered set. If the set of upper bounds for Y has a least element x, then x is called the *supremum* or *join* of Y, written $\bigsqcup Y$. Dually we have *infimum* or *meet* and write $\bigsqcap Y$. Note that a subset might not have supremum even if it has upper bounds; so is the case for infimum.

Example 2.2 Consider the set $X = \{0, 1, 2, 3\}$ with the binary relation $\sqsubseteq$ defined by $\sqsubseteq = Id_X \cup \{(0, 2), (0, 3), (1, 2), (1, 3)\}$. It forms the partially ordered set depicted in Fig. 2.1b. The subset $\{0, 1\}$ has two upper bounds, namely 2 and 3, but it has no supremum because the two upper bounds are incomparable in the sense that neither $2 \sqsubseteq 3$ nor $3 \sqsubseteq 2$ holds.

We call X a *lattice*, if suprema and infima exist for all the subsets of X with two elements, i.e. if $\bigsqcup\{x, y\}$ and $\bigsqcap\{x, y\}$ exist for any $x, y \in X$. We call X a *complete lattice* if suprema and infima exist for all the subsets of X.

Remark 2.1 A few characteristics of complete lattices are worth mentioning because they are useful in many applications.

1. It suffices to define complete lattices in terms of either suprema or infima only. For example, a candidate definition is to say X is a complete lattice if suprema exist for all subsets of X. Then the infimum of any subset Y also exists because

$$\bigsqcap Y = \bigsqcup\{x \in X \mid \forall y \in Y : x \sqsubseteq y\}$$

2. Let X be a complete lattice. By definition, X itself has a supremum, which is the greatest element of X, written $\top$, and an infimum, which is the least element of X, written $\bot$. It can be checked that the empty set $\emptyset$ has supremum $\bot$ and infimum $\top$.

3. If $(X, \sqsubseteq)$ is a complete lattice, then so is $(X, \sqsupseteq)$, where $\sqsupseteq$ is the inverse relation of $\sqsubseteq$, i.e. $x \sqsupseteq y$ iff $y \sqsubseteq x$ for any elements $x, y \in X$.

Example 2.3 The set $X = \{1, 2, 3, 5, 6, 10, 15, 30\}$ of all divisors of 30, partially ordered by divisibility, constitutes a complete lattice.

Example 2.4 The set of all natural numbers $\mathbb{N}$, with the usual relation $\leq$ on natural numbers, forms a lattice which is not a complete lattice.

Example 2.5 Let X be any set. Its powerset $\mathcal{P}(X) = \{Y \mid Y \subseteq X\}$ with the inclusion relation $\subseteq$ forms a complete lattice whose join and meet are set union and intersection, respectively.

Given a function $f : X \to Y$ and a set $Z \subseteq X$, we write $f(Z)$ for the image of Z under f, i.e. the set $\{f(z) \mid z \in Z\}$. Given a partially ordered set X and a function $f : X \to X$, we say $x \in X$ is a *fixed point* (resp. *prefixed point*, *postfixed point*) of f if $x = f(x)$ (resp. $f(x) \sqsubseteq x$, $x \sqsubseteq f(x)$).

Definition 2.2 Let X and Y be partially ordered sets. A function $f : X \to Y$ is called *monotone*, if $x \sqsubseteq y$ implies $f(x) \sqsubseteq f(y)$ for all $x, y \in X$.

Theorem 2.1 (Knaster–Tarski Fixed-Point Theorem) *If X is a complete lattice, then every monotone function f from X to X has a fixed point. The least of these is given by*

$$\mathit{lfp}(f) = \bigsqcap \{x \in X \mid f(x) \sqsubseteq x\}\ ,$$

and the greatest by

$$\mathit{gfp}(f) = \bigsqcup \{x \in X \mid x \sqsubseteq f(x)\}\ .$$

Proof Let $X' = \{x \in X \mid f(x) \sqsubseteq x\}$ and $x^* = \bigsqcap X'$. For each $x \in X'$ we have $x^* \sqsubseteq x$ and then by monotonicity $f(x^*) \sqsubseteq f(x) \sqsubseteq x$. Taking the infimum over x we get, $f(x^*) \sqsubseteq \bigsqcap f(X') \sqsubseteq \bigsqcap X' = x^*$, thus $x^* \in X'$ and is the least prefixed point. On the other hand, $x \in X'$ implies $f(x) \in X'$ by monotonicity. Applying this to x^* yields $f(x^*) \in X'$ which implies $x^* \sqsubseteq f(x^*)$. Therefore, we obtain that $x^* = f(x^*)$. In fact, x^* is the least fixed point because we have just shown that it is the least prefixed point.

The case for the greatest fixed point is dual and thus omitted. □

Definition 2.3 Given a complete lattice X, the function $f : X \to X$ is *continuous* if it preserves increasing chains, i.e. for all sequences $x_0 \sqsubseteq x_1 \sqsubseteq \ldots$ we have

$$f(\bigsqcup_{n \geq 0} x_n) \;=\; \bigsqcup_{n \geq 0} f(x_n).$$

Dually, f is *cocontinuous* if it preserves decreasing chains.

Notice that both continuity and cocontinuity imply monotonicity. For example, if f is continuous and $x \sqsubseteq y$, then from the increasing sequence $x \sqsubseteq y \sqsubseteq y \sqsubseteq \ldots$

we obtain that $f(\bigsqcup\{x, y\}) = f(y) = \bigsqcup\{f(x), f(y)\}$, which means $f(x) \sqsubseteq f(y)$. With continuity and cocontinuity we can construct, in a more tractable way, the least and greatest fixed point, respectively.

Proposition 2.1 *Let X be a complete lattice.*

1. *Every continuous function f on X has a least fixed point, given by* $\bigsqcup_{n\geq 0} f^n(\bot)$*, where* $\bot$ *is the bottom element of the lattice, and* $f^n(\bot)$ *is the nth iteration of f on* $\bot$: $f^0(\bot) := \bot$ *and* $f^{n+1}(\bot) := f(f^n(\bot))$ *for* $n \geq 0$.
2. *Every cocontinuous function f on X has a greatest fixed point, given by* $\bigsqcap_{n\geq 0} f^n(\top)$*, where* $\top$ *is the top element of the lattice.*

Proof We only prove the first clause, since the second one is dual.

We notice that $\bot \sqsubseteq f(\bot)$, and then monotonicity of f yields an increasing sequence:

$$\bot \sqsubseteq f(\bot) \sqsubseteq f^2(\bot) \sqsubseteq \dots$$

By continuity of f we have, $f(\bigsqcup_{n\geq 0} f^n(\bot)) = \bigsqcup_{n\geq 0} f^{n+1}(\bot)$ and the latter is equal to $\bigsqcup_{n\geq 0} f^n(\bot)$.

And in fact that limit is the least fixed point: for if some other x were also a fixed point, then we have $\bot \sqsubseteq x$ and moreover $f^n(\bot) \sqsubseteq x$ for all n by induction. So x is an upper bound of all $f^n(\bot)$. □

Let $(X, \sqsubseteq)$ be a partially ordered set. We say X is a *complete partially ordered set* (CPO) if it has suprema for all increasing chains. A CPO with *bottom* is a CPO with a least element $\bot$. The least fixed point of a continuous function f on the CPO can be characterised in the same way as in Proposition 2.1 (1), namely $\bigsqcup_{n\geq 0} f^n(\bot)$.

2.2 Induction and Coinduction

We observe that Theorem 2.1 provides two proof principles: the *induction principle* says that, to show $lfp(f) \sqsubseteq x$ it is enough to prove $f(x) \sqsubseteq x$; the *coinduction principle* says that, to show $x \sqsubseteq gfp(f)$ it suffices to prove $x \sqsubseteq f(x)$. Those are two important proof principles in concurrency theory. In this section, we briefly introduce inductive and coinductive definitions by rules and their associated proof techniques. For more detailed accounts, we refer the reader to [1, 2].

Given a set X, a *ground rule* on X is a pair $(S, x) \in \mathcal{P}(X) \times X$, meaning that, from the premises S we can derive the conclusion x. Usually, we write the rule (S, x) as

$$\frac{}{x} \text{ if } S = \emptyset \text{, and as } \frac{x_1, \dots, x_n}{x} \text{ if } X = \{x_1, \dots, x_n\}. \tag{2.1}$$

A rule without any premise, i.e. in the form $(\emptyset, x)$, is said to be an *axiom*.

Quite often we define a set of objectives by rules. For example, let Var be a set of variables ranged over by x. In the lambda calculus the set Λ of all lambda terms can be expressed by a BNF (Backus–Naur Form) grammar:

$$M, N ::= x \mid \lambda x.M \mid MN$$

which says that Λ is the least set satisfying the following three rules:

$$\frac{}{x} \qquad \frac{M}{\lambda x.M} \qquad \frac{M \quad N}{MN}$$

A set of ground rules R determines an operator $\hat{R} : X \to X$, which maps a set S to the set

$$\hat{R}(S) = \{x \mid \exists S' \subseteq S : (S', x) \in R\}$$

which includes all the elements derivable from those in S by using the rules in R.

Definition 2.4 A set S is *forward-closed* under R iff $\hat{R}(S) \subseteq S$; a set S is *backward-closed* under R iff $S \subseteq \hat{R}(S)$.
In other words, if S is forward-closed, then

$$\forall x \in X : (\exists S' \subseteq S : (S', x) \in R) \Rightarrow x \in S$$

If S is backward-closed, then

$$\forall x \in X : x \in S \Rightarrow (\exists S' \subseteq S : (S', x) \in R)$$

Given a set of rules R on X, we let

$$S_f = \bigcap\{S \subseteq X \mid \hat{R}(S) \subseteq S\} \qquad \text{and} \qquad S_b = \bigcup\{S \subseteq X \mid S \subseteq \hat{R}(S)\}.$$

So S_f is the intersection of all forward-closed sets and S_b is the union of all backward-closed sets. Note that $\hat{R}$ is monotone on the complete lattice $(\mathcal{P}(X), \subseteq)$. It follows from Theorem 2.1 that S_f is in fact the least fixed point of $\hat{R}$ and S_b is the greatest fixed point of $\hat{R}$. Usually, an object is defined *inductively* (resp. *coinductively*) if it is the *least* (resp. *greatest*) fixed point of a function. So the set S_f (resp. S_b) is inductively (resp. coinductively) defined by the set of rules R.

Example 2.6 The set of finite lists with elements from a set A is the set List inductively defined by the following two rules, i.e. the least set closed forward under these rules.

$$\frac{}{nil \in \mathsf{List}} \qquad \frac{a \in A \quad l \in \mathsf{List}}{a \cdot l \in \mathsf{List}} \tag{2.2}$$

In contrast, the set of all finite or infinite lists is coinductively defined by the same rules, i.e. the greatest set closed backward under the two rules above; the set of all infinite lists is the set coinductively defined by the second rule above, i.e. the greatest set closed backward under the second rule.

The two rules in (2.2) are not ground rules because they are not in the form shown in (2.1). However, we can convert them into ground rules. Take X to be the set of all (finite or infinite) strings with elements from $A \cup \{nil\}$. The ground rules determined by the two rules in (2.2) are

$$\frac{}{nil} \quad \text{and, for each } l \in X \text{ and } a \in A, \quad \frac{l}{a \cdot l}\,.$$

Let R be a set of ground rules on X. In general $\hat{R}$ is not necessarily continuous or cocontinuous on the complete lattice $(\mathcal{P}(X), \subseteq)$. For example, consider the rule

$$\frac{x_1 \;\ldots\; x_n \;\ldots}{x}$$

whose premise is an infinite set. For any $n \geq 1$, let $X_n = \{x_1, \ldots, x_n\}$. Then we have $x \in \hat{R}(\bigcup_n X_n)$, but $x \notin \bigcup_n \hat{R}(X_n)$. Thus, $\hat{R}$ is not continuous.

Below we impose some conditions on R to recover continuity and cocontinuity.

Definition 2.5 A set of rules R is *finite in the premises* (FP), if in each rule $(S, x) \in R$ the premise set S is finite. A set of rules R is *finite in the conclusions* (FC), if for each x, the set $\{S \mid (S, x) \in R\}$ is finite; that is, there are finitely many rules whose conclusion is x, though each premise set S may be infinite.

Proposition 2.2 *Let R be a set of ground rules.*

1. *If R is FP, then $\hat{R}$ is continuous;*
2. *If R is FC, then $\hat{R}$ is cocontinuous.*

Proof The two statements are dual, so we only prove the second one.

Let $S_0 \supseteq S_1 \supseteq \cdots \supseteq S_n \supseteq \cdots$ be a decreasing chain. We need to show that

$$\bigcap_{n \geq 0} \hat{R}(S_n) = \hat{R}(\bigcap_{n \geq 0} S_n)\,.$$

One direction is easy. Since $\bigcap_{n \geq 0} S_n \subseteq S_n$ and $\hat{R}$ is monotone, we have that

$$\hat{R}(\bigcap_{n \geq 0} S_n) \subseteq \bigcap_{n \geq 0} \hat{R}(S_n)\,.$$

The converse inclusion follows from the condition that R is FC. Let x be any element in $\bigcap_{n \geq 0} \hat{R}(S_n)$. Then $x \in \hat{R}(S_n)$ for each n, which means that for each $n \geq 0$ there is some $S'_n \subseteq S_n$ with $(S'_n, x) \in R$. Since R is FC, there exists some $k \geq 0$ such that $S'_n = S'_k$ for all $n \geq k$. Moreover, $S'_k \subseteq S_k \subseteq S_n$ for all $n \leq k$. Therefore, $S'_k \subseteq \bigcap_{n \geq 0} S_n$ and we have that $x \in \hat{R}(\bigcap_{n \geq 0} S_n)$. Thus, we have proved that

$$\bigcap_{n \geq 0} \hat{R}(S_n) \subseteq \hat{R}(\bigcap_{n \geq 0} S_n)\,.$$

□

Corollary 2.1 *Let R be a set of ground rules on X.*

1. *If R is FP, then* $lfp(\hat{R}) = \bigcup_{n\geq 0} \hat{R}^n(\emptyset)$;
2. *If R is FC, then* $gfp(\hat{R}) = \bigcap_{n\geq 0} \hat{R}^n(X)$.

Proof Combine Propositions 2.1 and 2.2. □

Intuitively, the set $\hat{R}^0(\emptyset) = \emptyset$; the set $\hat{R}^1(\emptyset) = \hat{R}(\emptyset)$ consisting of all the conclusions of instances of axioms. In general, the set $\hat{R}^{n+1}(\emptyset)$ contains all objects which immediately follow by ground rules with premises in $\hat{R}^n(\emptyset)$. The above corollary states that if R is FP, each element in the set $lfp(\hat{R})$ can be derived via a derivation tree of finite depth whose leaves are instances of axioms; if R is FC, elements in $gfp(\hat{R})$ can always be destructed as conclusions of ground rules whose premises can also be destructed similarly.

2.3 Topological Spaces

In this section, we review some fundamental concepts in general topology such as continuous functions, compact sets and some other related properties. They will be used in Chap. 6.

Definition 2.6 Let X be a nonempty set. A collection $\mathcal{T}$ of subsets of X is a *topology* on X iff $\mathcal{T}$ satisfies the following axioms.

1. X and $\emptyset$ belong to $\mathcal{T}$;
2. The union of any number of sets in $\mathcal{T}$ belongs to $\mathcal{T}$;
3. The intersection of any two sets in $\mathcal{T}$ belongs to $\mathcal{T}$.

The members of $\mathcal{T}$ are called *open sets*, and the pair $(X, \mathcal{T})$ is called a *topological space*.

Example 2.7 The collection of all open intervals in the real line $\mathbb{R}$ forms a topology, which is called the *usual topology* on $\mathbb{R}$.

Let $(X, \mathcal{T})$ be a topological space. A point $x \in X$ is an *accumulation point* or *limit point* of a subset Y of X iff every open set Z containing x must also contain a point of Y different from x, that is,

$$Z \text{ open}, \quad x \in Z \quad \text{implies} \quad (Z\backslash\{x\}) \cap Y \neq \emptyset \ .$$

Definition 2.7 A subset Y of a topological space $(X, \mathcal{T})$ is *closed* iff Y contains each of its limit points.

Definition 2.8 Let Y be a subset of a topological space $(X, \mathcal{T})$. The *closure* of Y is the set of all limit points of Y. We say Y is *dense* if the closure of Y is X, i.e. every point of X is a limit point of Y.

It is immediate that Y coincides with its closure if and only if Y is closed.

Definition 2.9 A topological space $(X, \mathcal{T})$ is said to be *separable* if it contains a countable dense subset; that is, if there exists a finite or denumerable subset Y of X such that the closure of Y is the entire space.

Definition 2.10 Let $(X, \mathcal{T})$ and $(X^\star, \mathcal{T}^\star)$ be topological spaces. A function f from X into $X^\star$ is *continuous* iff the inverse image $f^{-1}(Y)$ of every open subset Y of $X^\star$ is an open subset of X, i.e.

$$Y \in \mathcal{T}^\star \quad \text{implies} \quad f^{-1}(Y) \in \mathcal{T} .$$

Example 2.8 The projection mappings from the plane $\mathbb{R}^2$ into the line $\mathbb{R}$ are continuous with respect to the usual topologies. For example, consider the projection $\pi : \mathbb{R}^2 \to \mathbb{R}$ defined by $\pi(\langle x, y\rangle) = y$. The inverse of any open interval (a, b) is an infinite open strip parallel to the x-axis.

Continuous functions can also be characterised by their behaviour with respect to closed sets.

Theorem 2.2 *A function $f : X \to Y$ is continuous iff the inverse image of every closed subset of Y is a closed subset of X.* □

Let $\mathcal{Y} = \{Y_i\}$ be a class of subsets of X such that $Y \subseteq \cup_i Y_i$ for some $Y \subseteq X$. Then $\mathcal{Y}$ is called a *cover* of Y, and an *open cover* if each Y_i is open. Furthermore, if a finite subclass of $\mathcal{Y}$ is also a cover of Y, i.e. if

$$\exists Y_{i_1}, \ldots, Y_{i_m} \in \mathcal{Y}, \quad \text{such that} \quad Y \subseteq Y_{i_1} \cup \cdots \cup Y_{i_m}$$

then $\mathcal{Y}$ is said to contain a *finite subcover* of Y.

Example 2.9 The classical Heine–Borel theorem says that every open cover of a closed and bounded interval $[a, b]$ on the real line contains a finite subcover.

Definition 2.11 A subset Y of a topological space X is *compact* if every open cover of Y contains a finite subcover.

Example 2.10 By the Heine–Borel theorem, every closed and bounded interval $[a, b]$ on the real line $\mathbb{R}$ is compact.

Theorem 2.3 *Continuous images of compact sets are compact.* □

A set $\{X_i\}$ of sets is said to have the *finite intersection property* if every finite subset $\{X_{i_1}, \ldots, X_{i_m}\}$ has a nonempty intersection, i.e. $X_{i_1} \cap \cdots \cap X_{i_m} \neq \emptyset$.

Example 2.11 Consider the following class of open intervals:

$$X = \left\{(0, 1), \left(0, \frac{1}{2}\right), \left(0, \frac{1}{2^2}\right), \ldots\right\}$$

Clearly, it has the finite intersection property. Observe however that X has an empty intersection.

Theorem 2.4 *A topological space X is compact iff every class $\{X_i\}$ of closed subsets of X that satisfies the finite intersection property has, itself, a nonempty intersection.* □

2.4 Metric Spaces

The main subject in probabilistic concurrency theory is the quantitative analysis of system behaviour. So we meet metric spaces from time to time. This section gives some background knowledge on them.

Let $\mathbb{R}_{\geq 0}$ be the set of all nonnegative real numbers.

Definition 2.12 A *metric space* is a pair (X, d) consisting of a set X and a distance function $d : X \times X \to \mathbb{R}_{\geq 0}$ satisfying:

1. For all $x, y \in X$, $d(x, y) = 0$, iff $x = y$ (isolation);
2. For all $x, y \in X$, $d(x, y) = d(y, x)$ (symmetry);
3. For all $x, y, z \in X$, $d(x, z) \leq d(x, y) + d(y, z)$ (triangle inequality).

If we replace the first clause with $\forall x \in X : d(x, x) = 0$, we obtain the definition of *pseudometric space*.

A metric d is c-bounded if $\forall x, y \in X : d(x, y) \leq c$, where c is a positive real number.

Example 2.12 Let X be a set. The *discrete metric* $d : X \times X \longrightarrow [0, 1]$ is defined by

$$d(x, y) = \begin{cases} 0 & \text{if } x = y \\ 1 & \text{otherwise.} \end{cases}$$

Example 2.13 Let $\mathbb{R}^n$ denote the product set of n copies of the set $\mathbb{R}$ of real numbers, i.e. it consists of all n-tuples $\langle a_1, a_2, \ldots, a_n \rangle$ of real numbers. The function d is defined by

$$d(x, y) = \sqrt{(a_1 - b_1)^2 + \cdots + (a_n - b_n)^2}$$

where $x = \langle a_1, \ldots, a_n \rangle$ and $y = \langle b_1, \ldots, b_n \rangle$, is a metric, called the *Euclidean metric* on $\mathbb{R}^n$. This metric space $(\mathbb{R}^n, d)$ is called *Euclidean n-space*.

Let (X, d) be a metric space. For any point $x \in X$ in the space and any real number $\varepsilon > 0$, we let $S(x, \varepsilon)$ denote the set of points within a distance of ε from x:

$$S(x, \varepsilon) := \{y \mid d(x, y) < \varepsilon\}.$$

We call $S(x, \varepsilon)$ the *open sphere* with centre x and radius ε.

Theorem 2.5 *Let (X, d) be a metric space. The collection of open spheres in X generates a topology, called the metric topology, whose open sets are those open spheres in X.* □

Definition 2.13 A sequence (x_n) in a metric space (X, d) is *convergent* to $x \in X$, if for an arbitrary $\varepsilon > 0$ there exists $N \in \mathbb{N}$, such that $d(x_n, x) < \varepsilon$ whenever $n > N$.

Definition 2.14 A sequence (x_n) in a metric space (X, d) is called a *Cauchy sequence* if for an arbitrary $\varepsilon > 0$ there exists $N \in \mathbb{N}$, such that $d(x_m, x_n) < \varepsilon$ whenever $m, n > N$.

Definition 2.15 A metric space is *complete* if every Cauchy sequence is convergent.

For example, the space of real numbers with the usual metric is complete.

Example 2.14 Let X be a nonempty set and F denote the collection of functions from X to the interval $[0, 1]$. A metric is defined on F as follows:

$$d(f, g) := sup_{x \in X} |f(x) - g(x)| .$$

In fact, (F, d) is a complete metric space. Let (f_n) be a Cauchy sequence in F. Then for every $x \in X$, the sequence $(f_n(x))$ is Cauchy; and since $[0, 1]$ is complete, the sequence converges to some $a_x \in [0, 1]$. Let f be the function defined by $f(x) = a_x$. Thus (f_n) converges to f.

Example 2.15 Similar to Example 2.14, it can be seen that Euclidean n-space $\mathbb{R}^n$ is complete.

Let Y be a subset of a metric space (X, d) and let $\varepsilon > 0$. A finite set of points $Z = \{z_1, z_2, \ldots, z_m\}$ is called an *ε-net* for Y, if for every point $y \in Y$ there exists an $z \in Z$ with $d(y, z) < \varepsilon$.

Definition 2.16 A subset Y of a metric space X is *totally bounded* if Y possesses an ε-net for every $\varepsilon > 0$.

Theorem 2.6 *Let (X, d) be a complete metric space. Then $Y \subseteq X$ is compact iff Y is closed and totally bounded.* □

Definition 2.17 (Convex Set) A set $X \subseteq \mathbb{R}^n$ is *convex* if for every two points $x, y \in X$ the whole segment between x and y is also contained in X. In other words, for every $p \in [0, 1]$, the point $px + (1 - p)y$ belongs to X. We write $\updownarrow X$ for the *convex closure* of X, the smallest convex set containing X.

Given two n-dimensional vectors $x = \langle a_1, \ldots, a_n \rangle$ and $y = \langle b_1, \ldots, b_n \rangle$, we use the usual definition of dot-product $x \cdot y = \sum_{i=1}^{n} a_i \cdot b_i$.

A basic result about convex set is the separability of disjoint convex sets by a hyperplane [3].

Theorem 2.7 (Separation Theorem) *Let $X, Y \subseteq \mathbb{R}^n$ be convex sets with $X \cap Y = \emptyset$. Then there is a hyperplane whose normal is $h \in \mathbb{R}^n$ and a number $c \in \mathbb{R}$ such that*

$$\text{for all } x \in X \text{ and } y \in Y, \text{ we have } h \cdot x \leq c \leq h \cdot y$$

or

$$\text{for all } x \in X \text{ and } y \in Y, \text{ we have } h \cdot x \geq c \geq h \cdot y .$$

If X and Y are closed and at least one of them is bounded, they can be separated strictly, i.e. in such a way that

$$\text{for all } x \in X \text{ and } y \in Y, \text{ we have } h \cdot x < c < h \cdot y$$

or

$$\text{for all } x \in X \text{ and } y \in Y, \text{ we have } h \cdot x > c > h \cdot y.$$

□

Here the hyperplane is a set of the form $\{z \in \mathbb{R}^n \mid h \cdot z = c\}$.

Definition 2.18 Let (X, d) be a metric space. A function $f : X \to X$ is said to be a *contraction mapping* if there is a constant δ with $0 \leq \delta < 1$ such that

$$d(f(x), f(y)) \leq \delta \cdot d(x, y)$$

for all $x, y \in X$.

In the above definition, the constant δ is strictly less than 1. It entails the following property whose proof crucially relies on that constraint on δ and is worth writing down.

Theorem 2.8 (Banach Fixed-Point Theorem) *Every contraction on a complete metric space has a unique fixed point.*

Proof Let (X, d) be a complete metric space, and f be a contraction mapping on (X, d) with constant δ. For any $x_0 \in X$, define the sequence (x_n) by $x_{n+1} := f(x_n)$ for $n \geq 0$. Let $a := d(x_0, x_1)$. It is easy to show that

$$d(x_n, x_{n+1}) \leq \delta^n \cdot a$$

by repeated application of the property $d(f(x), f(y)) \leq \delta \cdot d(x, y)$. For any $\varepsilon > 0$, it is possible to choose a natural number N such that $\frac{\delta^n}{1-\delta} a < \varepsilon$ for all $n \geq N$. Now, for any $m, n \geq N$ with $m \leq n$,

$$\begin{aligned} d(x_m, x_n) &\leq d(x_m, x_{m+1}) + d(x_{m+1}, x_{m+2}) + \cdots + d(x_{n-1}, x_n) \\ &\leq \delta^m \cdot a + \delta^{m+1} \cdot a + \cdots + \delta^{n-1} \cdot a \\ &= \delta^m \frac{1-\delta^{n-m}}{1-\delta} a \\ &< \frac{\delta^m}{1-\delta} a < \varepsilon \end{aligned}$$

by repeated application of the triangle inequality. So the sequence (x_n) is a Cauchy sequence. Since (X, d) is complete, the sequence has a limit in (X, d). We define x^* to be this limit and show that it is a fixed point of f. Suppose it is not, i.e. $a^* := d(x^*, f(x^*)) > 0$. Since (x_n) converges to x^*, there exists some $N \in \mathbb{N}$ such that $d(x_n, x^*) < \frac{a^*}{2}$ for all $n \geq N$. Then

$$\begin{aligned} d(x^*, f(x^*)) &\leq d(x^*, x_{N+1}) + d(x_{N+1}, f(x^*)) \\ &\leq d(x^*, x_{N+1}) + \delta \cdot d(x_N, x^*) \\ &< \frac{a^*}{2} + \frac{a^*}{2} = a^*, \end{aligned}$$

which is a contradiction. So x^* is a fixed point of f. It is also unique. Otherwise, suppose there is another fixed point x'; we have $d(x', x^*) > 0$ since $x' \neq x^*$. But then we would have

$$d(x', x^*) = d(f(x'), f(x^*)) \leq \delta \cdot d(x', x^*) < d(x', x^*).$$

Therefore x^* is the unique fixed point of f.

2.5 Probability Spaces

In this section, we recall some basic concepts from probability and measure theory. More details can be found in many excellent textbooks, for example [4].

Definition 2.19 Let X be an arbitrary nonempty set and $\mathcal{X}$ a collection of subsets of X. We say that $\mathcal{X}$ is a *field* on X if

1. The empty set $\emptyset \in \mathcal{X}$;
2. Whenever $A \in \mathcal{X}$, then the complement $X \backslash A \in \mathcal{X}$;
3. Whenever $A, B \in \mathcal{X}$, then the union $A \cup B \in \mathcal{X}$.

A field $\mathcal{X}$ is a *σ-algebra*, if it is closed under countable union: whenever $A_i \in \mathcal{X}$ for $i \in \mathbb{N}$, then $\bigcup_{i \in \mathbb{N}} A_i \in \mathcal{X}$.

The elements of a σ-algebra are called *measurable sets*, and $(X, \mathcal{X})$ is called a *measurable space*. A measurable space $(X, \mathcal{X})$ is called *discrete* if $\mathcal{X}$ is the powerset $\mathcal{P}(X)$. A σ-algebra *generated* by a family of sets $\mathcal{X}$, denoted $\sigma(\mathcal{X})$, is the smallest σ-algebra that contains $\mathcal{X}$. The existence of $\sigma(\mathcal{X})$ is ensured by the following proposition.

Proposition 2.3 *For any nonempty set X and $\mathcal{X}$ a collection of subsets of X, there exists a unique smallest σ-algebra containing $\mathcal{X}$.* □

The *Borel σ-algebra* on a topological space $(X, \mathcal{X})$ is the smallest σ-algebra containing $\mathcal{X}$. The elements of the Borel σ-algebra are called *Borel sets*. If we have a topological space then we can always consider its Borel σ-algebra and regard $(X, \sigma(\mathcal{X}))$ as a measurable space.

Let $\mathcal{X}$ be a collection of subsets of a set X. We say $\mathcal{X}$ is a π-class if it is closed under finite intersections; $\mathcal{X}$ is a λ-class if it is closed under complementations and countable disjoint unions.

Theorem 2.9 (The π-λ theorem) *If $\mathcal{X}$ is a π-class, then $\sigma(\mathcal{X})$ is the smallest λ-class containing $\mathcal{X}$.* □

Definition 2.20 Let $(X, \mathcal{X})$ be a measurable space. A function $\mu : \mathcal{X} \to [-\infty, \infty]$ is a *measure* on $\mathcal{X}$ if it satisfies the following conditions:

1. $\mu(A) \geq 0$ for all $A \in \mathcal{X}$;
2. $\mu(\emptyset) = 0$;
3. If $A_1, A_2, \ldots$ are in $\mathcal{X}$, with $A_i \cap A_j = \emptyset$ for $i \neq j$, then $\mu(\bigcup_i A_i) = \sum_i \mu(A_i)$.

The triple $(X, \mathcal{X}, \mu)$ is called a *measure space*. A *Borel* measure is a measure on a Borel σ-algebra. If $\mu(X) = 1$, then the measure space $(X, \mathcal{X}, \mu)$ is called a *probability space*, μ a *probability measure*, also called *probability distribution*, the set X a *sample space*, and the elements of $\mathcal{X}$ *events*. If $\mu(X) \leq 1$, then we obtain a *subprobability measure*, also called *subprobability distribution* or simply *subdistribution*. A measure over a discrete measurable space $(X, \mathcal{P}(X))$ is called a *discrete measure* over X.

A (discrete) probability *subdistribution* over a set S can also be considered as a function $\Delta : S \to [0,1]$ with $\sum_{s \in S} \Delta(s) \leq 1$; the *support* of such a Δ is the set $\lceil \Delta \rceil := \{s \in S \mid \Delta(s) > 0\}$, and its *mass* $|\Delta|$ is $\sum_{s \in \lceil \Delta \rceil} \Delta(s)$. A subdistribution is a (total, or full) *distribution* if $|\Delta| = 1$. The *point distribution* $\overline{s}$ assigns probability 1 to s and 0 to all other elements of S, so that $\lceil \overline{s} \rceil = \{s\}$. With $\mathcal{D}_{sub}(s)$ we denote the set of subdistributions over S, and with $\mathcal{D}(S)$ its subset of full distributions. For $\Delta, \Theta \in \mathcal{D}_{sub}(S)$ we write $\Delta \leq \Theta$ iff $\Delta(s) \leq \Theta(s)$ for all $s \in S$.

Let $\{\Delta_k \mid k \in K\}$ be a set of subdistributions, possibly infinite. Then $\sum_{k \in K} \Delta_k$ is the real-valued function in $S \to \mathbb{R}$ defined by $(\sum_{k \in K} \Delta_k)(s) := \sum_{k \in K} \Delta_k(s)$. This is a partial operation on subdistributions because for some state s the sum of $\Delta_k(s)$ might exceed 1. If the index set is finite, say $\{1, \ldots, n\}$, we often write $\Delta_1 + \cdots + \Delta_n$. For a real number p from $[0,1]$ we use $p \cdot \Delta$ to denote the subdistribution given by $(p \cdot \Delta)(s) := p \cdot \Delta(s)$. Finally, we use ε to denote the everywhere-zero subdistribution that thus has empty support. These operations on subdistributions do not readily adapt themselves to distributions; yet if $\sum_{k \in K} p_k = 1$ for some collection of probabilities p_k, and the Δ_k are distributions, then so is $\sum_{k \in K} p_k \cdot \Delta_k$. In general when $0 \leq p \leq 1$, we write $x_p\!\oplus y$ for $p \cdot x + (1-p) \cdot y$ where that makes sense, so that for example $\Delta_1\,{}_p\!\oplus \Delta_2$ is always defined, and is full if Δ_1 and Δ_2 are. Finally, the *product* of two probability distributions Δ, Θ over S, T is the distribution $\Delta \times \Theta$ over $S \times T$ defined by $(\Delta \times \Theta)(s,t) := \Delta(s) \cdot \Theta(t)$.

2.6 Linear Programming

A linear programming problem is the problem of maximising or minimising a linear function subject to linear constraints. The constraints may be equalities or inequalities.

Example 2.16 (The Transportation Problem) There are m factories that supply n customers with a particular product. Maximum production at factory i is s_i. The demand for the product from customer j is r_j. Let c_{ij} be the cost of transporting one unit of the product from factory i to customer j. The problem is to determine the quantity of product to be supplied from each factory to each customer so as to minimise total costs.

Let y_{ij} be the quantity of product shipped from factory i to customer j. The total transportation cost is

$$\sum_{i=1}^{m}\sum_{j=1}^{n} y_{ij}c_{ij}. \tag{2.3}$$

The amount sent from factory i is $\sum_{j=1}^{n} y_{ij}$ and since the amount available at factory i is s_i, we must have

$$\sum_{j=1}^{n} y_{ij} \le s_i \qquad \text{for } i = 1,\dots,m. \tag{2.4}$$

The amount sent to customer j is $\sum_{i=1}^{m} y_{ij}$ and since the amount required there is r_j, we must have

$$\sum_{i=1}^{m} y_{ij} \ge r_j \qquad \text{for } j = 1,\dots,n. \tag{2.5}$$

It is assumed that we cannot send a negative amount from factory i to customer j, so we have also

$$y_{ij} \ge 0 \qquad \text{for } i = 1,\dots,m \text{ and } j = 1,\dots,n. \tag{2.6}$$

Our problem is to minimise (2.3) subject to (2.4), (2.5) and (2.6).

Two classes of problems, the *standard maximum problem* and the *standard minimum problem*, play a special role in linear programming. We are given an m-vector, $\mathbf{b} = (b_1,\dots,b_m)^T$, and an n-vector, $\mathbf{c} = (c_1,\dots,c_n)^T$ and an $m \times n$ matrix,

$$\mathbf{A} = \begin{pmatrix} a_{11} & a_{12} & \cdots & a_{1n} \\ a_{21} & a_{22} & \cdots & a_{2n} \\ \vdots & \vdots & \ddots & \vdots \\ a_{m1} & a_{m2} & \cdots & a_{mn} \end{pmatrix}$$

of real numbers.

The Standard Maximum Problem Find an n-vector, $\mathbf{x} = (x_1,\dots,x_n)^T$ that maximises

$$\mathbf{c}^T\mathbf{x} = c_1x_1 + \cdots + c_nx_n$$

subject to the constraints $\mathbf{Ax} \le \mathbf{b}$, i.e.

$$\begin{aligned} a_{11}x_1 + a_{12}x_2 + \cdots + a_{1n}x_n &\le b_1 \\ a_{21}x_1 + a_{22}x_2 + \cdots + a_{2n}x_n &\le b_2 \\ &\vdots \\ a_{m1}x_1 + a_{m2}x_2 + \cdots + a_{mn}x_n &\le b_m \end{aligned}$$

and $\mathbf{x} \geq \mathbf{0}$, i.e.

$$x_1 \geq 0,\ x_2 \geq 0,\ \ldots,\ x_n \geq 0.$$

The Standard Minimum Problem Find an m-vector, $\mathbf{y} = (y_1, \ldots, y_m)^T$ that minimises

$$\mathbf{b}^T\mathbf{y} = b_1 y_1 + \cdots + b_m y_m$$

subject to the constraints $\mathbf{A}^T\mathbf{y} \geq \mathbf{c}$, i.e.

$$\begin{aligned} a_{11}y_1 + a_{21}y_2 + \cdots + a_{m1}y_m &\geq c_1 \\ a_{12}y_1 + a_{22}y_2 + \cdots + a_{m2}y_m &\geq c_2 \\ &\vdots \\ a_{1n}y_1 + a_{2n}y_2 + \cdots + a_{mn}y_m &\geq c_n \end{aligned}$$

and $\mathbf{y} \geq \mathbf{0}$, i.e.

$$y_1 \geq 0,\ y_2 \geq 0,\ \ldots,\ y_m \geq 0.$$

In a linear programming problem, the function to be maximised or minimised is called the *objective function.* A vector, e.g. $\mathbf{x}$ for the standard maximum problem or $\mathbf{y}$ for the standard minimum problem, is said to be *feasible* if it satisfies the corresponding constraints. The set of feasible vectors is called the *constraint set.* A linear programming problem is said to be *feasible* if the constraint set is not empty; otherwise, it is said to be *infeasible.* A feasible maximum (resp. minimum) problem is said to be *unbounded* if the objective function can assume arbitrarily large positive (resp. negative) values at feasible vectors; otherwise, it is said to be *bounded.* The *value* of a bounded feasible maximum (resp, minimum) problem is the maximum (resp. minimum) value of the objective function as the variables range over the constraint set. A feasible vector at which the objective function achieves the value is called *optimal.*

Proposition 2.4 *All linear programming problems can be converted to a standard form.* □

Definition 2.21 The dual of the standard maximum problem

maximise $\mathbf{c}^T\mathbf{x}$
subject to the constraints $\mathbf{Ax} \leq \mathbf{b}$ and $\mathbf{x} \geq \mathbf{0}$

is defined to be the standard minimum problem

minimise $\mathbf{b}^T\mathbf{y}$
subject to the constraints $\mathbf{A}^T\mathbf{y} \geq \mathbf{c}$ and $\mathbf{y} \geq \mathbf{0}$

and vice versa.

Theorem 2.10 (The Duality Theorem) *If a standard linear programming problem is boundedfeasible, then so is its dual, their values are equal, and there exist optimal vectors for both problems.* □

References

1. Winskel, G.: The Formal Semantics of Programming Languages: an Introduction. The MIT Press, Cambridge (1993)
2. Sangiorgi, D.: Introduction to Bisimulation and Coinduction. Cambridge University Press, Cambridge (2012)
3. Matoušek, J.: Lectures on Discrete Geometry. Springer, New York (2002)
4. Billingsley, P.: Probability and Measure. Wiley, New York (1995)

Chapter 3
Probabilistic Bisimulation

Abstract We introduce the operational model of probabilistic labelled transition systems, where a state evolves into a distribution after performing some action. To define relations between distributions, we need to lift a relation on states to be a relation on distributions of states. There is a natural lifting operation that nicely corresponds to the Kantorovich metric, a fundamental concept used in mathematics to lift a metric on states to a metric on distributions of states, which is also related to the maximum flow problem in optimisation theory.

The lifting operation yields a neat notion of probabilistic bisimulation, for which we provide logical, metric and algorithmic characterisations. Specifically, we extend the Hennessy–Milner logic and the modal mu-calculus with a new modality, resulting in an adequate and an expressive logic for probabilistic bisimilarity. The correspondence of the lifting operation and the Kantorovich metric leads to a characterisation of bisimulations as pseudometrics that are postfixed points of a monotone function. Probabilistic bisimilarity also admits both partition refinement and "on-the-fly" decision algorithms; the latter exploits the close relationship between the lifting operation and the maximum flow problem.

Keywords Probabilistic labelled transition system · Lifting operation · Probabilistic bisimulation · Modal logic · Algorithm

3.1 Introduction

In recent years, probabilistic constructs have been proven useful for giving quantitative specifications of system behaviour. The first papers on probabilistic concurrency theory [1–3] proceed by *replacing* nondeterministic with probabilistic constructs. The reconciliation of nondeterministic and probabilistic constructs starts with [4] and has received a lot of attention in the literature [5–22]. It could be argued that it is one of the central problems of the area.

Also we shall work in a framework that features the coexistence of probability and nondeterminism. More specifically, we deal with *probabilistic labelled transition systems (pLTSs)* [19] that are an extension of the usual *labelled transition systems (LTSs)* so that a step of transition is in the form $s \xrightarrow{a} \Delta$, meaning that state s can perform action a and evolve into a distribution Δ over some successor states. In this

DOI 10.1007/978-3-662-45198-4_3

setting state s is related to state t by a relation $\mathcal{R}$, say probabilistic simulation, written $s\mathcal{R}t$, if for each transition $s \xrightarrow{a} \Delta$ from s there exists a transition $t \xrightarrow{a} \Theta$ from t such that Θ can somehow mimic the behaviour of Δ according to $\mathcal{R}$. To formalise the mimicking of Δ by Θ, we have to *lift* $\mathcal{R}$ to be a relation $\mathcal{R}^\dagger$ between distributions over states so that we can require $\Delta\mathcal{R}^\dagger\Theta$.

Various approaches of lifting relations have appeared in the literature; see e.g. [6, 19, 22–24]. We will show that although those approaches appear different, they can be reconciled. Essentially, there is only one lifting operation, which has been presented in different forms. Moreover, we argue that the lifting operation is interesting in itself. This is justified by its intrinsic connection with some fundamental concepts in mathematics, notably *the Kantorovich metric* [25]. For example, it turns out that our lifting of binary relations from states to distributions nicely corresponds to the lifting of metrics from states to distributions by using the Kantorovich metric. In addition, the lifting operation is closely related to *the maximum flow problem* in optimisation theory, as observed by Baier et al. [26].

A good scientific concept is often elegant, even seen from many different perspectives. Among the wealth of behavioural equivalences proposed in the traditional concurrency theory during the last three decades, *bisimilarity* [27, 28] is probably the most studied one as it admits a suitable semantics, a nice coinductive proof technique and efficient decision algorithms. In our opinion, bisimulation is a good concept because it can be characterised in a great many ways such as fixed-point theory, modal logics, game theory, coalgebras etc., and all the characterisations are very natural. We believe that probabilistic bisimulation is such a concept in probabilistic concurrency theory. As evidence, we will provide in this chapter three characterisations, from the perspectives of modal logics, metrics and decision algorithms.

1. Our logical characterisation of probabilistic bisimulation consists of two aspects: *adequacy* and *expressivity* [29]. A logic $\mathcal{L}$ is adequate when two states are bisimilar if and only if they satisfy exactly the same set of formulae in $\mathcal{L}$. The logic is expressive when each state s has a characteristic formula φ_s in $\mathcal{L}$ such that t is bisimilar to s if and only if t satisfies φ_s. We will introduce a probabilistic-choice modality to capture the behaviour of distributions. Intuitively, distribution Δ satisfies the formula $\bigoplus_{i\in I} p_i \cdot \varphi_i$ if there is a decomposition of Δ into a convex combination of some distributions, $\Delta = \sum_{i\in I} p_i \cdot \Delta_i$, and each Δ_i conforms to the property specified by φ_i. When the new modality is added to the Hennessy–Milner logic [30] we obtain an adequate logic for probabilistic bisimilarity; when it is added to the modal mu-calculus [31] we obtain an expressive logic.
2. By metric characterisation of probabilistic bisimulation, we mean to give a pseudometric[1] such that two states are bisimilar if and only if their distance is 0 when measured by the pseudometric. More specifically, we show that bisimulations correspond to pseudometrics that are postfixed points of a monotone function,

[1] We use a pseudometric rather than a proper metric because two *distinct* states can still be at distance zero if they are bisimilar.

and in particular bisimilarity corresponds to a pseudometric that is the greatest fixed point of the monotone function.

3. As to the algorithmic characterisation, we will see that a partition refinement algorithm can be used to check whether two states are bisimilar. We also propose an "on-the-fly" algorithm that checks whether two states are related by probabilistic bisimilarity. The schema of the algorithm is to approximate probabilistic bisimilarity by iteratively accumulating information about state pairs (s, t) where s and t are not bisimilar. In each iteration we dynamically construct a relation $\mathcal{R}$ as an approximant. Then we verify that every transition from one state can be matched by a transition from the other state, and that their resulting distributions are related by the lifted relation $\mathcal{R}^{\dagger}$. The latter involves solving the maximum flow problem of an appropriately constructed network, by taking advantage of the close relationship between our lifting operation and the above mentioned maximum flow problem.

3.2 Probabilistic Labelled Transition Systems

There are various ways of generalising the usual (nonprobabilistic) LTSs to a probabilistic setting. Here we choose one that is widely used in the literature.

Definition 3.1 A *probabilistic labelled transition system* (pLTS) is defined as a triple $\langle S, L, \rightarrow \rangle$, where

1. S is a set of states.
2. L is a set of transition actions.
3. Relation $\rightarrow$ is a subset of $S \times L \times \mathcal{D}(S)$.

A nonprobabilistic LTS may be viewed as a degenerate pLTS—one in which only point distributions are used. As with LTSs, we write $s \xrightarrow{\alpha} \Delta$ in place of $(s, \alpha, \Delta) \in \rightarrow$. If there exists some Δ with $s \xrightarrow{\alpha} \Delta$, then the predicate $s \xrightarrow{\alpha}$ holds. Similarly, $s \rightarrow$ holds if there is some α with $s \xrightarrow{\alpha}$. A pLTS is *finitely branching* if the set $\{\langle \alpha, \Delta \rangle \in L \times \mathcal{D}(S) \mid s \xrightarrow{\alpha} \Delta\}$ is finite for all states s; if moreover S is finite, then the pLTS is said to be *finitary*.

Convention All the pLTSs considered in this book are assumed to be finitary, unless otherwise stated.

In order to visualise pLTSs, we often draw them as directed graphs. Given that in a pLTS transitions go from states to distributions, we need to introduce additional edges to connect distributions back to states, thereby obtaining a bipartite graph. States are therefore represented by nodes of the form $\bullet$ and distributions by nodes of the form $\circ$. For any state s and distribution Δ with $s \xrightarrow{\alpha} \Delta$ we draw an edge from s to Δ, labelled with α. Consequently, the edges leaving a $\bullet$-node are all labelled with actions from L. For any distribution Δ and state s in $\lceil \Delta \rceil$, the support of Δ, we draw an edge from Δ to s, labelled with $\Delta(s)$. Consequently, the edges leaving a $\circ$-node

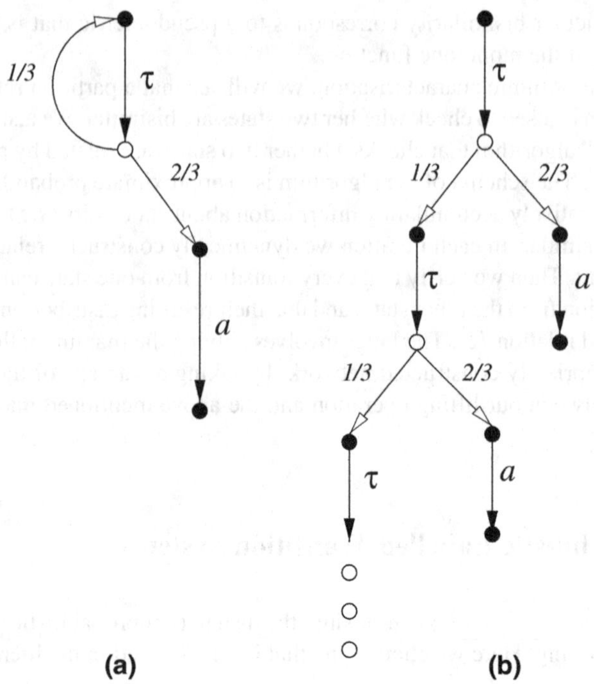

Fig. 3.1 Example pLTSs

are labelled with positive real numbers that sum to 1. Sometimes we partially unfold this graph by drawing the same nodes multiple times; in doing so, all outgoing edges of a given instance of a node are always drawn, but not necessarily all incoming edges. Edges labelled by probability 1 occur so frequently that it makes sense to omit them, together with the associated nodes $\circ$ representing point distributions.

Two example pLTSs are described this way in Fig. 3.1, where part (b) depicts the initial part of the pLTS obtained by unfolding the one in part (a).

For each state s, the outgoing transitions $s \xrightarrow{\alpha} \Delta$ represent the nondeterministic alternatives available in the state s. The nondeterministic choices provided by s are supposed to be resolved by the environment, which is formalised by a *scheduler* or an *adversary*. On the other hand, the probabilistic choices in the underlying distribution Δ are made by the system itself. Therefore, for each state s, the environment chooses some outgoing transition $s \xrightarrow{\alpha} \Delta$. Then the action α is performed, the system resolves the probabilistic choice, and subsequently with probability $\Delta(s')$ the system reaches state s'.

If we impose the constraint that for any state s and action α at most one outgoing transition from s is labelled α, then we obtain the special class of pLTSs called *reactive* (or *deterministic*) pLTSs that are the probabilistic counterpart to deterministic LTSs. Formally, a pLTS is reactive if for each $s \in S, \alpha \in L$ we have that $s \xrightarrow{\alpha} \Delta$ and $s \xrightarrow{\alpha} \Delta'$ imply $\Delta = \Delta'$.

3.3 Lifting Relations

In the probabilistic setting, formal systems are usually modelled as distributions over states. To compare two systems involves the comparison of two distributions. So we need a way of lifting relations on states to relations on distributions. This is used, for example, to define probabilistic bisimulation as we shall see in Sect. 3.5. A few approaches of lifting relations have appeared in the literature. We will take the one from [22], and show its coincidence with two other approaches.

Definition 3.2 Given two sets S and T and a binary relation $\mathcal{R} \subseteq S \times T$, the lifted relation $\mathcal{R}^\dagger \subseteq \mathcal{D}(S) \times \mathcal{D}(T)$ is the smallest relation that satisfies:

(1) $s\mathcal{R}t$ implies $\overline{s}\ \mathcal{R}^\dagger\ \overline{t}$.
(2) (Linearity) $\Delta_i\ \mathcal{R}^\dagger\ \Theta_i$ for all $i \in I$ implies $(\sum_{i\in I} p_i \cdot \Delta_i)\ \mathcal{R}^\dagger\ (\sum_{i\in I} p_i \cdot \Theta_i)$, where I is a finite index set and $\sum_{i\in I} p_i = 1$.

The are alternative presentations of Definition 3.2. The following is used in many proofs.

Proposition 3.1 *Let Δ and Θ be two distributions over S and T, respectively, and $\mathcal{R} \subseteq S \times T$. Then $\Delta\mathcal{R}^\dagger\Theta$ if and only if there are two collections of states, $\{s_i\}_{i\in I}$ and $\{t_i\}_{i\in I}$, and a collection of probabilities $\{p_i\}_{i\in I}$, for some finite index set I, such that $\sum_{i\in I} p_i = 1$ and Δ, Θ can be decomposed as follows:*

1. $\Delta = \sum_{i\in I} p_i \cdot \overline{s_i}$.
2. $\Theta = \sum_{i\in I} p_i \cdot \overline{t_i}$.
3. *For each $i \in I$ we have $s_i\ R\ t_i$.*

Proof ($\Leftarrow$) Suppose we can decompose Δ and Θ as follows: (i) $\Delta = \sum_{i\in I} p_i \cdot \overline{s_i}$, (ii) $\Theta = \sum_{i\in I} p_i \cdot \overline{t_i}$, and (iii) $s_i\mathcal{R}t_i$ for each $i \in I$. By (iii) and the first rule in Definition 3.2, we have $\overline{s_i}\ \mathcal{R}^\dagger\ \overline{t_i}$ for each $i \in I$. By the second rule in Definition 3.2 we obtain that $(\sum_{i\in I} p_i \cdot \overline{s_i})\ \mathcal{R}^\dagger\ (\sum_{i\in I} p_i \cdot \overline{t_i})$, that is $\Delta\ \mathcal{R}^\dagger\ \Theta$.

($\Rightarrow$) We proceed by rule induction.

- If $\Delta\mathcal{R}^\dagger\Theta$ because $\Delta = \overline{s}$, $\Theta = \overline{t}$ and $s\mathcal{R}t$, then we can simply take I to be the singleton set $\{i\}$ with $p_i = 1$ and $\Theta_i = \Theta$.
- If $\Delta\mathcal{R}^\dagger\Theta$ because of the conditions $\Delta = \sum_{i\in I} p_i \cdot \Delta_i$, $\Theta_i = \sum_{i\in I} p_i \cdot \Theta_i$ for some index set I, and $\Delta_i\ \mathcal{R}^\dagger\ \Theta_i$ for each $i \in I$, then by induction hypothesis there are index sets J_i such that $\Delta_i = \sum_{j\in J_i} p_{ij} \cdot \overline{s_{ij}}$, $\Theta_i = \sum_{j\in J_i} p_{ij} \cdot \overline{t_{ij}}$, and $s_{ij}\mathcal{R}t_{ij}$ for each $i \in I$ and $j \in J_i$. It follows that $\Delta = \sum_{i\in I}\sum_{j\in J_i} p_i p_{ij} \cdot \overline{s_{ij}}$, $\Theta = \sum_{i\in I}\sum_{j\in J_i} p_i p_{ij} \cdot \overline{t_{ij}}$, and $s_{ij}\mathcal{R}t_{ij}$ for each $i \in I$ and $j \in J_i$. Therefore, it suffices to take $\{ij \mid i \in I, j \in J_i\}$ to be the index set and $\{p_i p_{ij} \mid i \in I, j \in J_i\}$ be the collection of probabilities. □

An important point here is that in the decomposition of Δ into $\sum_{i\in I} p_i \cdot \overline{s_i}$, the states s_i are *not necessarily distinct*: that is, the decomposition is not in general unique. Thus when establishing the relationship between Δ and Θ, a given state s in Δ may play a number of different roles. This is reflected in the following property.

Proposition 3.2 *$\overline{s}\ \mathcal{R}^\dagger\ \Theta$ iff $s\ \mathcal{R}\ t$ for all $t \in \lceil\Theta\rceil$.* □

The lifting construction satisfies the following useful property.

Proposition 3.3 (Left-Decomposable) *Suppose $\mathcal{R} \subseteq S \times T$ and $\sum_{i\in I} p_i = 1$. If $(\sum_{i\in I} p_i \cdot \Delta_i)\ \mathcal{R}^\dagger\ \Theta$ then $\Theta = \sum_{i\in I} p_i \cdot \Theta_i$ for some set of distributions $\{\Theta_i\}_{i\in I}$ such that $\Delta_i\ \mathcal{R}^\dagger\ \Theta_i$ for each $i \in I$.*

Proof Suppose $\Delta = \sum_{i\in I} p_i \cdot \Delta_i$ and $\Delta\ \mathcal{R}^\dagger\ \Theta$. We have to find a family of Θ_i such that

(i) $\Delta_i\ \mathcal{R}^\dagger\ \Theta_i$ for each $i \in I$.
(ii) $\Theta = \sum_{i\in I} p_i \cdot \Theta_i$.

From the alternative characterisation of lifting, Proposition 3.1, we know that

$$\Delta = \sum_{j\in J} q_j \cdot \overline{s_j} \qquad s_j \mathcal{R} t_j \qquad \Theta = \sum_{j\in J} q_j \cdot \overline{t_j}$$

Define Θ_i to be

$$\sum_{s\in\lceil\Delta_i\rceil} \Delta_i(s) \cdot \left(\sum_{\{j\in J\,|\,s=s_j\}} \frac{q_j}{\Delta(s)} \cdot \overline{t_j} \right)$$

Note that $\Delta(s)$ can be written as $\sum_{\{j\in J\,|\,s=s_j\}} q_j$ and therefore

$$\Delta_i = \sum_{s\in\lceil\Delta_i\rceil} \Delta_i(s) \cdot \left(\sum_{\{j\in J\,|\,s=s_j\}} \frac{q_j}{\Delta(s)} \cdot \overline{s_j} \right)$$

Since $s_j \mathcal{R} t_j$ this establishes (i) above.

To establish (ii) above let us first abbreviate the sum $\sum_{\{j\in J\,|\,s=s_j\}} \frac{q_j}{\Delta(s)} \cdot \overline{t_j}$ to $X(s)$. Then $\sum_{i\in I} p_i \cdot \Theta_i$ can be written as

$$\begin{aligned}
&\sum_{s\in\lceil\Delta\rceil} \sum_{i\in I} p_i \cdot \Delta_i(s) \cdot X(s)\\
&= \sum_{s\in\lceil\Delta\rceil} \left(\sum_{i\in I} p_i \cdot \Delta_i(s) \right) \cdot X(s)\\
&= \sum_{s\in\lceil\Delta\rceil} \Delta(s) \cdot X(s)
\end{aligned}$$

The last equation is justified by the fact that $\Delta(s) = \sum_{i\in I} p_i \cdot \Delta_i(s)$.

Now $\Delta(s) \cdot X(s) = \sum_{\{j\in J\,|\,s=s_j\}} q_j \cdot \overline{t_j}$ and therefore we have

$$\begin{aligned}
\sum_{i\in I} p_i \cdot \Theta_i &= \sum_{s\in\lceil\Delta\rceil} \sum_{\{j\in J\,|\,s=s_j\}} q_j \cdot \overline{t_j}\\
&= \sum_{j\in J} q_j \cdot \overline{t_j}\\
&= \Theta
\end{aligned}$$

□

From Definition 3.2, the next two properties follow. In fact, they are sometimes used in the literature as definitions of lifting relations instead of being properties (see e.g. [6, 23]).

Theorem 3.1

1. *Let Δ and Θ be distributions over S and T, respectively. Then $\Delta \mathcal{R}^\dagger \Theta$ if and only if there is a probability distribution on $S \times T$, with support a subset of $\mathcal{R}$, such that Δ and Θ are its marginal distributions. In other words, there exists a weight function $w : S \times T \to [0, 1]$ such that*

   a) $\forall s \in S : \sum_{t \in T} w(s,t) = \Delta(s)$.
   b) $\forall t \in T : \sum_{s \in S} w(s,t) = \Theta(t)$.
   c) $\forall (s,t) \in S \times T : w(s,t) > 0 \Rightarrow s\mathcal{R}t$.

2. *Let Δ and Θ be distributions over S and $\mathcal{R}$ be an equivalence relation. Then $\Delta \mathcal{R}^\dagger \Theta$ if and only if $\Delta(C) = \Theta(C)$ for all equivalence classes $C \in S/\mathcal{R}$, where $\Delta(C)$ stands for the accumulation probability $\sum_{s \in C} \Delta(s)$.*

Proof

1. ($\Rightarrow$) Suppose $\Delta\ \mathcal{R}^\dagger\ \Theta$. By Proposition 3.1, we can decompose Δ and Θ such that $\Delta = \sum_{i \in I} p_i \cdot \overline{s_i}$, $\Theta = \sum_{i \in I} p_i \cdot \overline{t_i}$, and $s_i \mathcal{R} t_i$ for all $i \in I$. We define the weight function w by letting

$$w(s,t) = \sum \{p_i \mid s_i = s, t_i = t, i \in I\}$$

for any $s \in S, t \in T$. This weight function can be checked to meet our requirements.

   a) For any $s \in S$, we have

$$\begin{aligned}\sum_{t \in T} w(s,t) &= \sum\nolimits_{t \in T} \sum \{p_i \mid s_i = s, t_i = t, i \in I\} \\ &= \sum \{p_i \mid s_i = s, i \in I\} \\ &= \Delta(s).\end{aligned}$$

   b) Similarly, we have $\sum_{s \in S} w(s,t) = \Theta(t)$.
   c) For any $s \in S, t \in T$, if $w(s,t) > 0$ then there is some $i \in I$ such that $p_i > 0$, $s_i = s$, and $t_i = t$. It follows from $s_i \mathcal{R} t_i$ that $s\mathcal{R}t$.

($\Leftarrow$) Suppose there is a weight function w satisfying the three conditions in the hypothesis. We construct the index set $I = \{(s,t) \mid w(s,t) > 0, s \in S, t \in T\}$ and probabilities $p_{(s,t)} = w(s,t)$ for each $(s,t) \in I$.

a) We have $\Delta = \sum_{(s,t)\in I} p_{(s,t)} \cdot \overline{s}$ because, for any $s \in S$,

$$\begin{aligned}\left(\sum_{(s,t)\in I} p_{(s,t)} \cdot \overline{s}\right)(s) &= \sum_{(s,t)\in I} w(s,t)\\ &= \sum\{w(s,t) \mid w(s,t) > 0, t \in T\}\\ &= \sum\{w(s,t) \mid t \in T\}\\ &= \Delta(s).\end{aligned}$$

b) Similarly, we have $\Theta = \sum_{(s,t)\in I} w(s,t) \cdot \overline{t}$.

c) For each $(s,t) \in I$, we have $w(s,t) > 0$, which implies $s\mathcal{R}t$.

Hence, the above decompositions of Δ and Θ meet the requirement of the lifting $\Delta\ \mathcal{R}^\dagger\ \Theta$.

2. ($\Rightarrow$) Suppose $\Delta\ \mathcal{R}^\dagger\ \Theta$. By Proposition 3.1, we can decompose Δ and Θ such that $\Delta = \sum_{i\in I} p_i \cdot \overline{s_i}$, $\Theta = \sum_{i\in I} p_i \cdot \overline{t_i}$, and $s_i \mathcal{R} t_i$ for all $i \in I$. For any equivalence class $C \in S/\mathcal{R}$, we have that

$$\begin{aligned}\Delta(C) = \sum_{s\in C} \Delta(s) = \sum_{s\in C} \sum\{p_i \mid i \in I, s_i = s\} &\\ = \sum\{p_i \mid i \in I, s_i \in C\} &\\ = \sum\{p_i \mid i \in I, t_i \in C\} &\\ = \Theta(C) &\end{aligned}$$

where the equality in the third line is justified by the fact that $s_i \in C$ iff $t_i \in C$ since $s_i \mathcal{R} t_i$ and $C \in S/\mathcal{R}$.

($\Leftarrow$) Suppose $\Delta(C) = \Theta(C)$ for each equivalence class $C \in S/\mathcal{R}$. We construct the index set $I = \{(s,t) \mid s\mathcal{R}t \text{ and } s,t \in S\}$ and probabilities $p_{(s,t)} = \frac{\Delta(s)\Theta(t)}{\Delta([s])}$ for each $(s,t) \in I$, where $[s]$ stands for the equivalence class that contains s.

a) We have $\Delta = \sum_{(s,t)\in I} p_{(s,t)} \cdot \overline{s}$ because, for any $s' \in S$,

$$\begin{aligned}\left(\sum_{(s,t)\in I} p_{(s,t)} \cdot \overline{s}\right)(s') &= \sum_{(s',t)\in I} p_{(s',t)}\\ &= \sum\left\{\frac{\Delta(s')\Theta(t)}{\Delta([s'])} \mid s'\mathcal{R}t,\ t \in S\right\}\\ &= \sum\left\{\frac{\Delta(s')\Theta(t)}{\Delta([s'])} \mid t \in [s']\right\}\\ &= \frac{\Delta(s')}{\Delta([s'])} \sum\{\Theta(t) \mid t \in [s']\}\\ &= \frac{\Delta(s')}{\Delta([s'])} \Theta([s'])\end{aligned}$$

$$= \frac{\Delta(s')}{\Delta([s'])}\Delta([s'])$$
$$= \Delta(s').$$

b) Similarly, we have $\Theta = \sum_{(s,t)\in I} p_{(s,t)} \cdot \overline{t}$.
c) For each $(s,t) \in I$, we have $s\mathcal{R}t$.

Hence, the above decompositions of Δ and Θ meet the requirement of the lifting $\Delta\ \mathcal{R}^\dagger\ \Theta$. □

The lifting operation given in Definition 3.2 is monotone and also preserves the transitivity of the relation being lifted. Moreover, it is distributive with respect to the composition of relations.

Proposition 3.4

1. *If* $\mathcal{R}_1 \subseteq \mathcal{R}_2$ *then* ${\mathcal{R}_1}^\dagger \subseteq {\mathcal{R}_2}^\dagger$
2. $(\mathcal{R}_1 \cdot \mathcal{R}_2)^\dagger = {\mathcal{R}_1}^\dagger \cdot {\mathcal{R}_2}^\dagger$
3. *If* $\mathcal{R}$ *is a transitive relation, then so is* $\mathcal{R}^\dagger$.

Proof

1. By Definition 3.2, it is straightforward to show that if $\Delta_1\ {\mathcal{R}_1}^\dagger\ \Delta_2$ and $\mathcal{R}_1 \subseteq \mathcal{R}_2$ then $\Delta_1\ {\mathcal{R}_2}^\dagger\ \Delta_2$.
2. We first show that $(\mathcal{R}_1 \cdot \mathcal{R}_2)^\dagger \subseteq {\mathcal{R}_1}^\dagger \cdot {\mathcal{R}_2}^\dagger$. Suppose there are two distributions Δ_1, Δ_2 such that $\Delta_1\ (\mathcal{R}_1 \cdot \mathcal{R}_2)^\dagger\ \Delta_2$. Then we have that

$$\Delta_1 = \sum_{i\in I} p_i \cdot \overline{s_i}, \qquad s_i\ \mathcal{R}_1 \cdot \mathcal{R}_2\ t_i, \qquad \Delta_2 = \sum_{i\in I} p_i \cdot \overline{t_i}\ . \tag{3.1}$$

The middle part of (3.1) implies the existence of some states s_i' such that $s_i\mathcal{R}_1 s_i'$ and $s_i'\mathcal{R}_2 t_i$. Let Θ be the distribution $\sum_{i\in I} p_i \cdot \overline{s_i'}$. It is clear that $\Delta_1\ {\mathcal{R}_1}^\dagger\ \Theta$ and $\Theta\ {\mathcal{R}_2}^\dagger\ \Delta_2$. It follows that $\Delta_1\ {\mathcal{R}_1}^\dagger \cdot {\mathcal{R}_2}^\dagger\ \Delta_2$.
Then we show the inverse inclusion ${\mathcal{R}_1}^\dagger \cdot {\mathcal{R}_2}^\dagger \subseteq (\mathcal{R}_1 \cdot \mathcal{R}_2)^\dagger$. Given any three distributions Δ_1, Δ_2 and Δ_3, we show that if $\Delta_1\ {\mathcal{R}_1}^\dagger\ \Delta_2$ and $\Delta_2\ {\mathcal{R}_2}^\dagger\ \Delta_3$ then $\Delta_1\ (\mathcal{R}_1 \cdot \mathcal{R}_2)^\dagger\ \Delta_3$.
First $\Delta_1\ {\mathcal{R}_1}^\dagger\ \Delta_2$ means that

$$\Delta_1 = \sum_{i\in I} p_i \cdot \overline{s_i}, \qquad s_i\ \mathcal{R}_1\ s_i', \qquad \Delta_2 = \sum_{i\in I} p_i \cdot \overline{s_i'}\ ; \tag{3.2}$$

also $\Delta_2\ {\mathcal{R}_2}^\dagger\ \Delta_3$ means that

$$\Delta_2 = \sum_{j\in J} q_j \cdot \overline{t_j'}, \qquad t_j'\ \mathcal{R}_2\ t_j, \qquad \Delta_3 = \sum_{j\in J} q_j \cdot \overline{t_j}\ ; \tag{3.3}$$

and we can assume without loss of generality that all the coefficients p_i, q_j are nonzero. Now define $I_j = \{ i \in I \mid s_i' = t_j' \}$ and $J_i = \{ j \in J \mid t_j' = s_i' \}$, so that trivially

$$\{(i,j) \mid i \in I,\ j \in J_i\} \quad = \quad \{(i,j) \mid j \in J,\ i \in I_j\} \tag{3.4}$$

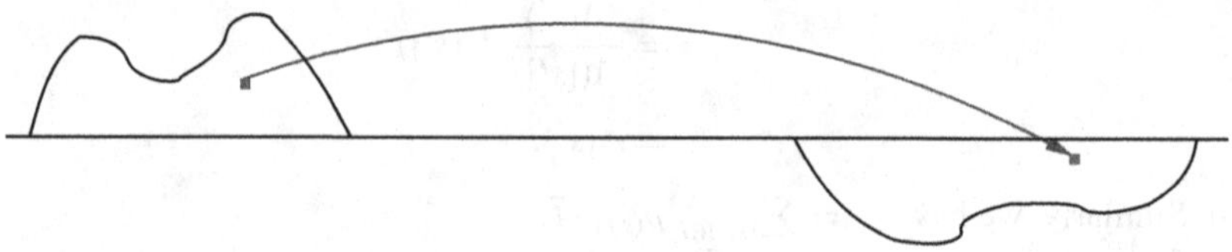

Fig. 3.2 The transportation problem

and note that

$$\Delta_2(s'_i) = \sum_{j \in J_i} q_j \qquad \text{and} \qquad \Delta_2(t'_j) = \sum_{i \in I_j} p_i. \tag{3.5}$$

Because of (3.5) we have

$$\begin{aligned} \Delta_1 = \textstyle\sum_{i \in I} p_i \cdot \overline{s_i} &= \textstyle\sum_{i \in I} p_i \cdot \sum_{j \in J_i} \frac{q_j}{\Delta_2(s'_i)} \cdot \overline{s_i} \\ &= \textstyle\sum_{i \in I} \sum_{j \in J_i} \frac{p_i \cdot q_j}{\Delta_2(s'_i)} \cdot \overline{s_i}. \end{aligned}$$

Similarly

$$\begin{aligned} \Delta_3 = \textstyle\sum_{j \in J} q_j \cdot \overline{t_j} &= \textstyle\sum_{j \in J} q_j \cdot \sum_{i \in I_j} \frac{p_i}{\Delta_2(t'_j)} \cdot \overline{t_j} \\ &= \textstyle\sum_{j \in J} \sum_{i \in I_j} \frac{p_i \cdot q_j}{\Delta_2(t'_j)} \cdot \overline{t_j} \\ &= \textstyle\sum_{i \in I} \sum_{j \in J_i} \frac{p_i \cdot q_j}{\Delta_2(t'_j)} \cdot \overline{t_j} \qquad \text{by (3.4).} \end{aligned}$$

Now for each j in J_i we know that in fact $t'_j = s'_i$, and so from the middle parts of (3.2) and (3.3), we obtain $\Delta_1 \; (\mathcal{R}_1 \cdot \mathcal{R}_2)^\dagger \; \Delta_3$.

3. By Clause 2 above we have $\mathcal{R}^\dagger \cdot \mathcal{R}^\dagger = (\mathcal{R} \cdot \mathcal{R})^\dagger$ for any relation $\mathcal{R}$. If $\mathcal{R}$ is transitive, then $\mathcal{R} \cdot \mathcal{R} \subseteq \mathcal{R}$. By Clause 1 above, we obtain $(\mathcal{R} \cdot \mathcal{R})^\dagger \subseteq \mathcal{R}^\dagger$. It follows that $\mathcal{R}^\dagger \cdot \mathcal{R}^\dagger \subseteq \mathcal{R}^\dagger$, thus $\mathcal{R}^\dagger$ is transitive. □

3.4 Justifying the Lifting Operation

The lifting operation given in Definition 3.2 is not only concise but also intrinsically related to some fundamental concepts in mathematics, notably the Kantorovich metric.

3.4.1 Justification by the Kantorovich Metric

We begin with some historical notes. The *transportation problem* plays an important role in linear programming due to its general formulation and methods of solution.

The original transportation problem, formulated by the French mathematician G. Monge in 1781 [32], consists of finding an optimal way of shovelling a pile of sand into a hole of the same volume; see Fig. 3.2. In the 1940s, the Russian mathematician and economist L. V. Kantorovich, who was awarded a Nobel prize in economics in 1975 for the theory of optimal allocation of resources, gave a relaxed formulation of the problem and proposed a variational principle for solving it [25]. Unfortunately, Kantorovich's work went unrecognised for a long time. The later-known *Kantorovich metric* has appeared in the literature under different names, because it has been rediscovered historically several times from different perspectives. Many metrics known in measure theory, ergodic theory, functional analysis, statistics etc. are special cases of the general definition of the Kantorovich metric [33]. The elegance of the formulation, the fundamental character of the optimality criterion as well as the wealth of applications that keep arising place the Kantorovich metric in a prominent position among the mathematical works of the twentieth century. In addition, this formulation can be computed in polynomial time [34], which is an appealing feature for its use in solving applied problems. For example, it is widely used to solve a variety of problems in business and economy such as market distribution, plant location, scheduling problems etc. In recent years the metric attracted the attention of computer scientists [35]: it has been used in various different areas in computer science such as probabilistic concurrency, image retrieval, data mining, bioinformatics etc.

Roughly speaking, the Kantorovich metric provides a way of measuring the distance between two distributions. Of course, this requires first a notion of distance between the basic elements that are aggregated into the distributions, which is often referred to as the *ground distance*. In other words, the Kantorovich metric defines a "lifted" distance between two distributions of mass in a space that is itself endowed with a ground distance. There is a host of metrics available in the literature (see e.g. [36]) to quantify the distance between probability measures; see [37] for a comprehensive review of metrics in the space of probability measures. The Kantorovich metric has an elegant formulation and a natural interpretation in terms of the transportation problem.

We now recall the mathematical definition of the Kantorovich metric. Let (X, m) be a *separable* metric space. (This condition will be used by Theorem 3.2 below.)

Definition 3.3 Given any two Borel probability measures Δ and Θ on X, the *Kantorovich distance* between Δ and Θ is defined by

$$K(\Delta, \Theta) = \sup \left\{ \left| \int f d\Delta - \int f d\Theta \right| \mid ||f|| \leq 1 \right\}.$$

where $|| \cdot ||$ is the *Lipschitz seminorm* defined by $||f|| = \sup_{x \neq y} \frac{|f(x)-f(y)|}{m(x,y)}$ for a function $f : X \to \mathbb{R}$.

The Kantorovich metric has an alternative characterisation. We denote by $\mathbf{P}(X)$ the set of all Borel probability measures on X such that for all $z \in X$, if $\Delta \in \mathbf{P}(X)$ then $\int_X m(x,z)\Delta(x) < \infty$. We write $M(\Delta, \Theta)$ for the set of all Borel probability measures on the product space $X \times X$ with marginal measures Δ and Θ, i.e. if $\Gamma \in M(\Delta, \Theta)$ then $\int_{y \in X} d\Gamma(x, y) = d\Delta(x)$ and $\int_{x \in X} d\Gamma(x, y) = d\Theta(y)$ hold.

Definition 3.4 For $\Delta, \Theta \in \mathbf{P}(X)$, we define the metric L as follows:

$$L(\Delta,\Theta) = \inf\left\{ \int m(x,y)d\Gamma(x,y) \;\mid\; \Gamma \in M(\Delta,\Theta) \right\}.$$

Lemma 3.1 *If (X, m) is a separable metric space then K and L are metrics on $\boldsymbol{P}(X)$.* □

The famous Kantorovich–Rubinstein duality theorem gives a dual representation of K in terms of L.

Theorem 3.2 (Kantorovich–Rubinstein [38]**)** *If (X, m) is a separable metric space then for any two distributions $\Delta, \Theta \in \boldsymbol{P}(X)$ we have $K(\Delta,\Theta) = L(\Delta,\Theta)$.* □

In view of the above theorem, many papers in the literature directly take Definition 3.4 as the definition of the Kantorovich metric. Here we keep the original definition, but it is helpful to understand K by using L. Intuitively, a probability measure $\Gamma \in M(\Delta,\Theta)$ can be understood as a *transportation* from one unit mass distribution Δ to another unit mass distribution Θ. If the distance $m(x, y)$ represents the cost of moving one unit of mass from location x to location y then the Kantorovich distance gives the optimal total cost of transporting the mass of Δ to Θ. We refer the reader to [39] for an excellent exposition on the Kantorovich metric and the duality theorem.

Many problems in computer science only involve finite state spaces, so discrete distributions with finite supports are sometimes more interesting than continuous distributions. For two discrete distributions Δ and Θ with finite supports $\{x_1, \ldots, x_n\}$ and $\{y_1, \ldots, y_l\}$, respectively, minimising the total cost of a discretised version of the transportation problem reduces to the following linear programming problem:

$$\begin{array}{ll} \text{Minimise} & \sum_{i=1}^{n}\sum_{j=1}^{l} m(x_i,y_j)\Gamma(x_i,y_j) \\ \text{subject to} & \bullet\ \forall 1 \le i \le n : \sum_{j=1}^{l} \Gamma(x_i,y_j) = \Delta(x_i) \\ & \bullet\ \forall 1 \le j \le l : \sum_{i=1}^{n} \Gamma(x_i,y_j) = \Theta(y_j) \\ & \bullet\ \forall 1 \le i \le n, 1 \le j \le l : \Gamma(x_i,y_j) \ge 0. \end{array} \tag{3.6}$$

Since (3.6) is a special case of the discrete mass transportation problem, some well-known polynomial time algorithms like [34] can be employed to solve it, which is an attractive feature for computer scientists.

Recall that a pseudometric is a function that yields a nonnegative real number for each pair of elements and satisfies the following: $m(s,s) = 0$, $m(s,t) = m(t,s)$, and $m(s,t) \le m(s,u) + m(u,t)$, for any $s, t \in S$. We say a pseudometric m is 1-bounded if $m(s,t) \le 1$ for any s and t. Let Δ and Θ be distributions over a finite set S of states. In [40] a 1-bounded pseudometric m on S is lifted to be a 1-bounded pseudometric $\hat{m}$ on $\mathcal{D}(S)$ by setting the distance $\hat{m}(\Delta,\Theta)$ to be the value of the following linear programming problem:

$$\begin{array}{ll} \text{Maximise} & \sum_{s \in S} (\Delta(s) - \Theta(s))x_s \\ \text{subject to} & \bullet\ \forall s,t \in S : x_s - x_t \leq m(s,t) \\ & \bullet\ \forall s \in S : 0 \leq x_s \leq 1. \end{array} \tag{3.7}$$

This problem can be dualised and then simplified to yield the following problem:

$$\begin{array}{ll} \text{Minimise} & \sum_{s,t \in S} m(s,t) y_{st} \\ \text{subject to} & \bullet\ \forall s \in S : \sum_{t \in S} y_{st} = \Delta(s) \\ & \bullet\ \forall t \in S : \sum_{s \in S} y_{st} = \Theta(t) \\ & \bullet\ \forall s,t \in S : y_{st} \geq 0. \end{array} \tag{3.8}$$

Now (3.8) is in exactly the same form as (3.6).

This way of lifting pseudometrics via the Kantorovich metric as given in (3.8) has an interesting connection with the lifting of binary relations given in Definition 3.2.

Theorem 3.3 *Let $\mathcal{R}$ be a binary relation and m a pseudometric on a state space S satisfying*

$$s\mathcal{R}t \quad \textit{iff} \quad m(s,t) = 0 \tag{3.9}$$

for any $s, t \in S$. Then it holds that

$$\Delta \mathcal{R}^{\dagger} \Theta \quad \textit{iff} \quad \hat{m}(\Delta, \Theta) = 0$$

for any distributions $\Delta, \Theta \in \mathcal{D}(S)$.

Proof Suppose $\Delta \mathcal{R}^{\dagger} \Theta$. From Theorem 3.1(1) we know there is a weight function w such that

1. $\forall s \in S : \sum_{t \in S} w(s,t) = \Delta(s)$.
2. $\forall t \in S : \sum_{s \in S} w(s,t) = \Theta(t)$.
3. $\forall s,t \in S : w(s,t) > 0 \Rightarrow s\mathcal{R}t$.

By substituting $w(s,t)$ for $y_{s,t}$ in (3.8), the three constraints there can be satisfied. For any $s, t \in S$ we distinguish two cases:

1. Either $w(s,t) = 0$;
2. or $w(s,t) > 0$. In this case we have $s\mathcal{R}t$, which implies $m(s,t) = 0$ by (3.9).

Therefore, we always have $m(s,t)w(s,t) = 0$ for any $s, t \in S$. Consequently, we get $\sum_{s,t \in S} m(s,t)w(s,t) = 0$ and the optimal value of the problem in (3.8) must be 0, i.e. $\hat{m}(\Delta, \Theta) = 0$, and the optimal solution is determined by w.

The above reasoning can be reversed to show that the optimal solution of (3.8) determines a weight function, thus $\hat{m}(\Delta, \Theta) = 0$ implies $\Delta \mathcal{R}^{\dagger} \Theta$. □

The above property will be used in Sect. 3.7 to give a metric characterisation of probabilistic bisimulation (cf. Theorem 3.11).

In the remainder of this subsection we do a sanity check and show that (3.7) and (3.8) do entail the same metric on $\mathcal{D}(S)$. Given a metric m on S, we write $m^\star$ for the lifted metric by using the linear programming problem in (3.8).

Proposition 3.5 *Let m be a metric over S. Then $\hat{m}$ is a metric over $\mathcal{D}(S)$.*

Proof We verify that $(\mathcal{D}(S), \hat{m})$ satisfies the definition of pseudometric space.

1. It is clear that $\hat{m}(\Delta, \Delta) = 0$.
2. We observe that

$$\sum_{s\in S}(\Delta(s) - \Delta'(s))x_s = \sum_{s\in S}(\Delta'(s) - \Delta(s))(1 - x_s) + \sum_{s\in S}\Delta(s) - \sum_{s\in S}\Delta'(s)$$
$$= \sum_{s\in S}(\Delta'(s) - \Delta(s))(1 - x_s).$$

Now $x'_s = 1 - x_s$ also satisfy the constraints on x_s in (3.7), hence the symmetry of $\hat{m}$ can be shown.
3. Let $\Delta_1, \Delta_2, \Delta_3 \in \mathcal{D}(S)$, we have

$$\sum_{s\in S}(\Delta_1(s) - \Delta_3(s))x_s = \sum_{s\in S}(\Delta_1(s) - \Delta_2(s))x_s + \sum_{s\in S}(\Delta_2(s) - \Delta_3(s))x_s.$$

By taking the maximum over the x_s for the left hand side, we obtain

$$\hat{m}(\Delta_1, \Delta_3) \le \hat{m}(\Delta_1, \Delta_2) + \hat{m}(\Delta_2, \Delta_3).$$

□

Proposition 3.6 *Let m be a metric over S. Then $m^\star$ is a metric over $\mathcal{D}(S)$.*

Proof We verify that $(\mathcal{D}(S), m^\star)$ satisfies the definition of pseudometric space.

1. It is easy to see that $m^\star(\Delta, \Delta) = 0$, by letting $y_{s,s} = \Delta(s)$ and $y_{s,t} = 0$ for all $s \neq t$ in (3.8).
2. Suppose $m^\star(\Delta, \Delta')$ is obtained by using some real numbers $y_{s,t}$ in (3.8). Let $y'_{s,t} = y_{t,s}$ for all $s, t \in S$. We have that

$$\sum_{s,t\in S} m(s,t) \cdot y_{s,t} = \sum_{s,t\in S} m(s,t) \cdot y'_{t,s} = \sum_{t,s\in S} m(t,s) \cdot y'_{t,s}.$$

Observe that $\sum_{s\in S} y'_{t,s} = \sum_{s\in S} y_{s,t} = \Delta'(t)$ for all $t \in S$, and similarly we have the dual $\sum_{t\in S} y'_{t,s} = \Delta(s)$ for all $s \in S$. It follows that $m^\star(\Delta, \Delta') = m^\star(\Delta', \Delta)$.
3. Suppose $m^\star(\Delta_1, \Delta_2)$ is obtained by using some real numbers $x_{s,t}$ in (3.8), and $m^\star(\Delta_2, \Delta_3)$ is obtained by using $y_{s,t}$. Let $z_{s,t} = \sum_{r\in\lceil\Delta_2\rceil} x_{s,r} \cdot \frac{y_{r,t}}{\Delta_2(r)}$. Note that if $\Delta_2(r) = 0$ for some $r \in S$, then we must have that $x_{s,r} = y_{r,t} = 0$ for any

$s, t \in S$. For any $s \in S$, we have that

$$\begin{aligned}
\sum_{t\in S} z_{s,t} &= \sum_{t\in S}\sum_{r\in\lceil\Delta_2\rceil} x_{s,r}\cdot\frac{y_{r,t}}{\Delta_2(r)}\\
&= \sum_{r\in\lceil\Delta_2\rceil}\frac{x_{s,r}}{\Delta_2(r)}\cdot\left(\sum_{t\in S} y_{r,t}\right)\\
&= \sum_{r\in\lceil\Delta_2\rceil}\frac{x_{s,r}}{\Delta_2(r)}\cdot\Delta_2(r)\\
&= \sum_{r\in\lceil\Delta_2\rceil} x_{s,r}\\
&= \sum_{r\in S} x_{s,r}\\
&= \Delta_1(s)
\end{aligned}$$

and for any $t \in S$, we have that

$$\begin{aligned}
\sum_{s\in S} z_{s,t} &= \sum_{s\in S}\sum_{r\in\lceil\Delta_2\rceil} x_{s,r}\cdot\frac{y_{r,t}}{\Delta_2(r)}\\
&= \sum_{r\in\lceil\Delta_2\rceil}\sum_{s\in S} x_{s,r}\cdot\frac{y_{r,t}}{\Delta_2(r)}\\
&= \sum_{r\in\lceil\Delta_2\rceil}\left(\sum_{s\in S} x_{s,r}\right)\cdot\frac{y_{r,t}}{\Delta_2(r)}\\
&= \sum_{r\in\lceil\Delta_2\rceil}\Delta_2(r)\cdot\frac{y_{r,t}}{\Delta_2(r)}\\
&= \sum_{r\in\lceil\Delta_2\rceil} y_{r,t}\\
&= \sum_{r\in S} y_{r,t}\\
&= \Delta_3(t).
\end{aligned}$$

Therefore, the real numbers $z_{s,t}$ satisfy the constraints in (3.8) and we obtain that

$$\begin{aligned}
&m^\star(\Delta_1,\Delta_3)\\
&\le \sum_{s,t\in S} m(s,t)\cdot z_{s,t}\\
&= \sum_{s,t\in S} m(s,t)\cdot\sum_{r\in\lceil\Delta_2\rceil} x_{s,r}\cdot\frac{y_{r,t}}{\Delta_2(r)}\\
&= \sum_{s,t\in S}\sum_{r\in\lceil\Delta_2\rceil} m(s,t)\cdot x_{s,r}\cdot\frac{y_{r,t}}{\Delta_2(r)}\\
&\le \sum_{s,t\in S}\sum_{r\in\lceil\Delta_2\rceil}(m(s,r)+m(r,t))\cdot x_{s,r}\cdot\frac{y_{r,t}}{\Delta_2(r)}\\
&= \sum_{s,t\in S}\sum_{r\in\lceil\Delta_2\rceil} m(s,r)\cdot x_{s,r}\cdot\frac{y_{r,t}}{\Delta_2(r)}+\sum_{s,t\in S}\sum_{r\in\lceil\Delta_2\rceil} m(r,t)\cdot x_{s,r}\cdot\frac{y_{r,t}}{\Delta_2(r)}\\
&= \sum_{s\in S}\sum_{r\in\lceil\Delta_2\rceil} m(s,r)\cdot x_{s,r}\cdot\frac{\sum_{t\in S} y_{r,t}}{\Delta_2(r)}+\sum_{s,t\in S}\sum_{r\in\lceil\Delta_2\rceil} m(r,t)\cdot x_{s,r}\cdot\frac{y_{r,t}}{\Delta_2(r)}\\
&= \sum_{s\in S}\sum_{r\in\lceil\Delta_2\rceil} m(s,r)\cdot x_{s,r}\cdot\frac{\Delta_2(r)}{\Delta_2(r)}+\sum_{s,t\in S}\sum_{r\in\lceil\Delta_2\rceil} m(r,t)\cdot x_{s,r}\cdot\frac{y_{r,t}}{\Delta_2(r)}\\
&= \sum_{s\in S}\sum_{r\in S} m(s,r)\cdot x_{s,r}+\sum_{s,t\in S}\sum_{r\in\lceil\Delta_2\rceil} m(r,t)\cdot x_{s,r}\cdot\frac{y_{r,t}}{\Delta_2(r)}\\
&= m^\star(\Delta_1,\Delta_2)+\sum_{s,t\in S}\sum_{r\in\lceil\Delta_2\rceil} m(r,t)\cdot x_{s,r}\cdot\frac{y_{r,t}}{\Delta_2(r)}
\end{aligned}$$

$$
\begin{aligned}
&= m^\star(\Delta_1,\Delta_2) + \sum\nolimits_{t\in S}\sum\nolimits_{r\in\lceil\Delta_2\rceil} m(r,t)\cdot(\sum\nolimits_{s\in S} x_{s,r})\cdot\frac{y_{r,t}}{\Delta_2(r)}\\
&= m^\star(\Delta_1,\Delta_2) + \sum\nolimits_{t\in S}\sum\nolimits_{r\in\lceil\Delta_2\rceil} m(r,t)\cdot\Delta_2(r)\cdot\frac{y_{r,t}}{\Delta_2(r)}\\
&= m^\star(\Delta_1,\Delta_2) + \sum\nolimits_{t\in S}\sum\nolimits_{r\in\lceil\Delta_2\rceil} m(r,t)\cdot y_{r,t}\\
&= m^\star(\Delta_1,\Delta_2) + \sum\nolimits_{t\in S}\sum\nolimits_{r\in S} m(r,t)\cdot y_{r,t}\\
&= m^\star(\Delta_1,\Delta_2) + m^\star(\Delta_2,\Delta_3).
\end{aligned}
$$

□

Proposition 3.7 *$\hat{m}$ coincides with $m^\star$.*

Proof The problem in (3.7) can be rewritten in the following form:

$$
\begin{array}{ll}
\text{Maximise} & \sum_{s\in S}\Delta(s)\cdot x_s - \sum_{s\in S}\Delta'(s)y_s\\
\text{subject to} & \bullet\ \forall s,t\in S: x_s - y_t \le m(s,t)\\
& \bullet\ \forall s\in S: y_s - x_s \le 0\\
& \bullet\ \forall s\in S: x_s\le 1,\ y_s\le 1\\
& \bullet\ \forall s\in S: x_s\ge 0,\ y_s\ge 0.
\end{array}
\tag{3.10}
$$

Dualising (3.10) yields:

$$
\begin{array}{ll}
\text{Minimise} & \sum_{s,t\in S} m(s,t)z_{s,t} + \alpha_s + \beta_s\\
\text{subject to} & \bullet\ \forall t\in S: \sum_{t\in S} z_{s,t} - \alpha_s + \beta_s \ge \Delta(s)\\
& \bullet\ \forall s\in S: \sum_{t\in S} z_{t,s} - \alpha_s - \beta_s \le \Delta'(s)\\
& \bullet\ \forall s,t\in S: z_{s,t}\ge 0,\ \alpha_s\ge 0,\ \beta_s\ge 0.
\end{array}
\tag{3.11}
$$

The duality theorem of linear programming (Theorem 2.10) tells us that the dual problem has the same value as the primal problem. We now observe that in the optimal vector of (3.11) it must be the case that $\alpha_s = \beta_s = 0$ for all $s \in S$. In addition, we have $\sum_{s\in S}\Delta(s) = \sum_{s\in S}\Delta'(s) = 1$. So we can simplify (3.11) to (3.8). □

3.4.2 *Justification by Network Flow*

The lifting operation discussed in Sect. 3.3 is also related to the maximum flow problem in optimisation theory. This was already observed by Baier et al. in [26].

We briefly recall the basic definitions of networks. More details can be found in e.g. [41]. A *network* is a tuple $\mathcal{N} = (N, E, s_\perp, s_\top, c)$ where (N, E) is a finite directed graph (i.e. N is a set of nodes and $E \subseteq N \times N$ is a set of edges) with two special nodes $s_\perp$ (the *source*) and $s_\top$ (the *sink*) and a *capacity* c, i.e. a function that assigns to each edge $(v, w) \in E$ a nonnegative number $c(v, w)$. A *flow function* f for $\mathcal{N}$ is a function that assigns to edge e a real number $f(e)$ such that

- $0 \leq f(e) \leq c(e)$ for all edges e.
- Let $in(v)$ be the set of incoming edges to node v and $out(v)$ the set of outgoing edges from node v. Then, for each node $v \in N \backslash \{s_\perp, s_\top\}$,

$$\sum_{e \in in(v)} f(e) = \sum_{e \in out(v)} f(e).$$

The *flow* $F(f)$ of f is given by

$$F(f) = \sum_{e \in out(s_\perp)} f(e) - \sum_{e \in in(s_\perp)} f(e).$$

The *maximum flow* in $\mathcal{N}$ is the supremum (maximum) over the flows $F(f)$, where f is a flow function in $\mathcal{N}$.

We will see that the question whether $\Delta \ \mathcal{R}^\dagger \ \Theta$ holds can be reduced to a maximum flow problem in a suitably chosen network. Suppose $\mathcal{R} \subseteq S \times S$ and $\Delta, \Theta \in \mathcal{D}(S)$. Let $S' = \{s' \mid s \in S\}$ where s' are pairwise distinct new states, i.e. $(\_)' : S \to S'$ is a bijective function. We create two states $s_\perp$ and $s_\top$ not contained in $S \cup S'$ with $s_\perp \neq s_\top$. We associate with the pair (Δ, Θ) the following network $\mathcal{N}(\Delta, \Theta, \mathcal{R})$.

- The nodes are $N = S \cup S' \cup \{s_\perp, s_\top\}$.
- The edges are $E = \{(s, t') \mid (s, t) \in \mathcal{R}\} \cup \{(s_\perp, s) \mid s \in S\} \cup \{(s', s_\top) \mid s \in S\}$.
- The capability c is defined by $c(s_\perp, s) = \Delta(s)$, $c(t', s_\top) = \Theta(t)$ and $c(s, t') = 1$ for all $s, t \in S$.

The network is depicted in Fig. 3.3.

The next lemma appeared as Lemma 5.1 in [26].

Lemma 3.2 *Let S be a finite set, $\Delta, \Theta \in \mathcal{D}(S)$ and $\mathcal{R} \subseteq S \times S$. The following statements are equivalent.*

(i) There exists a weight function w for (Δ, Θ) with respect to $\mathcal{R}$.
(ii) The maximum flow in $\mathcal{N}(\Delta, \Theta, \mathcal{R})$ is 1. □

Proof (i) $\Longrightarrow$ (ii): Let w be a weight function for (Δ, Θ) with respect to $\mathcal{R}$. We define a flow function f as follows: $f(s_\perp, s) = \Delta(s)$, $f(t', s_\top) = \Theta(t)$ and moreover $f(s, t') = w(s, t)$ for all $s, t \in S$. Then $F(f) = \sum_{s \in S} f(s_\perp, s) = \sum_{s \in S} \Delta(s) = 1$. For each outgoing edge $(s_\perp, s)$ from node $s_\perp$, its maximum capacity is reached, so the maximum flow of $\mathcal{N}(\Delta, \Theta, \mathcal{R})$ is 1.

(ii) $\Longrightarrow$ (i): Let f be a flow function such that $F(f) = 1$. We observe that $f(s_\perp, s) \leq c(s_\perp, s) = \Delta(s)$ and

$$\sum_{s \in S} f(s_\perp, s) = F(f) = 1 = \sum_{s \in S} \Delta(s),$$

so it must be the case that $f(s_\perp, s) = \Delta(s)$ for all $s \in S$. Similarly, we have the dual $f(t', s_\top) = \Theta(t)$ for all $t \in S$. Let w be the weight function defined by letting

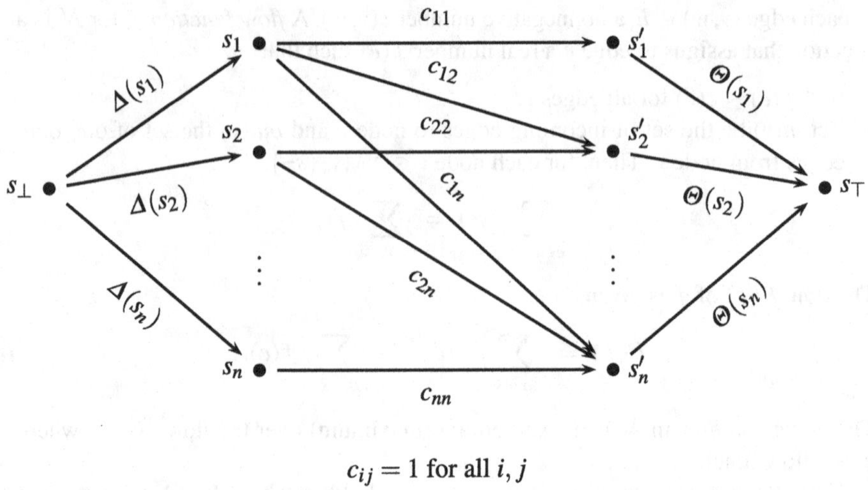

Fig. 3.3 The network $\mathcal{N}(\Delta, \Theta, \mathcal{R})$

$w(s,t) = f(s,t')$ for all $(s,t) \in \mathcal{R}$ and $w(s,t) = 0$ if $(s,t) \notin \mathcal{R}$. We can check that

$$\sum_{t \in S} w(s,t) = \sum_{t \in S} f(s,t') = f(s_\bot, s) = \Delta(s)$$

and similarly, $\sum_{s \in S} w(s,t) = \Theta(t)$. So w is a weight function for (Δ, Θ) with respect to $\mathcal{R}$. □

Since the lifting operation given in Definition 3.2 can also be stated in terms of weight functions, we obtain the following characterisation using network flow.

Theorem 3.4 *Let S be a finite set, $\Delta, \Theta \in \mathcal{D}(S)$ and $\mathcal{R} \subseteq S \times S$. Then $\Delta \; \mathcal{R}^\dagger \; \Theta$ if and only if the maximum flow in $\mathcal{N}(\Delta, \Theta, \mathcal{R})$ is* 1.

Proof Combine Theorem 3.1(1) and Lemma 3.2. □

The above property will play an important role in Sect. 3.8.2 to give an "on the fly" algorithm for checking probabilistic bisimilarity.

Besides Theorems 3.1 and 3.4, there are other equivalent ways of lifting relations. Given a binary relation $\mathcal{R} \subseteq S \times T$ and a set $A \subseteq S$, we write $\mathcal{R}(A)$ for the set $\{t \in T \mid \exists s \in A : s\mathcal{R}t\}$. A set A is $\mathcal{R}$-closed if $\mathcal{R}(A) \subseteq A$.

Theorem 3.5 *Let Δ and Θ be distributions over finite sets S and T, respectively.*

1. *$\Delta \; \mathcal{R}^\dagger \; \Theta$ if and only if $\Delta(A) \leq \Theta(\mathcal{R}(A))$ for all $A \subseteq S$.*
2. *If $\mathcal{R}$ is a preorder, then $\Delta \; \mathcal{R}^\dagger \; \Theta$ if and only if $\Delta(A) \leq \Theta(A)$ for each $\mathcal{R}$-closed set $A \subseteq S$.*

Proof

1. ($\Rightarrow$) Since $\Delta \ \mathcal{R}^{\dagger} \ \Theta$, by Proposition 3.1 we can decompose Δ and Θ as follows.

$$\Delta = \sum_{i \in I} p_i \cdot \overline{s_i} \qquad s_i \ \mathcal{R} \ t_i \qquad \Theta = \sum_{i \in I} p_i \cdot \overline{t_i}.$$

   Note that $\{s_i\}_{i \in I} = \lceil \Delta \rceil$. We define an index set $J := \{i \in I \mid s_i \in A\}$. Then $\Delta(A) = \{p_j\}_{j \in J}$. For each $j \in J$ we have $s_j \ \mathcal{R} \ t_j$, i.e. $t_j \in \mathcal{R}(s_j)$. It follows that $\{t_j\}_{j \in J} \subseteq \mathcal{R}(A)$. Therefore, we can infer that

$$\Delta(A) \ = \ \Delta(\{s_j\}_{j \in J}) \ = \sum_{j \in J} p_j \ \leq \ \Theta(\{t_j\}_{j \in J}) \ \leq \ \Theta(\mathcal{R}(A)).$$

   ($\Leftarrow$) In view of Theorem 3.4 it suffices to show that the maximum flow of the network $\mathcal{N}(\Delta, \Theta, \mathcal{R})$ is 1. According to the maximum flow minimum cut theorem by Ford and Fulkerson [42], the maximum flow equals the capacity of a minimal cut. We show that $\{s_\perp\}$ is a minimal cut of capacity 1. Clearly this cut has capacity 1. To see the minimality, let $C \neq \{s_\perp\}$ be some minimal cut. The capacity of C is $c(C) = \sum\{c(s, t') \mid s \in C, \ t' \notin C\}$. If the cut C involves an edge of capacity 1, i.e. $(s, t) \in \mathcal{R}$ with $s \in C$ and $t' \notin C$, then $\{s_\perp\}$ is a cut of smaller or equal capacity since its capacity is 1. Let $B = C \cap S$, and thus we can assume that if $s \in B$ then $\{t' \mid t \in \mathcal{R}(s)\} \subseteq C$. Hence the capacity of C is $c(C) = \Delta(S \backslash B) + \Theta(\mathcal{R}(B))$. Since $\Delta(B) \leq \Theta(\mathcal{R}(B))$, we have

$$c(C) \ \geq \ \Delta(S \backslash B) + \Delta(B) \ = \ \Delta(S) \ = \ 1.$$

   Therefore, the capacity of C is greater than or equal to 1, which means that the minimum cut has capacity 1.
2. When $\mathcal{R}$ is a preorder, we can show that the following two conditions are equivalent: (i) $\Delta(A) \leq \Theta(\mathcal{R}(A))$ for all $A \subseteq S$, (ii) $\Delta(A) \leq \Theta(A)$ for each $\mathcal{R}$-closed set $A \subseteq S$. For one direction, suppose (i) holds and A is a $\mathcal{R}$-closed set. Then $\mathcal{R}(A) \subseteq A$, and thus $\Theta(\mathcal{R}(A)) \leq \Theta(A)$. Combining this with (i) we see that $\Delta(A) \leq \Theta(A)$. For the other direction, suppose (ii) holds and pick up any $A \subseteq S$. Since $\mathcal{R}$ is reflexive, we have $A \subseteq \mathcal{R}(A)$, and thus $\Delta(A) \leq \Delta(\mathcal{R}(A))$. By the transitivity of $\mathcal{R}$ we know that $\mathcal{R}(A)$ is $\mathcal{R}$-closed. By (ii) we get $\Delta(\mathcal{R}(A)) \leq \Theta(\mathcal{R}(A))$. Combining the previous two inequalities, we obtain $\Delta(A) \leq \Theta(\mathcal{R}(A))$. □

Remark 3.1 Note that in Theorem 3.5(2) it is important to require $\mathcal{R}$ to be a preorder, which is used in showing the equivalence of the two conditions (i) and (ii) in the above proof. If $\mathcal{R}$ is not a preorder, the implication from (ii) to (i) is invalid in general. For example, let $S = \{s, t\}$, $\mathcal{R} = \{(s, t)\}$, $\Delta = \frac{1}{2}\overline{s} + \frac{1}{2}\overline{t}$ and $\Theta = \frac{1}{3}\overline{s} + \frac{2}{3}\overline{t}$. There are only two nonempty $\mathcal{R}$-closed sets: $\{t\}$ and S. Clearly, we have both $\Delta(\{t\}) \leq \Theta(\{t\})$ and $\Delta(S) \leq \Theta(\{t\})$. However,

$$\Delta(\{t\}) = \frac{1}{2} \not\leq 0 = \Theta(\emptyset) = \Theta(\mathcal{R}(\{t\})).$$

Note also that Theorem 3.5 can be generalised to countable state spaces. The proof is almost the same except that the maximum flow minimum cut theorem for countable networks [43] is used. See [44] for more details.

3.5 Probabilistic Bisimulation

With a solid base of the lifting operation, we can proceed to define a probabilistic version of bisimulation.

Let s and t be two states in a pLTS. We say t can simulate the behaviour of s if whenever the latter can exhibit action a and lead to distribution Δ then the former can also perform a and lead to a distribution, say Θ, which then in turn can mimic Δ in successor states. We are interested in a relation between two states, but it is expressed by invoking a relation between two distributions. To formalise the mimicking of one distribution by the other, we make use of the lifting operation investigated in Sect. 3.3.

Definition 3.5 A relation $\mathcal{R} \subseteq S \times S$ is a *probabilistic simulation* if $s \ \mathcal{R}\ t$ implies

- if $s \xrightarrow{a} \Delta$ then there exists some Θ such that $t \xrightarrow{a} \Theta$ and $\Delta \mathcal{R}^\dagger \Theta$.

If both $\mathcal{R}$ and $\mathcal{R}^{-1}$ are probabilistic simulations, then $\mathcal{R}$ is a *probabilistic bisimulation*. The largest probabilistic bisimulation, denoted by $\sim$, is called *probabilistic bisimilarity*.

As in the nonprobabilistic setting, probabilistic bisimilarity can be approximated by a family of inductively defined relations.

Definition 3.6 Let S be the state set of a pLTS. We define:

- $\sim_0 := S \times S$
- $s \sim_{n+1} t$, for $n \geq 0$, if
  1. whenever $s \xrightarrow{a} \Delta$, there exists some Θ such that $t \xrightarrow{a} \Theta$ and $\Delta\ (\sim_n)^\dagger\ \Theta$;
  2. whenever $t \xrightarrow{a} \Theta$, there exists some Δ such that $s \xrightarrow{a} \Delta$ and $\Delta\ (\sim_n)^\dagger\ \Theta$.
- $\sim_\omega := \bigcap_{n\geq 0} \sim_n$

In general, $\sim$ is a strictly finer relation than $\sim_\omega$. However, the two relations coincide when limited to *image-finite* pLTSs where for any state s and action a, the set $\{\Delta \in \mathcal{D}(S) \mid s \xrightarrow{a} \Delta\}$ is finite.

Proposition 3.8 *On image-finite pLTSs,* $\sim_\omega$ *coincides with* $\sim$.

Proof It is trivial to show by induction that $s \sim t$ implies $s \sim_n t$ for all $n \geq 0$, and thus that $s \sim_\omega t$.

Now we show that $\sim_\omega$ is a bisimulation. Suppose $s \sim_\omega t$ and $s \xrightarrow{a} \Delta$. By assumption for all $n \geq 0$ there exists some Θ_n with $t \xrightarrow{a} \Theta_n$ and $\Delta\ (\sim_n)^\dagger\ \Theta_n$. Since we are considering image-finite pLTSs, there are only finitely many different Θ_n's. Then for at least one of them, say Θ, we have $\Delta\ (\sim_n)^\dagger\ \Theta$ for infinitely many different n's. By a straightforward induction we can show that $s \sim_n t$ implies $s \sim_m t$ for all $m, n \geq 0$ with $n > m$. It follows that $\Delta\ (\sim_n)^\dagger\ \Theta$ for all $n \geq 0$. We now claim that

$$\Delta\ (\sim_\omega)^\dagger\ \Theta \tag{3.12}$$

To see this, let A be any subset of the whole state space that may be countable. For any $n \geq 0$, since $\Delta\ (\sim_n)^\dagger\ \Theta$ we know from Theorem 3.5(1) that $\Delta(A) \leq \Theta(\sim_n(A))$.

Therefore,

$$\Delta(A) \leq \inf_{n\geq 0} \Theta(\sim_n (A)) = \Theta((\bigcap_{n\geq 0} \sim_n)(A)) = \Theta(\sim_\omega (A)).$$

Using Theorem 3.5(1) again, we obtain (3.12).

By symmetry we also have that if $t \xrightarrow{a} \Theta$ then there is some Δ with $s \xrightarrow{a} \Delta$ and $\Delta\ (\sim_\omega)^\dagger\ \Theta$. □

Let $\prec$ be the largest probabilistic simulation, called probabilistic similarity, and $\asymp$ be the kernel of probabilistic similarity, i.e. $\prec \cap \prec^{-1}$, called *simulation equivalence*. In general, simulation equivalence is coarser than bisimilarity. However, for reactive pLTSs, the two relations do coincide. Recall that in a reactive pLTS, for each state s and action a there is at most one distribution Δ with $s \xrightarrow{a} \Delta$. To prove that result, we need a technical lemma.

Lemma 3.3 *Let $\mathcal{R}$ be a preorder on a set S and $\Delta, \Theta \in \mathcal{D}(S)$. If $\Delta\ \mathcal{R}^\dagger\ \Theta$ and $\Theta\ \mathcal{R}^\dagger\ \Delta$ then $\Delta(C) = \Theta(C)$ for all equivalence classes C with respect to the kernel $\mathcal{R} \cap \mathcal{R}^{-1}$ of $\mathcal{R}$.*

Proof Let us write $\equiv$ for $\mathcal{R} \cap \mathcal{R}^{-1}$. For any $s \in S$, let $[s]_\equiv$ be the equivalence class that contains s. Let A_s be the set $\{t \in S \mid s\ \mathcal{R}\ t\ \wedge\ t\ \not\mathcal{R}\ s\}$. It is easy to see that

$$\begin{aligned} \mathcal{R}(s) &= \{t \in S \mid s\ \mathcal{R}\ t\} \\ &= \{t \in S \mid s\ \mathcal{R}\ t\ \wedge\ t\ \mathcal{R}\ s\} \uplus \{t \in S \mid s\ \mathcal{R}\ t\ \wedge\ t\ \not\mathcal{R}\ s\} \\ &= [s]_\equiv\ \uplus\ A_s \end{aligned}$$

where $\uplus$ stands for a disjoint union. Therefore, we have

$$\Delta(\mathcal{R}(s)) = \Delta([s]_\equiv) + \Delta(A_s) \qquad \text{and} \qquad \Theta(\mathcal{R}(s)) = \Theta([s]_\equiv) + \Theta(A_s). \quad (3.13)$$

We now check that both $\mathcal{R}(s)$ and A_s are $\mathcal{R}$-closed sets, that is $\mathcal{R}(\mathcal{R}(s)) \subseteq \mathcal{R}(s)$ and $\mathcal{R}(A_s) \subseteq A_s$. Suppose $u \in \mathcal{R}(\mathcal{R}(s))$. Then there exists some $t \in \mathcal{R}(s)$ such that $t\ \mathcal{R}\ u$, which means that $s\ \mathcal{R}\ t$ and $t\ \mathcal{R}\ u$. As a preorder $\mathcal{R}$ is a transitive relation. So we have $s\ \mathcal{R}\ u$ that implies $u \in \mathcal{R}(s)$. Therefore we can conclude that $\mathcal{R}(\mathcal{R}(s)) \subseteq \mathcal{R}(s)$.

Suppose $u \in \mathcal{R}(A_s)$. Then there exists some $t \in A_s$ such that $t\ \mathcal{R}\ u$, which means that $s\ \mathcal{R}\ t$, $t\ \not\mathcal{R}\ s$ and $t\ \mathcal{R}\ u$. As a preorder $\mathcal{R}$ is a transitive relation. So we have $s\ \mathcal{R}\ u$. Note that we also have $u\ \not\mathcal{R}\ s$. Otherwise we would have $u\ \mathcal{R}\ s$, which means, together with $t\ \mathcal{R}\ u$ and the transitivity of $\mathcal{R}$, that $t\ \mathcal{R}\ s$, a contradiction to the hypothesis $t\ \not\mathcal{R}\ s$. It then follows that $u \in A_s$ and then we conclude that $\mathcal{R}(A_s) \subseteq A_s$.

We have verified that $\mathcal{R}(s)$ and A_s are $\mathcal{R}$-closed sets. Since $\Delta\ \mathcal{R}^\dagger\ \Theta$ and $\Theta\ \mathcal{R}^\dagger\ \Delta$, we apply Theorem 3.5(2) and obtain that $\Delta(\mathcal{R}(s)) \leq \Theta(\mathcal{R}(s))$ and $\Theta(\mathcal{R}(s)) \leq \Delta(\mathcal{R}(s))$, that is

$$\Delta(\mathcal{R}(s)) = \Theta(\mathcal{R}(s)). \quad (3.14)$$

Similarly, using the fact that A_s is $\mathcal{R}$-closed we obtain that

$$\Delta(A_s) = \Theta(A_s) \quad (3.15)$$

It follows from (3.13)–(3.15) that

$$\Delta([s]_{\equiv}) = \Theta([s]_{\equiv}).$$

as we have desired. □

Remark 3.2 Note that in the above proof the equivalence classes $[s]_{\equiv}$ are not necessarily $\mathcal{R}$-closed. For example, let $S = \{s,t\}$, $Id_S = \{(s,s), (t,t)\}$ and the relation $\mathcal{R} = Id_S \cup \{(s,t)\}$. The kernel of $\mathcal{R}$ is $\equiv\, = \mathcal{R} \cap \mathcal{R}^{-1} = Id_S$ and then $[s]_{\equiv} = \{s\}$. We have $\mathcal{R}(s) = S \not\subseteq [s]_{\equiv}$. So a more direct attempt to apply Theorem 3.5(2) to those equivalence classes would not work.

Theorem 3.6 *For reactive pLTSs, $\asymp$ coincides with $\sim$.*

Proof It is obvious that $\sim$ is included in $\asymp$. For the other direction, we show that $\asymp$ is a bisimulation. Let $s,t \in S$ and $s \asymp t$. Suppose that $s \xrightarrow{a} \Delta$. There exists a transition $t \xrightarrow{a} \Theta$ with $\Delta\ (\prec)^{\dagger}\ \Theta$. Since we are considering reactive probabilistic systems, the transition $t \xrightarrow{a} \Theta$ from t must be matched by the unique outgoing transition $s \xrightarrow{a} \Delta$ from s, with $\Theta\ (\prec)^{\dagger}\ \Delta$. Note that by using Proposition 3.4 it is easy to show that $\prec$ is a preorder on S. It follows from Lemma 3.3 that $\Delta(C) = \Theta(C)$ for any $C \in S/\asymp$. By Theorem 3.1(2) we see that $\Delta\ (\asymp)^{\dagger}\ \Theta$. Therefore, $\asymp$ is indeed a probabilistic bisimulation relation. □

3.6 Logical Characterisation

Let $\mathcal{L}$ be a logic and $\langle S, L, \rightarrow \rangle$ be a pLTS. For any $s \in S$, we use the notation $\mathcal{L}(s)$ to stand for the set of formulae that state s satisfies. This induces an equivalence relation on states: $s \ =^{\mathcal{L}}\ t$ iff $\mathcal{L}(s) = \mathcal{L}(t)$. Thus, two states are equivalent when they satisfy exactly the same set of formulae.

In this section we consider two kinds of logical characterisations of probabilistic bisimilarity.

Definition 3.7 (Adequacy and Expressivity)

1. $\mathcal{L}$ is *adequate* with respect to $\sim$ if for any states s and t,

$$s =^{\mathcal{L}} t \text{ iff } s \sim t.$$

2. $\mathcal{L}$ is *expressive* with respect to $\sim$ if for each state s there exists a *characteristic formula* $\varphi_s \in \mathcal{L}$ such that, for any state t,

$$t \text{ satisfies } \varphi_s \text{ iff } s \sim t.$$

We will propose a probabilistic extension of the Hennessy–Milner logic, showing its adequacy, and then a probabilistic extension of the modal mu-calculus, showing its expressivity. In general the latter is more expressive than the former because it has fixed-point operators to describe infinite behaviour. But for finite processes where no infinite behaviour occurs, an adequate logic will also be expressive for those processes.

3.6.1 An Adequate Logic

We extend the Hennessy–Milner logic by adding a probabilistic-choice modality to express the behaviour of distributions.

Definition 3.8 The class $\mathcal{L}_{\mathsf{hm}}$ of modal formulae over L, ranged over by φ, is defined by the following grammar:

$$\varphi := \top \mid \varphi_1 \wedge \varphi_2 \mid \neg\varphi \mid \langle a \rangle \psi$$
$$\psi := \varphi_1 \,{}_p\!\oplus \varphi_2$$

where $p \in [0,1]$. We call φ a *state formula* and ψ a *distribution formula*. Note that a distribution formula ψ only appears as the continuation of a diamond modality $\langle a \rangle \psi$. We sometimes use the finite conjunction $\bigwedge_{i \in I} \varphi_i$ as syntactic sugar.

The *satisfaction relation* $\models \subseteq S \times \mathcal{L}_{\mathsf{hm}}$ is defined by

- $s \models \top$ for all $s \in S$.
- $s \models \varphi_1 \wedge \varphi_2$ if $s \models \varphi_i$ for $i = 1, 2$.
- $s \models \neg\varphi$ if it is not the case that $s \models \varphi$.
- $s \models \langle a \rangle \psi$ if for some $\Delta \in \mathcal{D}(S)$, $s \xrightarrow{a} \Delta$ and $\Delta \models \psi$.
- $\Delta \models \varphi_1 \,{}_p\!\oplus \varphi_2$ if there are $\Delta_1, \Delta_2 \in \mathcal{D}(S)$ with $\Delta = p \cdot \Delta_1 + (1-p) \cdot \Delta_2$ and for each $i = 1, 2,\ t \in \lceil \Delta_i \rceil$ we have $t \models \varphi_i$.

With a slight abuse of notation, we write $\Delta \models \psi$ above to mean that Δ satisfies the distribution formula ψ. For any $\varphi \in \mathcal{L}_{\mathsf{hm}}$, the set of states that satisfies φ is denoted by $[\![\varphi]\!]$, i.e. $[\![\varphi]\!] = \{s \in S \mid s \models \varphi\}$.

Example 3.1 Let Δ be a distribution and φ be a state formula. We consider the distribution formula

$$\psi := \varphi \,{}_p\!\oplus \top$$

where $p = \Delta([\![\varphi]\!])$. It is not difficult to see that $\Delta \models \psi$ holds. In the case $p = 0$ the formula ψ degenerates to $\top$. In the case $p > 0$ the distribution Δ can be decomposed as follows.

$$\Delta = \sum_{s \in S} \Delta(s) \cdot \overline{s} = p \cdot \sum_{s \in [\![\varphi]\!]} \frac{\Delta(s)}{p} \cdot \overline{s} + (1-p) \cdot \sum_{s \notin [\![\varphi]\!]} \frac{\Delta(s)}{1-p} \cdot \overline{s}.$$

So Δ can be written as the linear combination of two distributions; the first one has as support all the states that can satisfy φ. In general, for any $\Theta \in \mathcal{D}(S)$ we have $\Theta \models \psi$ if and only if $\Theta([\![\varphi]\!]) \geq p$.

It turns out that $\mathcal{L}_{\mathsf{hm}}$ is adequate with respect to probabilistic bisimilarity.

Theorem 3.7 (Adequacy) *Let s and t be any two states in a finite-state pLTS. Then $s \sim t$ if and only if $s =^{\mathcal{L}_{hm}} t$.*

Proof ($\Rightarrow$) Suppose $s \sim t$. We show that $s \models \varphi \Leftrightarrow t \models \varphi$ by structural induction on φ.

- Let $s \models \top$. Then we clearly have $t \models \top$.
- Let $s \models \varphi_1 \wedge \varphi_2$. Then $s \models \varphi_i$ for $i = 1, 2$. So by induction $t \models \varphi_i$, and we have $t \models \varphi_1 \wedge \varphi_2$. By symmetry we also have $t \models \varphi_1 \wedge \varphi_2$ implies $s \models \varphi_1 \wedge \varphi_2$.
- Let $s \models \neg\varphi$. So $s \not\models \varphi$, and by induction we have $t \not\models \varphi$. Thus $t \models \neg\varphi$. By symmetry we also have $t \not\models \varphi$ implies $s \not\models \varphi$.
- Let $s \models \langle a \rangle(\varphi_1\,{}_p\!\oplus \varphi_2)$. Then $s \xrightarrow{a} \Delta$ and $\Delta \models \varphi_1\,{}_p\!\oplus \varphi_2$ for some Δ. So we have $\Delta = p \cdot \Delta_1 + (1-p) \cdot \Delta_2$ and for all $i = 1, 2$ and $s' \in \lceil \Delta_i \rceil$ we have $s' \models \varphi_i$. Since $s \sim t$, there is some Θ with $t \xrightarrow{a} \Theta$ and $\Delta \sim^\dagger \Theta$. By Proposition 3.3 we have that $\Theta = p \cdot \Theta_1 + (1-p) \cdot \Theta_2$ and $\Delta_i \sim^\dagger \Theta_i$ for $i = 1, 2$. It follows that for each $t' \in \lceil \Theta_i \rceil$ there is some $s' \in \lceil \Delta_i \rceil$ with $s' \sim t'$. So by induction we have $t' \models \varphi_i$ for all $t' \in \lceil \Theta_i \rceil$ with $i = 1, 2$. Therefore, we have $\Theta \models \varphi_1\,{}_p\!\oplus \varphi_2$. It follows that $t \models \langle a \rangle(\varphi_1\,{}_p\!\oplus \varphi_2)$. By symmetry we also have $t \models \langle a \rangle(\varphi_1\,{}_p\!\oplus \varphi_2)$ implies $s \models \langle a \rangle(\varphi_1\,{}_p\!\oplus \varphi_2)$.

($\Leftarrow$) We show that the relation $=^{\mathcal{L}_{\mathsf{hm}}}$ is a probabilistic bisimulation. Obviously $=^{\mathcal{L}_{\mathsf{hm}}}$ is an equivalence relation. Let $E = \{U_i \mid i \in I\}$ be the set of all equivalence classes of $=^{\mathcal{L}_{\mathsf{hm}}}$. We first claim that, for any equivalence class U_i, there exists a formula φ_i satisfying $[\![\varphi_i]\!] = U_i$. This can be proved as follows:

- If E contains only one equivalence class U_1, then $U_1 = S$. So we can take the required formula to be $\top$ because $[\![\top]\!] = S$.
- If E contains more than one equivalence class, then for any $i, j \in I$ with $i \neq j$, there exists a formula φ_{ij} such that $s_i \models \varphi_{ij}$ and $s_j \not\models \varphi_{ij}$ for any $s_i \in U_i$ and $s_j \in U_j$. Otherwise, for any formula φ, $s_i \models \varphi$ implies $s_j \models \varphi$. Since the negation exists in the logic $\mathcal{L}_{\mathsf{hm}}$, we also have $s_i \models \neg\varphi$ implies $s_j \models \neg\varphi$, which means $s_j \models \varphi$ implies $s_i \models \varphi$. Then $s_i \models \varphi \Leftrightarrow s_j \models \varphi$ for any $\varphi \in \mathcal{L}_{\mathsf{hm}}$, which contradicts the fact that s_i and s_j are taken from different equivalence classes. For each index $i \in I$, define $\varphi_i = \bigwedge_{j \neq i} \varphi_{ij}$, then by construction $[\![\varphi_i]\!] = U_i$. Let us check the last equality. On one hand, if $s' \in [\![\varphi_i]\!]$, then $s' \models \varphi_i$ which means that $s' \models \varphi_{ij}$ for all $j \neq i$. That is, $s' \notin U_j$ for all $j \neq i$, and this in turn implies that $s' \in U_i$. On the other hand, if $s' \in U_i$ then $s' \models \varphi_i$ as $s_i \models \varphi_i$, which means that $s' \in [\![\varphi_i]\!]$.

This completes the proof of the claim that for each equivalence U_i we can find a formula φ_i with $[\![\varphi_i]\!] = U_i$.

Now let $s =^{\mathcal{L}_{\mathsf{hm}}} t$ and $s \xrightarrow{a} \Delta$. By Theorem 3.1(2) it remains to show that $t \xrightarrow{a} \Theta$ for some Θ with

$$\Delta(U_i) = \Theta(U_i) \text{ for any } i \in I. \tag{3.16}$$

Let

$$\varphi := \langle a \rangle \bigwedge_{i \in I} (\varphi_i \,{}_{p_i}\!\oplus \top)$$

where $p_i = \Delta([\![\varphi_i]\!])$. It is easy to see that $s \models \varphi$, which implies that $t \models \varphi$. Therefore, there exists a distribution Θ with $t \xrightarrow{a} \Theta$ and $\Theta \models \bigwedge_{i \in I} (\varphi_i \,{}_{p_i}\!\oplus \top)$. Then for each $i \in I$ we have $\Theta \models \varphi_i \,{}_{p_i}\!\oplus \top$, implying that $\Theta(U_i) = \Theta([\![\varphi_i]\!]) \geq p_i = \Delta([\![\varphi_i]\!]) = \Delta(U_i)$. Note that $\sum_{i \in I} p_i = 1$. Thus we have $\Delta(U_i) = \Theta(U_i)$ for each $i \in I$, the goal set in (3.16).

By symmetry each transition of t can be matched by some transition of s. □

The above theorem still holds if the pLTS has a countable state space but is image-finite. The proof is a bit subtle; see [45] for more details.

When restricted to reactive pLTSs, probabilistic bisimulations can be characterised by simpler forms of logics [23, 46, 47] or a simple test language [48]. Most notably, in the absence of nondeterminism, there is no need of negation to characterise probabilistic bisimulations.

Let us fix a reactive pLTS $\langle S, L, \rightarrow \rangle$ where the state space S may be countable. Let $\mathcal{L}_{\mathsf{rhm}}$ be the sublogic of $\mathcal{L}_{\mathsf{hm}}$ generated by the following grammar:

$$\varphi \;:=\; \top \mid \varphi_1 \wedge \varphi_2 \mid \langle a \rangle (\varphi_1 \,{}_{p}\!\oplus \top)$$

where p is a *rational number* in the unit interval $[0, 1]$. Recall that $s \models \langle a \rangle (\varphi \,{}_{p}\!\oplus \top)$ iff $s \xrightarrow{a} \Delta$ and $\Delta([\![\varphi]\!]) \geq p$ (cf. Example 3.1). The logic above induces a logical equivalence relation $=^{\mathcal{L}_{\mathsf{rhm}}}$ between states.

The following lemma says that the transition probabilities to sets of the form $[\![\varphi]\!]$ are completely determined by the formulae. It has appeared as Lemma 7.7.6 in [49].

Lemma 3.4 *If $s =^{\mathcal{L}_{\mathsf{rhm}}} t$ and $s \xrightarrow{a} \Delta$, then some Θ exists with $t \xrightarrow{a} \Theta$, and for any formula $\varphi \in \mathcal{L}_{\mathsf{rhm}}$ we have $\Delta([\![\varphi]\!]) = \Theta([\![\varphi]\!])$.*

Proof First of all, the existence of Θ is obvious because otherwise the formula $\langle a \rangle (\top \,{}_{1}\!\oplus \top)$ would be satisfied by s but not by t.

Let us assume, without loss of generality, that there exists a formula φ such that $\Delta([\![\varphi]\!]) < \Theta([\![\varphi]\!])$. Then we can always squeeze in a rational number p with $\Delta([\![\varphi]\!]) < p \leq \Theta([\![\varphi]\!])$. It follows that $t \models \langle a \rangle (\varphi \,{}_{p}\!\oplus \top)$ but $s \not\models \langle a \rangle (\varphi \,{}_{p}\!\oplus \top)$, which contradicts the hypothesis that $s =^{\mathcal{L}_{\mathsf{rhm}}} t$. □

We will show that the logic $\mathcal{L}_{\mathsf{rhm}}$ can characterise bisimulation for reactive pLTSs. The completeness proof of the characterisation crucially relies on the π-λ theorem (cf. Theorem 2.9). The next proposition is a typical application of that theorem [50], which tells us that when two probability distributions agree on a π-class they also agree on the generated σ-algebra.

Proposition 3.9 *Let $\mathcal{A}_0 = \{[\![\varphi]\!] \mid \varphi \in \mathcal{L}_{\mathsf{rhm}}\}$ and $\mathcal{A} = \sigma(\mathcal{A}_0)$. For any two distributions $\Delta, \Theta \in \mathcal{D}(S)$, if $\Delta(A) = \Theta(A)$ for any $A \in \mathcal{A}_0$, then $\Delta(B) = \Theta(B)$ for any $B \in \mathcal{A}$.*

Proof Let $\mathcal{X} = \{A \in \mathcal{A} \mid \Delta(A) = \Theta(A)\}$. Then $\mathcal{X}$ is closed under countable disjoint unions because probability distributions are σ-additive. Since Δ and Θ are distributions, we have $\Delta(S) = \Theta(S) = 1$. It follows that if $A \in \mathcal{X}$ then $\Delta(S\backslash A) = \Delta(S) - \Delta(A) = \Theta(S) - \Theta(A) = \Theta(S\backslash A)$, i.e. $S\backslash A \in \mathcal{X}$. Thus $\mathcal{X}$ is closed under complementation as well. It follows that $\mathcal{X}$ is a λ-class. Note that $\mathcal{A}_0$ is a π-class in view of the equation $[\![\varphi_1 \wedge \varphi_2]\!] = [\![\varphi_1]\!] \cap [\![\varphi_2]\!]$. Since $\mathcal{A}_0 \subseteq \mathcal{X}$, we can apply the π-λ Theorem to obtain that $\mathcal{A} = \sigma(\mathcal{A}_0) \subseteq \mathcal{X} \subseteq \mathcal{A}$, i.e. $\mathcal{A} = \mathcal{X}$. Therefore, $\Delta(B) = \Theta(B)$ for any $B \in \mathcal{A}$. □

The following theorem has appeared in [51], which is obtained by simplifying the results of [46] in reactive pLTSs with countable state spaces.

Theorem 3.8 *Let s and t be two states in a reactive pLTS. Then* $s \sim t$ *iff* $s =^{\mathcal{L}_{rhm}} t$.

Proof The proof of soundness is carried out by a routine induction on the structure of formulae. Below we focus on the completeness. It suffices to show that $=^{\mathcal{L}_{\mathrm{rhm}}}$ is a bisimulation. Note that $=^{\mathcal{L}_{\mathrm{rhm}}}$ is clearly an equivalence relation. For any $u \in S$ the equivalence class in $S/_{=^{\mathcal{L}_{\mathrm{rhm}}}}$ that contains u is

$$[u] = \bigcap\{[\![\varphi]\!] \mid u \models \varphi\} \cap \bigcap\{S\backslash[\![\varphi]\!] \mid u \not\models \varphi\}. \tag{3.17}$$

In (3.17) only countable intersections are used because the set of all the formulae in the logic $\mathcal{L}_{\mathrm{rhm}}$ is countable. Let $\mathcal{A}_0$ be defined as in Proposition 3.9. Then each equivalence class of $S/_{=^{\mathcal{L}_{\mathrm{rhm}}}}$ is a member of $\sigma(\mathcal{A}_0)$.

On the other hand, $s =^{\mathcal{L}_{\mathrm{rhm}}} t$ and $s \xrightarrow{a} \Delta$ implies that some distribution Θ exists with $t \xrightarrow{a} \Theta$ and for any $\varphi \in \mathcal{L}_{\mathrm{rhm}}$, $\Delta([\![\varphi]\!]) = \Theta([\![\varphi]\!])$ by Lemma 3.4. Thus by Proposition 3.9 we have

$$\Delta([u]) = \Theta([u]) \tag{3.18}$$

where $[u]$ is any equivalence class of $S/_{=^{\mathcal{L}_{\mathrm{rhm}}}}$. Then it follows from Theorem 3.1(2) that $\Delta\ (=^{\mathcal{L}_{\mathrm{rhm}}})^{\dagger}\ \Theta$. Symmetrically, any transition of t can be mimicked by a transition from s. Therefore, the relation $=^{\mathcal{L}_{\mathrm{rhm}}}$ is a bisimulation. □

Theorem 3.8 tells us that $\mathcal{L}_{\mathrm{rhm}}$ can characterise bisimulation for reactive pLTSs, and this logic has neither negation nor infinite conjunction. Moreover, the above result holds for general reactive pLTSs that may have countable state space and are not necessarily finitely branching.

3.6.2 An Expressive Logic

In this section we add the probabilistic-choice modality introduced in Sect. 3.6.1 to the modal mu-calculus, and show that the resulting probabilistic mu-calculus is expressive with respect to probabilistic bisimilarity.

$$
\begin{aligned}
[\![\top]\!]_\rho &= S\\
[\![\bot]\!]_\rho &= \emptyset\\
[\![\varphi_1 \wedge \varphi_2]\!]_\rho &= [\![\varphi_1]\!]_\rho \cap [\![\varphi_2]\!]_\rho\\
[\![\varphi_1 \vee \varphi_2]\!]_\rho &= [\![\varphi_1]\!]_\rho \cup [\![\varphi_2]\!]_\rho\\
[\![\langle a\rangle\psi]\!]_\rho &= \{s \in S \mid \exists\Delta : s \xrightarrow{a} \Delta \wedge \Delta \in [\![\psi]\!]_\rho\}\\
[\![[a]\psi]\!]_\rho &= \{s \in S \mid \forall\Delta : s \xrightarrow{a} \Delta \Rightarrow \Delta \in [\![\psi]\!]_\rho\}\\
[\![X]\!]_\rho &= \rho(X)\\
[\![\mu X.\varphi]\!]_\rho &= \bigcap\{V \subseteq S \mid [\![\varphi]\!]_{\rho[X\mapsto V]} \subseteq V\}\\
[\![\nu X.\varphi]\!]_\rho &= \bigcup\{V \subseteq S \mid [\![\varphi]\!]_{\rho[X\mapsto V]} \supseteq V\}\\
[\![\bigoplus_{i\in I} p_i \cdot \varphi_i]\!]_\rho &= \{\Delta \in \mathscr{D}(S) \mid \Delta = \bigoplus_{i\in I} p_i \cdot \Delta_i \wedge \forall i \in I, \forall t \in \lceil\Delta_i\rceil : t \in [\![\varphi_i]\!]_\rho\}
\end{aligned}
$$

Fig. 3.4 Semantics of probabilistic modal mu-calculus

3.6.2.1 Probabilistic Modal Mu-Calculus

Let Var be a countable set of variables. We define a set $\mathcal{L}_\mu$ of modal formulae in positive normal form given by the following grammar:

$$
\begin{aligned}
\varphi &:= \top \mid \bot \mid \langle a\rangle\psi \mid [a]\psi \mid \varphi_1 \wedge \varphi_2 \mid \varphi_1 \vee \varphi_2 \mid X \mid \mu X.\varphi \mid \nu X.\varphi\\
\psi &:= \bigoplus_{i\in I} p_i \cdot \varphi_i
\end{aligned}
$$

where $a \in L$, I is a finite index set and $\sum_{i\in I} p_i = 1$. Here we still write φ for a state formula and ψ a distribution formula. Sometimes we also use the finite conjunction $\bigwedge_{i\in I} \varphi_i$ and disjunction $\bigvee_{i\in I} \varphi_i$. As usual, we have $\bigwedge_{i\in\emptyset} \varphi_i = \top$ and $\bigvee_{i\in\emptyset} \varphi_i = \bot$.

The two fixed-point operators μX and νX bind the respective variable X. We apply the usual terminology of free and bound variables in a formula and write $fv(\varphi)$ for the set of free variables in φ.

We use *environments*, which bind free variables to sets of states, in order to give semantics to formulae. We fix a finitary pLTS and let S be its state set. Let

$$\mathsf{Env} = \{\rho \mid \rho : \mathsf{Var} \rightarrow \mathcal{P}(S)\}$$

be the set of all environments ranged over by ρ. For a set $V \subseteq S$ and a variable $X \in \mathsf{Var}$, we write $\rho[X \mapsto V]$ for the environment that maps X to V and Y to $\rho(Y)$ for all $Y \neq X$.

The semantics of a formula φ can be given as the set of states satisfying it. This entails a semantic function $[\![]\!] : \mathcal{L}_\mu \rightarrow \mathsf{Env} \rightarrow \mathcal{P}(S)$ defined inductively in Fig. 3.4, where we also apply $[\![]\!]$ to distribution formulae and $[\![\psi]\!]$ is interpreted as the set of distributions that satisfy ψ. As the meaning of a closed formula φ does not depend on the environment, we write $[\![\varphi]\!]$ for $[\![\varphi]\!]_\rho$ where ρ is an arbitrary environment.

The semantics of probabilistic modal mu-calculus is the same as that of the modal mu-calculus [31] except for the probabilistic-choice modality that is used to represent decompositions of distributions. The characterisation of *least fixed-point formula* $\mu X.\varphi$ and *greatest fixed-point formula* $\nu X.\varphi$ follows from the well-known Knaster–Tarski fixed-point theorem [52] (cf. Theorem 2.1).

We shall consider (closed) *equation systems* of formulae of the form

$$\begin{aligned} E: X_1 &= \varphi_1 \\ &\vdots \\ X_n &= \varphi_n \end{aligned}$$

where $X_1, \ldots, X_n$ are mutually distinct variables and $\varphi_1, \ldots, \varphi_n$ are formulae having at most $X_1, \ldots, X_n$ as free variables. Here E can be viewed as a function $E : \mathsf{Var} \to \mathcal{L}_\mu$ defined by $E(X_i) = \varphi_i$ for $i = 1, \ldots, n$ and $E(Y) = Y$ for other variables $Y \in \mathsf{Var}$.

An environment ρ is a *solution* of an equation system E if $\forall i : \rho(X_i) = [\![\varphi_i]\!]_\rho$. The existence of solutions for an equation system can be seen from the following arguments. The set Env, which includes all candidates for solutions, together with the partial order $\leq$ defined by

$$\rho \leq \rho' \text{ iff } \forall X \in \mathsf{Var} : \rho(X) \subseteq \rho'(X)$$

forms a complete lattice. The *equation function* $\mathcal{E} : \mathsf{Env} \to \mathsf{Env}$ given in the λ-calculus notation by

$$\mathcal{E} := \lambda\rho.\lambda X.[\![E(X)]\!]_\rho$$

is monotone. Thus, the Knaster–Tarski fixed-point theorem guarantees existence of solutions, and the largest solution

$$\rho_E := \bigsqcup \{\rho \mid \rho \leq \mathcal{E}(\rho)\}$$

3.6.2.2 Characteristic Equation Systems

As studied in [53], the behaviour of a process can be characterised by an equation system of modal formulae. Below we show that this idea also applies in the probabilistic setting.

Definition 3.9 Given a finitary pLTS, its *characteristic equation system* consists of one equation for each state $s_1, \ldots, s_n \in S$.

$$\begin{aligned} E: X_{s_1} &= \varphi_{s_1} \\ &\vdots \\ X_{s_n} &= \varphi_{s_n} \end{aligned}$$

where

$$\varphi_s := \left(\bigwedge_{s \xrightarrow{a} \Delta} \langle a \rangle X_\Delta \right) \wedge \left(\bigwedge_{a \in L} [a] \bigvee_{s \xrightarrow{a} \Delta} X_\Delta \right) \tag{3.19}$$

with $X_\Delta := \bigoplus_{s\in\lceil\Delta\rceil} \Delta(s)\cdot X_s$.

Theorem 3.9 *Suppose E is a characteristic equation system. Then $s \sim t$ if and only if $t \in \rho_E(X_s)$.*

Proof ($\Leftarrow$) Let $\mathcal{R} = \{(s,t) \mid t \in \rho_E(X_s)\}$. We first show that

$$\Theta \in [\![X_\Delta]\!]_{\rho_E} \text{ implies } \Delta\ \mathcal{R}^\dagger\ \Theta. \tag{3.20}$$

Let $\Delta = \bigoplus_{i\in I} p_i \cdot \overline{s_i}$, then $X_\Delta = \bigoplus_{i\in I} p_i \cdot X_{s_i}$. Suppose $\Theta \in [\![X_\Delta]\!]_{\rho_E}$. We have that $\Theta = \bigoplus_{i\in I} p_i \cdot \Theta_i$ and, for all $i \in I$ and $t' \in \lceil\Theta_i\rceil$, that $t' \in [\![X_{s_i}]\!]_{\rho_E}$, i.e. $s_i \mathcal{R} t'$. It follows that $\overline{s_i}\ \mathcal{R}^\dagger\ \Theta_i$ and thus $\Delta\mathcal{R}^\dagger\Theta$.

Now we show that $\mathcal{R}$ is a bisimulation.

1. Suppose $s\mathcal{R}t$ and $s \xrightarrow{a} \Delta$. Then $t \in \rho_E(X_s) = [\![\varphi_s]\!]_{\rho_E}$. It follows from (3.19) that $t \in [\![\langle a\rangle X_\Delta]\!]_{\rho_E}$. So there exists some Θ such that $t \xrightarrow{a} \Theta$ and $\Theta \in [\![X_\Delta]\!]_{\rho_E}$. Now we apply (3.20).
2. Suppose $s\mathcal{R}t$ and $t \xrightarrow{a} \Theta$. Then $t \in \rho_E(X_s) = [\![\varphi_s]\!]_{\rho_E}$. It follows from (3.19) that $t \in [\![[a] \bigvee_{s\xrightarrow{a}\Delta} X_\Delta]\!]$. Notice that it must be the case that s can enable action a, otherwise, $t \in [\![[a]\bot]\!]_{\rho_E}$ and thus t cannot enable a either, in contradiction with the assumption $t \xrightarrow{a} \Theta$. Therefore, $\Theta \in [\![\bigvee_{s\xrightarrow{a}\Delta} X_\Delta]\!]_{\rho_E}$, which implies $\Theta \in [\![X_\Delta]\!]_{\rho_E}$ for some Δ with $s \xrightarrow{a} \Delta$. Now we apply (3.20).

($\Rightarrow$) We define the environment $\rho_\sim$ by

$$\rho_\sim(X_s) := \{t \mid s \sim t\}.$$

It suffices to show that $\rho_\sim$ is a postfixed point of $\mathcal{E}$, i.e.

$$\rho_\sim \le \mathcal{E}(\rho_\sim) \tag{3.21}$$

because in that case we have $\rho_\sim \le \rho_E$, thus $s \sim t$ implies $t \in \rho_\sim(X_s)$ that in turn implies $t \in \rho_E(X_s)$.

We first show that

$$\Delta \sim^\dagger \Theta \text{ implies } \Theta \in [\![X_\Delta]\!]_{\rho_\sim}. \tag{3.22}$$

Suppose that $\Delta \sim^\dagger \Theta$. Using Proposition 3.1 we have that (i) $\Delta = \bigoplus_{i\in I} p_i \cdot \overline{s_i}$, (ii) $\Theta = \bigoplus_{i\in I} p_i \cdot \overline{t_i}$, (iii) $s_i \sim t_i$ for all $i \in I$. We know from (iii) that $t_i \in [\![X_{s_i}]\!]_{\rho_\sim}$. Using (ii) we have that $\Theta \in [\![\bigoplus_{i\in I} p_i \cdot X_{s_i}]\!]_{\rho_\sim}$. Using (i) we obtain $\Theta \in [\![X_\Delta]\!]_{\rho_\sim}$.

Now we are in a position to show (3.21). Suppose $t \in \rho_\sim(X_s)$. We must prove that $t \in [\![\varphi_s]\!]_{\rho_\sim}$, i.e.

$$t \in \Big(\bigcap_{s\xrightarrow{a}\Delta} [\![\langle a\rangle X_\Delta]\!]_{\rho_\sim}\Big) \cap \Big(\bigcap_{a\in L} [\![[a] \bigvee_{s\xrightarrow{a}\Delta} X_\Delta]\!]_{\rho_\sim}\Big)$$

by (3.19). This can be done by showing that t belongs to each of the two parts of this intersection.

1. Rule 1: $E \to F$
2. Rule 2: $E \to G$
3. Rule 3: $E \to H$ if $X_n \notin fv(\varphi_1, ..., \varphi_n)$

$$\begin{array}{cccc} E: X_1 = \varphi_1 & F: X_1 = \varphi_1 & G: X_1 = \varphi_1[\varphi_n/X_n] & H: X_1 = \varphi_1 \\ \vdots & \vdots & \vdots & \vdots \\ X_{n-1} = \varphi_{n-1} & X_{n-1} = \varphi_{n-1} & X_{n-1} = \varphi_{n-1}[\varphi_n/X_n] & X_{n-1} = \varphi_{n-1} \\ X_n = \varphi_n & X_n = \nu X_n.\varphi_n & X_n = \varphi_n & \end{array}$$

Fig. 3.5 Transformation rules

1. In the first case, we assume that $s \xrightarrow{a} \Delta$. Since $s \sim t$, there exists some Θ such that $t \xrightarrow{a} \Theta$ and $\Delta \sim^\dagger \Theta$. By (3.22), we get $\Theta \in [\![X_\Delta]\!]_{\rho_\sim}$. It follows that $t \in [\![\langle a \rangle X_\Delta]\!]_{\rho_\sim}$.
2. In the second case, we suppose $t \xrightarrow{a} \Theta$ for any action $a \in L$ and distribution Θ. Then by $s \sim t$ there exists some Δ such that $s \xrightarrow{a} \Delta$ and $\Delta \sim^\dagger \Theta$. By (3.22), we get $\Theta \in [\![X_\Delta]\!]_{\rho_\sim}$. As a consequence, $t \in [\![[a] \bigvee_{s \xrightarrow{a} \Delta} X_\Delta]\!]_{\rho_\sim}$. Since this holds for arbitrary action a, our desired result follows. □

3.6.2.3 Characteristic Formulae

So far we know how to construct the characteristic equation system for a finitary pLTS. As introduced in [54], the three transformation rules in Fig. 3.5 can be used to obtain from an equation system E a formula whose interpretation coincides with the interpretation of X_1 in the greatest solution of E. The formula thus obtained from a characteristic equation system is called a *characteristic formula*.

Theorem 3.10 *Given a characteristic equation system E, there is a characteristic formula φ_s such that $\rho_E(X_s) = [\![\varphi_s]\!]$ for any state s.* □

The above theorem, together with the results in Sect. 3.6.2.2, gives rise to the following corollary.

Corollary 3.1 *For each state s in a finitary pLTS, there is a characteristic formula φ_s such that $s \sim t$ iff $t \in [\![\varphi_s]\!]$.* □

3.7 Metric Characterisation

In the bisimulation game, probabilities are treated as labels since they are matched only when they are identical. One might argue that this does not provide a robust relation: Processes that differ with a very small probability, for instance, would be considered just as different as processes that perform completely different actions. This is particularly relevant to security systems where specifications can be given as perfect, but impractical processes and other practical processes are considered safe if they only differ from the specification with a negligible probability.

To find a more flexible way to differentiate processes, we borrow from pure mathematics the notion of metric[2] and measure the difference between two processes that are not quite bisimilar. Since different processes may behave the same, they will be given distance zero in our metric semantics. So we are more interested in pseudometrics than metrics.

In the rest of this section, we fix a finite-state pLTS $(S, L, \longrightarrow)$ and provide the set of pseudometrics on S with the following partial order.

Definition 3.10 The relation $\preceq$ for the set $\mathcal{M}$ of 1-bounded pseudometrics on S is defined by

$$m_1 \preceq m_2 \text{ if } \forall s, t : m_1(s,t) \geq m_2(s,t).$$

Here we reverse the ordering with the purpose of characterising bisimilarity by a *greatest* fixed point (cf. Corollary 3.2).

Lemma 3.5 *$(\mathcal{M}, \preceq)$ is a complete lattice.*

Proof The top element is given by $\forall s, t : \top(s,t) = 0$; the bottom element is given by $\bot(s,t) = 1$ if $s \neq t$, 0 otherwise. Greatest lower bounds are given by

$$\left(\bigsqcap X\right)(s,t) = \sup\{m(s,t) \mid m \in X\}$$

for any $X \subseteq \mathcal{M}$. Finally, least upper bounds are given by

$$\bigsqcup X = \bigsqcap \{m \in \mathcal{M} \mid \forall m' \in X : m' \preceq m\}.$$ □

In order to define the notion of state-metrics (that will correspond to bisimulations) and the monotone transformation on metrics, we need to lift metrics from S to $\mathcal{D}(S)$. Here we use the lifting operation based on the Kantorovich metric [25] on probability measures (cf. Sect. 3.4.1), which has been used by van Breugel and Worrell for defining metrics on fully probabilistic systems [40] and reactive probabilistic systems [55]; and by Desharnais et al. for labelled Markov chains [56] and labelled concurrent Markov chains [57], respectively.

Definition 3.11 $m \in \mathcal{M}$ is a *state-metric* if, for all $\varepsilon \in [0, 1)$, $m(s,t) \leq \varepsilon$ implies:

- If $s \xrightarrow{a} \Delta$ then there exists some Δ' such that $t \xrightarrow{a} \Delta'$ and $\hat{m}(\Delta, \Delta') \leq \varepsilon$.

Note that if m is a state-metric then it is also a metric. By $m(s,t) \leq \varepsilon$ we have $m(t,s) \leq \varepsilon$, which implies

- If $t \xrightarrow{a} \Delta'$ then there exists some Δ such that $s \xrightarrow{a} \Delta$ and $\hat{m}(\Delta', \Delta) \leq \varepsilon$.

---

[2] For simplicity, in this section we use the term metric to denote both metric and pseudometric. All the results are based on pseudometrics.

In the above definition, we prohibit ε from being 1 because we use 1 to represent the distance between any two incomparable states including the case where one state may perform a transition and the other may not.

The greatest state-metric is defined as

$$m_{max} = \bigsqcup \{m \in \mathcal{M} \mid m \text{ is a state-metric}\}.$$

It turns out that state-metrics correspond to bisimulations and the greatest state-metric corresponds to bisimilarity. To make the analogy closer, in what follows we will characterise m_{max} as a fixed point of a suitable monotone function on $\mathcal{M}$. First we recall the definition of Hausdorff distance.

Definition 3.12 Given a 1-bounded metric d on Z, the *Hausdorff distance* between two subsets X, Y of Z is defined as follows:

$$H_d(X, Y) = max\{sup_{x\in X} inf_{y\in Y} d(x, y), sup_{y\in Y} inf_{x\in X} d(y, x)\}$$

where $inf\ \emptyset = 1$ and $sup\ \emptyset = 0$.

Next we define a function F on $\mathcal{M}$ by using the Hausdorff distance.

Definition 3.13 Let $der(s, a) = \{\Delta \mid s \xrightarrow{a} \Delta\}$. For any $m \in \mathcal{M}$, $F(m)$ is a pseudometric given by

$$F(m)(s, t) = max_{a\in L}\{H_{\hat{m}}(der(s, a), der(t, a))\}.$$

Thus we have the following property.

Lemma 3.6 *For all $\varepsilon \in [0, 1)$, $F(m)(s, t) \leq \varepsilon$ if and only if:*

- *If $s \xrightarrow{a} \Delta$ then there exists some Δ' such that $t \xrightarrow{a} \Delta'$ and $\hat{m}(\Delta, \Delta') \leq \varepsilon$.*
- *If $t \xrightarrow{a} \Delta'$ then there exists some Δ such that $s \xrightarrow{a} \Delta$ and $\hat{m}(\Delta', \Delta) \leq \varepsilon$.*

□

The above lemma can be proved by directly checking the definition of F, as can the next lemma.

Lemma 3.7 *m is a state-metric iff $m \preceq F(m)$.* □

Consequently we have the following characterisation:

$$m_{max} = \bigsqcup \{m \in \mathcal{M} \mid m \preceq F(m)\}.$$

Lemma 3.8 *F is monotone on $\mathcal{M}$.* □

Because of Lemmas 3.5 and 3.8, we can apply Theorem 2.1, which tells us that m_{max} is the greatest fixed point of F. Furthermore, by Lemma 3.7 we know that m_{max} is indeed a state-metric, and it is the greatest state-metric.

In addition, if our pLTS is *image-finite*, i.e. for all $a \in L, s \in S$ the set $der(s, a)$ is finite, the closure ordinal of F is ω. Therefore one can proceed in a standard way to show that

$$m_{max} = \bigsqcap \{F^i(\top) \mid i \in \mathbb{N}\}$$

where $\top$ is the top metric in $\mathcal{M}$ and $F^0(\top) = \top$.

Lemma 3.9 *For image-finite pLTSs, the closure ordinal of F is ω.*

Proof Let $m_{max}(s,t) \leq \varepsilon$ and $s \xrightarrow{a} \Delta$. For each $m_i = F^i(\top)$ there is a Θ_i such that $t \xrightarrow{a} \Theta_i$ and $\hat{m}_i(\Delta, \Theta_i) \leq \varepsilon$. Since the pLTSs are image-finite, there is a Θ such that for all but finitely many i, $t \xrightarrow{a} \Theta$ and $\hat{m}_i(\Delta, \Theta) \leq \varepsilon$. □

We now show the correspondence between our state-metrics and bisimulations.

Theorem 3.11 *Given a binary relation $\mathcal{R}$ and a pseudometric $m \in \mathcal{M}$ on a finite-state pLTS such that*

$$m(s,t) = \begin{cases} 0 & \text{if } s\mathcal{R}t \\ 1 & \text{otherwise.} \end{cases} \tag{3.23}$$

Then $\mathcal{R}$ is a probabilistic bisimulation if and only if m is a state-metric.

Proof The result can be proved by using Theorem 3.3, which in turn relies on Theorem 3.1(1). Below we give an alternative proof that uses Theorem 3.1(2) instead.

Given two distributions Δ, Δ' over S, let us consider how to compute $\hat{m}(\Delta, \Delta')$ if $\mathcal{R}$ is an equivalence relation. Since S is finite, we may assume that $V_1, \ldots, V_n \in S/\mathcal{R}$ are all the equivalence classes of S under $\mathcal{R}$. If $s, t \in V_i$ for some $i \in 1, \ldots, n$, then $m(s,t) = 0$, which implies $x_s = x_t$ by the first constraint of (3.7). So for each $i \in 1, \ldots, n$ there exists some x_i such that $x_i = x_s$ for all $s \in V_i$. Thus, some summands of (3.7) can be grouped together and we have the following linear program:

$$\sum_{i \in 1,\ldots,n} (\Delta(V_i) - \Delta'(V_i))x_i \tag{3.24}$$

with the constraint $x_i - x_j \leq 1$ for any $i, j \in 1, \ldots, n$ with $i \neq j$. Briefly speaking, if $\mathcal{R}$ is an equivalence relation then $\hat{m}(\Delta, \Delta')$ is obtained by maximising the linear program in (3.24).

($\Rightarrow$) Suppose $\mathcal{R}$ is a bisimulation and $m(s,t) = 0$. From the assumption in (3.23) we know that $\mathcal{R}$ is an equivalence relation. By the definition of m we have $s\mathcal{R}t$. If $s \xrightarrow{a} \Delta$ then $t \xrightarrow{a} \Delta'$ for some Δ' such that $\Delta \ \mathcal{R}^\dagger \ \Delta'$. To show that m is a state-metric it suffices to prove $\hat{m}(\Delta, \Delta') = 0$. We know from $\Delta \ \mathcal{R}^\dagger \ \Delta'$ and Theorem 3.1 (2) that $\Delta(V_i) = \Delta'(V_i)$, for each $i \in 1, \ldots, n$. It follows that (3.24) is maximised to be 0, thus $\hat{m}(\Delta, \Delta') = 0$.

($\Leftarrow$) Suppose m is a state-metric and has the relation in (3.23). Notice that $\mathcal{R}$ is an equivalence relation. We show that it is a bisimulation. Suppose $s\mathcal{R}t$, which means $m(s,t) = 0$. If $s \xrightarrow{a} \Delta$ then $t \xrightarrow{a} \Delta'$ for some Δ' such that $\hat{m}(\Delta, \Delta') = 0$. To ensure that $\hat{m}(\Delta, \Delta') = 0$, in (3.24) the following two conditions must be satisfied.

1. No coefficient is positive. Otherwise, if $\Delta(V_i) - \Delta'(V_i) > 0$ then (3.24) would be maximised to a value not less than $(\Delta(V_i) - \Delta'(V_i))$, which is greater than 0.

2. It is not the case that at least one coefficient is negative and the other coefficients are either negative or 0. Otherwise, by summing up all the coefficients, we would get

$$\Delta(S) - \Delta'(S) < 0$$

that contradicts the assumption that Δ and Δ' are distributions over S.

Therefore the only possibility is that all the coefficients in (3.24) are 0, that is $\Delta(V_i) = \Delta'(V_i)$ for any equivalence class $V_i \in S/\mathcal{R}$. It follows from Theorem 3.1 (2) that $\Delta\ \mathcal{R}^\dagger\ \Delta'$. So we have shown that $\mathcal{R}$ is indeed a bisimulation. □

Corollary 3.2 *Let s and t be two states in a finite state pLTS. Then $s \sim t$ if and only if $m_{max}(s,t) = 0$.*

Proof (⇒) Since $\sim$ is a bisimulation, by Theorem 3.11 there exists some state-metric m such that $s \sim t$ iff $m(s,t) = 0$. By the definition of m_{max} we have $m \preceq m_{max}$. Therefore $m_{max}(s,t) \leq m(s,t) = 0$.

(⇐) From m_{max} we construct a pseudometric m as follows.

$$m(s,t) = \begin{cases} 0 & \text{if } m_{max}(s,t) = 0 \\ 1 & \text{otherwise.} \end{cases}$$

Since m_{max} is a state-metric, it is easy to see that m is also a state-metric. Now we construct a binary relation $\mathcal{R}$ such that $\forall s, s' : s\mathcal{R}s'$ iff $m(s,s') = 0$. It follows from Theorem 3.11 that $\mathcal{R}$ is a bisimulation. If $m_{max}(s,t) = 0$, then $m(s,t) = 0$ and thus $s\mathcal{R}t$. Therefore we have the required result $s \sim t$ because $\sim$ is the largest bisimulation. □

3.8 Algorithmic Characterisation

Bisimulation is useful for verifying formal systems and is the foundation of state-aggregation algorithms that compress models by merging bisimilar states. State aggregation is routinely used as a preprocessing step before model checking [58]. In this section we present two algorithms for computing probabilistic bisimulation.

3.8.1 A Partition Refinement Algorithm

We first introduce an algorithm that, given a pLTS $(S, L, \rightarrow)$ where both S and L are finite, iteratively computes bisimilarity. The idea is originally proposed by Kanellakis and Smolka [59] for computing nonprobabilistic bisimulation and commonly known as a *partition refinement* algorithm (see Fig. 3.6). The point is to represent the state space as a set of *blocks*, where a block is set of states standing for an equivalence class, and the equivalence of two given states can be tested by checking whether

$\mathscr{B} := \{S\}$
repeat
  $\mathscr{B}_{old} := \mathscr{B}$
  $\mathscr{B} :=$ **Refine**$(\mathscr{B})$
until $\mathscr{B}_{old} = \mathscr{B}$
return $\mathscr{B}$

Fig. 3.6 Schema for the partition refinement algorithm

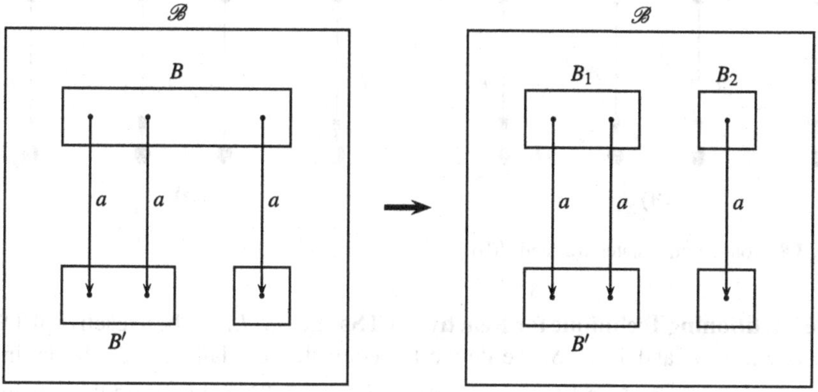

Fig. 3.7 Splitting a block

they belong to a same block. The blocks form a partition of the state space. Starting from the partition $\{S\}$, the algorithm iteratively refines the partition by splitting each block into two smaller blocks if two states in one block are found to exhibit different behaviour. Eventually, when no further refinement is possible for a partition, the algorithm terminates and all states in a block of the partition are bisimilar.

Let $\mathcal{B} = \{B_1, \ldots, B_n\}$ be a partition consisting of a set of blocks. The algorithm tries to refine the partition by splitting each block. A *splitter* for a block $B \in \mathcal{B}$ is the block $B' \in \mathcal{B}$ such that some states in B have a-transitions, for $a \in L$, into B' and others do not. In this case B' splits B with respect to a into two blocks B_1 and B_2, where $B_1 = \{s \in B \mid \exists s' \in B' : s \xrightarrow{a} s'\}$ and $B_2 = B - B_1$. This is illustrated in Fig. 3.7.

The refinement operator **Refine**$(\mathcal{B})$ yields the partition

$$\textbf{Refine}(\mathcal{B}) = \bigcup_{B \in \mathcal{B}, a \in L} \textbf{Split}(B, a, \mathcal{B}) \tag{3.25}$$

where **Split**$(B, a, \mathcal{B})$ is the splitting procedure that detects whether the partition $\mathcal{B}$ contains a splitter for a given block $B \in \mathcal{B}$ with respect to action $a \in L$. If such splitter exists, B is split into two blocks B_1 and B_2. Otherwise, B itself is returned.

We will introduce a partition refinement algorithm for pLTSs that was presented in [26]. Before that we briefly sketch how the above partitioning technique can be modified for reactive pLTSs and explain why this method fails for general pLTSs.

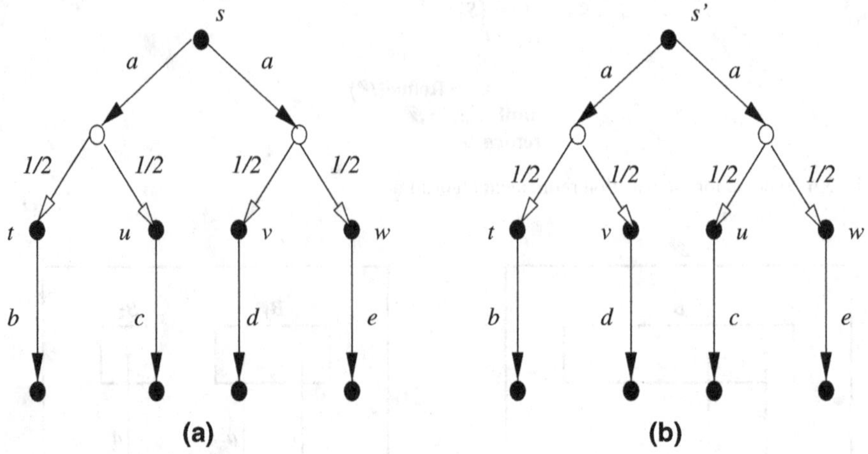

Fig. 3.8 Nonbisimilar states s(**a**) and s'(**b**)

The Partitioning Technique for Reactive pLTSs Let $\langle S, L, \rightarrow\rangle$ be a reactive pLTS. For any $a \in L$ and $B \subseteq S$, we define the equivalence relation $\sim_{(a,B)}$ by letting $s \sim_{(a,B)} t$ if $s \xrightarrow{a} \Delta$ and $t \xrightarrow{a} \Theta$ with $\Delta(B) = \Theta(B)$. We still use the schema shown in Fig. 3.6 and the refinement operator in (3.25), but change the splitting procedure as follows

$$\mathbf{Split}(B, a, \mathcal{B}) = \bigcup_{C \in \mathcal{B}} \mathbf{Split}(B, a, C) \text{ where } \mathbf{Split}(B, a, C) = C / \sim_{(a,B)} . \quad (3.26)$$

An implementation of the algorithm using some tricks on data structures yields the following complexity result.

Theorem 3.12 *The bisimulation equivalence classes of a reactive pLTS with n states and m transitions can be computed in time $\mathcal{O}(mn \log n)$ and space $\mathcal{O}(mn)$.* □

The splitter technique in (3.26) does not work for general pLTSs when we use the obvious modification of the equivalence relation $\sim_{(a,B)}$ where $s \sim_{(a,B)} t$ iff for any transition $s \xrightarrow{a} \Delta$ there is a transition $t \xrightarrow{a} \Theta$ with $\Delta(B) = \Theta(B)$ and vice versa.

Example 3.2 Consider the pLTS described in Fig. 3.8, we have $s \not\sim s'$ because the transition $s \xrightarrow{a} \frac{1}{2}\overline{t} + \frac{1}{2}\overline{u}$ cannot be matched by any transition from s'. However, s and s' cannot be distinguished by using the above partitioning technique. The problem is that after one round of refinement, we obtain the blocks

$$\{s, s'\}, \{t\}, \{u\}, \{v\}, \{w\}$$

and then no further refinement can split the block $\{s, s'\}$.

The Partitioning Technique for pLTSs To compute bisimulation equivalence classes in general pLTSs, we can keep the schema sketched in Fig. 3.6 but use two partitions: a partition $\mathcal{B}$ for states and a partition $\mathcal{M}$ for transitions. By a *transition* *partition* we mean a set $\mathcal{M}$ consisting of pairs (a, M) where $a \in L$ and $M \subseteq M_a$ with $M_a = \bigcup_{s \in S}\{\Delta \mid s \xrightarrow{a} \Delta\}$ such that, for any action a, the set $\{M \mid (a, M) \in \mathcal{M}\}$ is a partition of the set M_a.

The algorithm works as follows. We skip the first refinement step and start with the state partition

$$\mathcal{B}_{init} = S/\sim_L \text{ where } s \sim_L t \text{ iff } \{a \mid s \xrightarrow{a}\} = \{a \mid t \xrightarrow{a}\}$$

that identifies those states that can perform the same actions immediately. The initial transition partition

$$\mathcal{M}_{init} = \{(a, M_a) \mid a \in L\}$$

identifies all transitions with the same label. In each iteration, we try to refine the state partition $\mathcal{B}$ according to an equivalence class (a, M) of $\mathcal{M}$ or the transition partition $\mathcal{M}$ according to a block $B \in \mathcal{B}$. The refinement of $\mathcal{B}$ by (a, M) is done by the operation **Split**$(M, a, \mathcal{B})$ that divides each block B of $\mathcal{B}$ into two subblocks $B_{(a,M)} = \{s \in B \mid s \xrightarrow{a} M\}$ and $B \backslash B_{(a,M)}$. In other words,

$$\mathbf{Split}(M, a, \mathcal{B}) = \bigcup_{B \in \mathcal{B}} \mathbf{Split}(M, a, B)$$

where $\mathbf{Split}(M, a, B) = \{B_{(a,M)}, B \backslash B_{(a,M)}$ and $B_{(a,M)} = \{s \in B \mid s \xrightarrow{a} M\}$. The refinement of $\mathcal{M}$ by B is done by the operation **Split**$(B, \mathcal{M})$ that divides any block $(a, M) \in \mathcal{M}$ by the subblocks $(a, M_1), \ldots, (a, M_n)$ where $\{M_1, \ldots, M_n\} = M/\sim_B$ and $\Delta \sim_B \Theta$ iff $\Delta(B) = \Theta(B)$. Formally,

$$\mathbf{Split}(B, \mathcal{M}) = \bigcup_{(a,M) \in \mathcal{M}} \mathbf{Split}(B, (a, M))$$

where $\mathbf{Split}(B, (a, M)) = \{(a, M') \mid M' \in M/\sim_B\}$. If no further refinement of $\mathcal{B}$ and $\mathcal{M}$ is possible then we have $\mathcal{B} = S/\sim$. The algorithm is sketched in Fig. 3.9. See [26] for the correctness proof of the algorithm and the suitable data structures used to obtain the following complexity result.

Theorem 3.13 *The bisimulation equivalence classes of a pLTS with n states and m transitions can be decided in time $\mathcal{O}(mn(\log m + \log n))$ and space $\mathcal{O}(mn)$.* □

3.8.2 An "On-the-Fly" Algorithm

We now propose an "on-the-fly" algorithm for checking whether two states in a finitary pLTS are bisimilar.

An important ingredient of the algorithm is to check whether two distributions are related via a lifted relation. Fortunately, Theorem 3.4 already provides us a method

$\mathscr{B} := S/\sim_L$ where $s \sim_L t$ iff $\{a \mid s \xrightarrow{a}\} = \{a \mid t \xrightarrow{a}\}$
$\mathscr{M} := \{(a, M_a) \mid a \in L\}$ where $M_a = \bigcup_{s \in S}\{\Delta \mid s \xrightarrow{a} \Delta\}$
As long as $\mathscr{B}$ or $\mathscr{M}$ can be modified perform one of the following steps:
• either choose some $B \in \mathscr{B}$ and put $\mathscr{M} :=$ Split$(B, \mathscr{M})$
• or choose some $(a, M) \in \mathscr{M}$ and put $\mathscr{B} :=$ Split$(M, a, \mathscr{B})$
Return $\mathscr{B}$

Fig. 3.9 Computing bisimulation equivalence classes in pLTSs

for deciding whether $\Delta\ \mathcal{R}^\dagger\ \Theta$, for two given distributions Δ, Θ and a relation $\mathcal{R}$. We construct the network $\mathcal{N}(\Delta, \Theta, \mathcal{R})$ and compute the maximum flow with well-known methods, as sketched in the procedure **Check** shown in Fig. 3.10.

As shown in [60], computing the maximum flow in a network can be done in time $O(n^3/\log n)$ and space $O(n^2)$, where n is the number of nodes in the network. So we immediately have the following result.

Lemma 3.10 *The test whether $\Delta\ \mathcal{R}^\dagger\ \Theta$ can be done in time $O(n^3/\log n)$ and space $O(n^2)$.* □

We now present a bisimilarity-checking algorithm by adapting the algorithm proposed in [61] for value-passing processes, which in turn was inspired by [62].

The main procedure in the algorithm is **Bisim**(s, t) shown in Fig. 3.10. It starts with the initial state pair (s, t), trying to find the smallest bisimulation relation containing the pair by matching transitions from each pair of states it reaches. It uses three auxiliary data structures:

- *NotBisim* collects all state pairs that have already been detected as not bisimilar.
- *Visited* collects all state pairs that have already been visited.
- *Assumed* collects all state pairs that have already been visited and assumed to be bisimilar.

The core procedure, **Match**, is called from function **Bis** inside the main procedure **Bisim**. Whenever a new pair of states is encountered it is inserted into *Visited*. If two states fail to match each other's transitions then they are not bisimilar and the pair is added to *NotBisim*. If the current state pair has been visited before, we check whether it is in *NotBisim*. If this is the case, we return $false$. Otherwise, a loop has been detected and we make assumption that the two states are bisimilar, by inserting the pair into *Assumed*, and return $true$. Later on, if we find that the two states are not bisimilar after finishing searching the loop, then the assumption is wrong, so we first add the pair into *NotBisim* and then raise the exception *WrongAssumption*, which forces the function **Bis** to run again, with the new information that the two states in this pair are not bisimilar. In this case, the size of *NotBisim* has been increased by at least one. Hence, **Bis** can only be called for finitely many times. Therefore, the procedure **Bisim**(s, t) will terminate. If it returns $true$, then the set of state pairs *(Visited - NotBisim)* constitutes a bisimulation relation containing the pair (s, t).

The main difference from the algorithm of checking nonprobabilistic bisimilarity in [61] is the introduction of the procedure **MatchDistribution**(Δ, Θ), where we

```
Bisim(s,t)

Bisim(s,t) = {
NotBisim := {}
fun Bis(s,t)={
    Visited := {}
    Assumed := {}
    Match(s,t)}
} handle WrongAssumption ⇒ Bis(s,t)
return Bis(s,t)

Match(s,t) =
Visited := Visited ∪ {(s,t)}
b = ⋀_{a∈L} MatchAction(s,t,a)
if b = false then
    NotBisim := NotBisim ∪ {(s,t)}
    if (s,t) ∈ Assumed then
        raise WrongAssumption
    end if
end if
return b

MatchAction(s,t,a) =
for all s -a-> Δ_i do
    for all t -a-> Θ_j do
        b_ij = MatchDistribution(Δ_i, Θ_j)
    end for
end for
return (⋀_i(⋁_j b_ij)) ∧ (⋀_j(⋁_i b_ij))

MatchDistribution(Δ, Θ) =
Assume ⌈Δ⌉ = {s_1, ..., s_n} and ⌈Θ⌉ = {t_1, ..., t_m}
ℛ := {(s_i, t_j) | Close(s_i, t_j) = true}
return Check(Δ, Θ, ℛ)

Close(s,t) =
if (s,t) ∈ NotBisim then
    return false
else if (s,t) ∈ Visited then
    Assumed := Assumed ∪ {(s,t)}
    return true
else
    return Match(s,t)
end if

Check(Δ, Θ, ℛ) =
Construct the network 𝒩(Δ, Θ, ℛ)
Compute the maximum flow F in 𝒩(Δ, Θ, ℛ)
return (F = 1)
```

Fig. 3.10 Check whether s is bisimilar to t. ©2009 IEEE. Reprinted, with permission, from [63].

approximate $\sim$ by a binary relation $\mathcal{R}$ that is coarser than $\sim$ in general, and we check the validity of $\Delta\ \mathcal{R}^\dagger\ \Theta$. If $\Delta\ \mathcal{R}^\dagger\ \Theta$ does not hold, then $\Delta \sim^\dagger \Theta$ is invalid either and **MatchDistribution**(Δ, Θ) returns *false* correctly. Otherwise, the two distributions Δ and Θ are considered equivalent with respect to $\mathcal{R}$ and we move on to match other pairs of distributions. The correctness of the algorithm is stated in the following theorem.

Theorem 3.14 *Given two states s_0, t_0 in a finitary pLTS, the function* ***Bisim***(s_0, t_0) *terminates, and it returns true if and only if $s_0 \sim t_0$.*

Proof Let $\mathbf{Bis}_i$ stand for the ith execution of the function **Bis**. Let $Assumed_i$ and $NotBisim_i$ be the set *Assumed* and *NotBisim* at the end of $\mathbf{Bis}_i$. When $\mathbf{Bis}_i$ is finished, either a *WrongAssumption* is raised or no *WrongAssumption* is raised. In the former case, $Assumed_i \cap NotBisim_i \neq \emptyset$; in the latter case, the execution of the function **Bisim** is completed. By examining function **Close** we see that $Assumed_i \cap NotBisim_{i-1} = \emptyset$. Now it follows from the fact $NotBisim_{i-1} \subseteq NotBisim_i$ that $NotBisim_{i-1} \subset NotBisim_i$. Since we are considering finitary pLTSs, there is some j such that $NotBisim_{j-1} = NotBisim_j$, when all the nonbisimilar state pairs reachable from s_0 and t_0 have been found and **Bisim** must terminate.

For the correctness of the algorithm, we consider the relation

$$\mathcal{R}_i = Visited_i - NotBisim_i,$$

where $Visited_i$ is the set *Visited* at the end of $\mathbf{Bis}_i$. Let $\mathbf{Bis}_k$ be the last execution of **Bis**. For each $i \leq k$, the relation $\mathcal{R}_i$ can be regarded as an approximation of $\sim$, as far as the states appeared in $\mathcal{R}_i$ are concerned. Moreover, $\mathcal{R}_i$ is a coarser approximation because if two states s, t are revisited but their relation is unknown, they are assumed to be bisimilar. Therefore, if $\mathbf{Bis}_k(s_0, t_0)$ returns $false$, then $s_0 \not\sim t_0$. On the other hand, if $\mathbf{Bis}_k(s_0, t_0)$ returns $true$, then $\mathcal{R}_k$ constitutes a bisimulation relation containing the pair (s_0, t_0). This follows because **Match**$(s_0, t_0) = true$, which basically means that whenever $s \mathcal{R}_k t$ and $s \xrightarrow{a} \Delta$ there exists some transition $t \xrightarrow{a} \Theta$ such that **Check**$(\Delta, \Theta, \mathcal{R}_k) = true$, i.e. $\Delta\ \mathcal{R}_k{}^\dagger\ \Theta$. Indeed, this rules out the possibility that $s_0 \not\sim t_0$ as otherwise we would have $s_0 \not\sim_\omega t_0$ by Proposition 3.8, that is $s_0 \not\sim_n t_0$ for some $n > 0$. The latter means that some transition $s_0 \xrightarrow{a} \Delta$ exists such that for all $t_0 \xrightarrow{a} \Theta$ we do not have $\Delta\ (\sim_{n-1})^\dagger\ \Theta$, or symmetrically with the roles of s_0 and t_0 exchanged, i.e. Δ and Θ can be distinguished at level n, so a contradiction arises.

Below we consider the time and space complexities of the algorithm.

Theorem 3.15 *Let s and t be two states in a pLTS with n states in total. The function* ***Bisim***(s, t) *terminates in time $O(n^7/\log n)$ and space $O(n^2)$.*

Proof The number of state pairs is bounded by n^2. In the worst case, each execution of the function **Bis**(s, t) only yields one new pair of states that are not bisimilar. The number of state pairs examined in the first execution of **Bis**(s, t) is at most $O(n^2)$, in the second execution is at most $O(n^2 - 1), \ldots$. Therefore, the total number of state pairs examined is at most $O(n^2 + (n^2 - 1) + \cdots + 1) = O(n^4)$. When a state pair (s, t)

is examined, each transition of s is compared with all transitions of t labelled with the same action. Since the pLTS is finitely branching, we could assume that each state has at most c outgoing transitions. Therefore, for each state pair, the number of comparisons of transitions is bound by c^2. As a comparison of two transitions calls the function **Check** once, which requires time $O(n^3/\log n)$ by Lemma 3.10. As a result, examining each state pair takes time $O(c^2n^3/\log n)$.

Finally, the worst case time complexity of executing **Bisim**(s,t) is $O(n^7/\log n)$.

The space requirement of the algorithm is easily seen to be $O(n^2)$, in view of Lemma 3.10. □

Remark 3.3 With mild modification, the above algorithm can be adapted to check probabilistic similarity. We simply remove the underlined part in the function **MatchAction**; the rest of the algorithm remains unchanged. Similar to the analysis in Theorems 3.14 and 3.15, the new algorithm can be shown to correctly check probabilistic similarity over finitary pLTSs; its worst case time and space complexities are still $O(n^7/\log n)$ and $O(n^2)$, respectively.

3.9 Bibliographic Notes

3.9.1 Probabilistic Models

Models for probabilistic concurrent systems have been studied for a long time [64–67]. One of the first models obtained as a simple adaptation of the traditional LTSs from concurrency theory appears in [23]. Their *probabilistic transition systems* are classical LTSs, where in addition every transition is labelled with a probability, a real number in the interval [0,1], such that for every state s and every action a, the probabilities of all a-labelled transitions leaving s sum to either 0 or 1.

In [1] a similar model is proposed, but where the probabilities of *all* transitions leaving s sum to either 0 or 1. Van Glabbeek et al. [68] propose the terminology *reactive* for the type of model studied in [23], and *generative* for the type of model studied in [1]. In a generative model, a process can be considered to spontaneously generate actions, unless restricted by the environment; in generating actions, a probabilistic choice is made between all transitions that can be taken from a given state, even if they have different labels. In a reactive model, on the other hand, processes are supposed to perform actions only in response to requests by the environment. The choice between two different actions is therefore not under the control of the process itself. When the environment requests a specific action, a probabilistic choice is made between all transitions (if any) that are labelled with the requested action.

In the aforementioned models, the nondeterministic choice that can be modelled by nonprobabilistic LTSs is *replaced* by a probabilistic choice (and in the generative model also a deterministic choice, a choice between different actions, is made probabilistic). Hence reactive and generative probabilistic transition systems do not generalise nonprobabilistic LTSs. A model, or rather a calculus, which features both nondeterministic and reactive probabilistic choices is proposed in [4]. It is slightly reformulated in [6] under the name *simple probabilistic automata*, which is akin to our pLTS model. Note that essentially the same model has appeared in the literature under different names such as *NP-systems* [69], *probabilistic processes* [70], *probabilistic transition systems* [14] etc. Furthermore, there are strong structural similarities with *Markov decision processes* [71].

Following the classification above, our model is reactive rather than generative. The reactive model of [23] can be reformulated by saying that a state s has at most one outgoing transition for any action a, and this transition ends in a probability distribution over its successor states. The generalisation of [6], that we use here as well, is that a state can have multiple outgoing transitions with the same label, each ending in a probability distribution. Simple probabilistic automata are a special case of the *probabilistic automata* of [6], that also generalise the generative models of probabilistic processes to a setting with nondeterministic choice.

3.9.2 Probabilistic Bisimulation

Probabilistic bisimulation was first introduced by Larsen and Skou [23]. Later on, it has been investigated in a great many probabilistic models. An adequate logic for probabilistic bisimulation in a setting similar to our pLTSs has been studied in [45, 47] and then used in [72] to characterise a notion of bisimulation defined on an LTS whose states are distributions. It is also based on a probabilistic extension of the Hennessy–Milner logic. The main difference from our logic in Sect. 3.6.1 is the introduction of the operator $[\cdot]_p$. Intuitively, a distribution Δ satisfies the formula $[\varphi]_p$ when the set of states satisfying φ is measured by Δ with probability at least p. So the formula $[\varphi]_p$ can be expressed by our logic in terms of the probabilistic choice $\varphi_p \oplus \top$, as we have seen in Example 3.1.

An expressive logic for nonprobabilistic bisimulation has been proposed in [53]. In Sect. 3.6.2 we partially extend the results of [53] to a probabilistic setting that admits both probabilistic and nondeterministic choice. We present a probabilistic extension of the modal mu-calculus [31], where a formula is interpreted as the set of states satisfying it. This is in contrast to the probabilistic semantics of the mu-calculus as studied in [11, 12, 18] where formulae denote lower bounds of probabilistic evidence of properties, and the semantics of the generalised probabilistic logic of [16] where a mu-calculus formula is interpreted as a set of deterministic trees that satisfy it.

The probabilistic bisimulation studied in this chapter (cf. Definition 3.5) is a relation between states. In [73] a notion of probabilistic bisimulation is directly defined on distributions, and it is shown that the distribution-based bisimulations correspond to a lifting of state-based bisimulations with combined transitions. By combined transitions we mean that if a state can perform an action and then nondeterministically evolve into two different distributions, it can also evolve into any linear combination of the two distributions. The logical characterisation of the distribution-based bisimilarity in [73] is similar to ours in Sect. 3.6.1 but formulae are interpreted on distributions rather than on states. Another notion of distribution-based bisimulation is given in [74]. Unlike the other bisimulations we have discussed so far, it is essentially a linear-time semantics rather than a branching-time semantics [75].

The Kantorovich metric has been used by van Breugel et al. for defining behavioural pseudometrics on fully probabilistic systems [40, 76, 77] and reactive probabilistic systems [55, 78–80]; and by Desharnais et al. for labelled Markov chains [81, 57] and labelled concurrent Markov chains [56]; and later on by Ferns et al. for Markov decision processes [82, 83]; and by Deng et al. for action-labelled quantitative transition systems [84]. In this chapter we are mainly interested in the correspondence of our lifting operation to the Kantorovich metric. The metric characterisation of probabilistic bisimulation in Sect. 3.7 is merely a direct consequence of this correspondence. In [85] a general family of bisimulation metrics is proposed, which can be used to specify some properties in security and privacy. In [86] a different kind of approximate reasoning technique is proposed that measures the distances of labels used in (nonprobabilistic) bisimulation games rather than the distances of states in probabilistic bisimulation games.

Decision algorithms for probabilistic bisimilarity and similarity are first investigated by Baier et al. in [26]. In [87] Zhang et al. have improved the original algorithm of computing probabilistic simulation by exploiting the parametric maximum flow algorithm [88] to compute the maximum flows of networks and resulted in an algorithm that computes strong simulation in time $\mathcal{O}(m^2n)$ and in space $\mathcal{O}(m^2)$. In [89] a framework based on abstract interpretation [90] is proposed to design algorithms that compute bisimulation and simulation. It entails the bisimulation algorithm by Baier et al. as an instance and leads to a simulation algorithm that improves the one for pLTSs by Zhang et al. All those algorithms above are global in the sense that a whole state space has to be fully generated in advance. In contrast, "on-the-fly" algorithms are local in the sense that the state space is dynamically generated, which is often more efficient to determine that one state fails to be related to another. Our algorithm in Sect. 3.8.2 is inspired by [26] because we also reduce the problem of checking whether two distributions are related by a lifted relation to the maximum flow problem of a suitable network. We generalise the local algorithm of checking nonprobabilistic bisimilarity [61, 62] to the probabilistic setting.

An important line of research is on denotational semantics of probabilistic processes, where bisimulations are interpreted in analytic spaces. This topic is well covered in [50, 91], and thus omitted from this book.

References

1. Giacalone, A., Jou, C.C., Smolka, S.A.: Algebraic reasoning for probabilistic concurrent systems. Proceedings of IFIP TC2 Working Conference on Programming Concepts and Methods, Sea of Galilee, Israel (1990)
2. Christoff, I.: Testing equivalences and fully abstract models for probabilistic processes. Proceedings of the 1st International Conference on Concurrency Theory. Lecture Notes in Computer Science, vol. 458, pp. 126–140. Springer, Heidelberg (1990)

3. Larsen, K.G., Skou, A.: Compositional verification of probabilistic processes. Proceedings of the 3rd International Conference on Concurrency Theory. Lecture Notes in Computer Science, vol. 630, pp. 456–471. Springer, Heidelberg (1992)
4. Hansson, H., Jonsson, B.: A calculus for communicating systems with time and probabilities. Proceedings of IEEE Real-Time Systems Symposium, pp. 278–287. IEEE Computer Society Press (1990)
5. Yi, W., Larsen, K.G.: Testing probabilistic and nondeterministic processes. Proceedings of the IFIP TC6/WG6.1 12th International Symposium on Protocol Specification, Testing and Verification, IFIP Transactions, vol. C-8, pp. 47–61. North-Holland (1992)
6. Segala, R., Lynch, N.: Probabilistic simulations for probabilistic processes. Proceedings of the 5th International Conference on Concurrency Theory. Lecture Notes in Computer Science, vol. 836, pp. 481–496. Springer, Heidelberg (1994)
7. Lowe, G.: Probabilistic and prioritized models of timed CSP. Theor. Comput. Sci. **138**, 315–352 (1995)
8. Segala, R.: Modeling and verification of randomized distributed real-time systems. Tech. Rep. MIT/LCS/TR-676, PhD thesis, MIT, Dept. of EECS (1995)
9. Morgan, C.C., McIver, A.K., Seidel, K.: Probabilistic predicate transformers. ACM Trans. Progr. Lang. Syst. **18**(3), 325–353 (1996)
10. He, J., Seidel, K., McIver, A.K.: Probabilistic models for the guarded command language. Sci. Comput. Program. **28**, 171–192 (1997)
11. Huth, M., Kwiatkowska, M.: Quantitative analysis and model checking. Proceedings of the 12th Annual IEEE Symposium on Logic in Computer Science, pp. 111–122. IEEE Computer Society (1997)
12. McIver, A., Morgan, C.: An expectation-based model for probabilistic temporal logic. Tech. Rep. PRG-TR-13-97, Oxford University Computing Laboratory (1997)
13. Bandini, E., Segala, R.: Axiomatizations for probabilistic bisimulation. Proceedings of the 28th International Colloquium on Automata, Languages and Programming, Lecture Notes in Computer Science, vol. 2076, pp. 370–381. Springer (2001)
14. Jonsson, B., Yi, W.: Testing preorders for probabilistic processes can be characterized by simulations. Theor. Comput. Sci. **282**(1), 33–51 (2002)
15. Mislove, M.M., Ouaknine, J., Worrell, J.: Axioms for probability and nondeterminism. Electron. Notes Theor. Comput. Sci. **96**, 7–28 (2004)
16. Cleaveland, R., Iyer, S.P., Narasimha, M.: Probabilistic temporal logics via the modal mu-calculus. Theor. Comput. Sci. **342**(2–3), 316–350 (2005)
17. Tix, R., Keimel, K., Plotkin, G.: Semantic domains for combining probability and non-determinism. Electron. Notes Theor. Comput. Sci. **129**, 1–104 (2005)
18. McIver, A., Morgan, C.: Results on the quantitative mu-calculus. ACM Trans. Comput. Logic **8**(1) (2007)
19. Deng, Y., van Glabbeek, R., Hennessy, M., Morgan, C.C., Zhang, C.: Remarks on testing probabilistic processes. Electron Notes Theor. Comput. Sci. **172**, 359–397 (2007)
20. Deng, Y., van Glabbeek, R., Morgan, C.C., Zhang, C.: Scalar outcomes suffice for finitary probabilistic testing. Proceedings of the 16th European Symposium on Programming. Lecture Notes in Computer Science, vol. 4421, pp. 363–378. Springer, Heidelberg (2007)
21. Deng, Y., van Glabbeek, R., Hennessy, M., Morgan, C.C.: Characterising testing preorders for finite probabilistic processes. Log. Methods Comput. Sci. **4**(4), 1–33 (2008)
22. Deng, Y., van Glabbeek, R., Hennessy, M., Morgan, C.: Testing finitary probabilistic processes (extended abstract). Proceedings of the 20th International Conference on Concurrency Theory. Lecture Notes in Computer Science, vol. 5710, pp. 274–288. Springer, Heidelberg (2009)
23. Larsen, K.G., Skou, A.: Bisimulation through probabilistic testing. Inf. Comput. **94**(1), 1–28 (1991)
24. Deng, Y., Du, W.: Probabilistic barbed congruence. Electron. Notes Theor. Comput. Sci. **190**(3), 185–203 (2007)
25. Kantorovich, L.: On the transfer of masses (in Russian). Dokl. Akademii Nauk **37**(2), 227–229 (1942)

26. Baier, C., Engelen, B., Majster-Cederbaum, M.E.: Deciding bisimilarity and similarity for probabilistic processes. J. Comput. Syst. Sci. **60**(1), 187–231 (2000)
27. Milner, R.: Communication and Concurrency. Prentice Hall, Upper Saddle River (1989)
28. Park, D.: Concurrency and automata on infinite sequences. Proceedings of the 5th GI-Conference on Theoretical Computer Science. Lecture Notes in Computer Science, vol. 104, pp. 167–183. Springer, Heidelberg (1981)
29. Pnueli, A.: Linear and branching structures in the semantics and logics of reactive systems. Proceedings of the 12th International Colloquium on Automata, Languages and Programming. Lecture Notes in Computer Science, vol. 194, pp. 15–32. Springer, Heidelberg (1985)
30. Hennessy, M., Milner, R.: Algebraic laws for nondeterminism and concurrency. J. the ACM **32**(1), 137–161 (1985)
31. Kozen, D.: Results on the propositional mu-calculus. Theor. Comput. Sci. **27**, 333–354 (1983)
32. Monge, G.: Mémoire sur la théorie des déblais et des remblais. Histoire de l'Academie des Science de Paris p. 666 (1781)
33. Vershik, A.: Kantorovich metric: Initial history and little-known applications. J. Math. Sci. **133**(4), 1410–1417 (2006)
34. Orlin, J.B.: A faster strongly polynomial minimum cost flow algorithm. Proceedings of the 20th ACM Symposium on the Theory of Computing, pp. 377–387. ACM (1988)
35. Deng, Y., Du, W.: Kantorovich metric in computer science: A brief survey. Electron. Notes Theor. Comput. Sci. **353**(3), 73–82 (2009)
36. Gibbs, A.L., Su, F.E.: On choosing and bounding probability metrics. Int. Stat. Rev. **70**(3), 419–435 (2002)
37. Rachev, S.: Probability Metrics and the Stability of Stochastic Models. Wiley, New York (1991)
38. Kantorovich, L.V., Rubinshtein, G.S.: On a space of totally additive functions. Vestn Len. Univ. **13**(7), 52–59 (1958)
39. Villani, C.: Topics in Optimal Transportation, Graduate Studies in Mathematics, vol. 58. American Mathematical Society (2003)
40. van Breugel, F., Worrell, J.: An algorithm for quantitative verification of probabilistic transition systems. Proceedings of the 12th International Conference on Concurrency Theory. Lecture Notes in Computer Science, vol. 2154, pp. 336–350. Springer, Heidelberg (2001)
41. Even, S.: Graph Algorithms. Computer Science Press, Potomac (1979)
42. Ford, L., Fulkerson, D.: Flows in Networks. Princeton University Press, Princeton (2010)
43. Aharonia, R., Bergerb, E., Georgakopoulosc, A., Perlsteina, A., Sprüssel, P.: The max-flow min-cut theorem for countable networks. J. Comb. Theory, Ser. B **101**(1), 1–17 (2011)
44. Sack, J., Zhang, L.: A general framework for probabilistic characterizing formulae. Proceedings of the 13th International Conference on Verification, Model Checking, and Abstract Interpretation. Lecture Notes in Computer Science, vol. 7148, pp. 396–411. Springer, Heidelberg (2012)
45. Hermanns, H., Parma, A. et al.: Probabilistic logical characterization. Inf. Comput. **209**(2), 154–172 (2011)
46. Desharnais, J., Edalat, A., Panangaden, P.: A logical characterization of bisimulation for labelled Markov processes. Proceedings of the 13th Annual IEEE Symposium on Logic in Computer Science, pp. 478–489. IEEE Computer Society Press (1998)
47. Parma, A., Segala, R.: Logical characterizations of bisimulations for discrete probabilistic systems. Proceedings of the 10th International Conference on Foundations of Software Science and Computational Structures. Lecture Notes in Computer Science, vol. 4423, pp. 287–301. Springer, Heidelberg (2007)
48. van Breugel, F., Mislove, M., Ouaknine, J., Worrell, J.: Domain theory, testing and simulation for labelled Markov processes. Theor. Comput. Sci. **333**(1–2), 171–197 (2005)
49. Sangiorgi, D., Rutten, J. (eds.): Advanced Topics in Bisimulation and Coinduction. Cambridge University Press, New York (2011)
50. Doberkat, E.E.: Stochastic Coalgebraic Logic. Springer, Heidelberg (2010)
51. Deng, Y., Wu, H.: Modal Characterisations of Probabilistic and Fuzzy Bisimulations. Proceedings of the 16th International Conference on Formal Engineering Methods. Lecture Notes in Computer Science, vol. 8829, pp. 123–138. Springer, Heidelberg (2014)

52. Tarski, A.: A lattice-theoretical fixpoint theorem and its application. Pac. J. Math. **5**, 285–309 (1955)
53. Steffen, B., Ingólfsdóttir, A.: Characteristic formulae for processes with divergence. Inf. Comput. **110**, 149–163 (1994)
54. Müller-Olm, M.: Derivation of characteristic formulae. Electron. Notes Theor. Comput. Sci. **18**, 159–170 (1998)
55. van Breugel, F., Worrell, J.: Towards quantitative verification of probabilistic transition systems. Proceedings of the 28th International Colloquium on Automata, Languages and Programming. Lecture Notes in Computer Science, vol. 2076, pp. 421–432. Springer, Heidelberg (2001)
56. Desharnais, J., Jagadeesan, R., Gupta, V., Panangaden, P.: The metric analogue of weak bisimulation for probabilistic processes. Proceedings of the 17th Annual IEEE Symposium on Logic in Computer Science, pp. 413–422. IEEE Computer Society (2002)
57. Desharnais, J., Jagadeesan, R., Gupta, V., Panangaden, P.: Metrics for labelled markov processes. Theor. Comput. Sci. **318**(3), 323–354 (2004)
58. Baier, C., Katoen, J.P.: Principles of Model Checking. The MIT Press, Cambridge (2008)
59. Kanellakis, P., Smolka, S.A.: CCS expressions, finite state processes, and three problems of equivalence. Inf. Comput. **86**(1), 43–65 (1990)
60. Cheriyan, J., Hagerup, T., Mehlhorn, K.: Can a maximum flow be computed on O(nm) time? Proceedings of the 17th International Colloquium on Automata, Languages and Programming. Lecture Notes in Computer Science, vol. 443, pp. 235–248. Springer, Heidelberg (1990)
61. Lin, H.: "On-the-fly" instantiation of value-passing processes. Proceedings of FORTE'98, IFIP Conference Proceedings, vol. 135, pp. 215–230. Kluwer (1998)
62. Fernandez, J.C., Mounier, L.: Verifying bisimulations "on the fly". Proceedings of the 3rd International Conference on Formal Description Techniques for Distributed Systems and Communication Protocols, pp. 95–110. North-Holland (1990)
63. Deng, Y., Du, W.: A local algorithm for checking probabilistic bisimilarity. Proceedings of the 4th International Conference on Frontier of Computer Science and Technology, pp. 401–409. IEEE Computer Society (2009)
64. Rabin, M.O.: Probabilistic automata. Inf. Control **6**, 230–245 (1963)
65. Derman, C.: Finite State Markovian Decision Processes. Academic, New York (1970)
66. Vardi, M.: Automatic verification of probabilistic concurrent finite-state programs. Proceedings 26th Annual Symposium on Foundations of Computer Science, pp. 327–338 (1985)
67. Jones, C., Plotkin, G.: A probabilistic powerdomain of evaluations. Proceedings of the 4th Annual IEEE Symposium on Logic in Computer Science, pp. 186–195. Computer Society Press (1989)
68. van Glabbeek, R., Smolka, S.A., Steffen, B., Tofts, C.: Reactive, generative, and stratified models of probabilistic processes. Proceedings of the 5th Annual IEEE Symposium on Logic in Computer Science, pp. 130–141. Computer Society Press (1990)
69. Jonsson, B., Ho-Stuart, C., Yi, W.: Testing and refinement for nondeterministic and probabilistic processes. Proceedings of the 3rd International Symposium on Formal Techniques in Real-Time and Fault-Tolerant Systems. Lecture Notes in Computer Science, vol. 863, pp. 418–430. Springer, Heidelberg (1994)
70. Jonsson, B., Yi, W.: Compositional testing preorders for probabilistic processes. Proceedings of the 10th Annual IEEE Symposium on Logic in Computer Science, pp. 431–441. Computer Society Press (1995)
71. Puterman, M.L.: Markov Decision Processes. Wiley, New York (1994)
72. Crafa, S., Ranzato, F.: A spectrum of behavioral relations over LTSs on probability distributions. Proceedings the 22nd International Conference on Concurrency Theory. Lecture Notes in Computer Science, vol. 6901, pp. 124–139. Springer, Heidelberg (2011)
73. Hennessy, M.: Exploring probabilistic bisimulations, part I. Form. Asp. Comput. **24**(4–6), 749–768 (2012)
74. Feng, Y., Zhang, L.: When equivalence and bisimulation join forces in probabilistic automata. Proceedings of the 19th International Symposium on Formal Methods. Lecture Notes in Computer Science, vol. 8442, pp. 247–262. Springer, Heidelberg (2014)

75. van Glabbeek, R.: The linear time-branching time spectrum I; the semantics of concrete, sequential processes. Handbook of Process Algebra, Chapter 1, pp. 3–99. Elsevier (2001)
76. van Breugel, F., Worrell, J.: Approximating and computing behavioural distances in probabilistic transition systems. Theor. Comput Sci. **360**(1–3), 373–385 (2006)
77. van Breugel, F., Sharma, B., Worrell, J.: Approximating a behavioural pseudometric without discount for probabilistic systems. Proceedings of the 10th International Conference on Foundations of Software Science and Computational Structures. Lecture Notes in Computer Science, vol. 4423, pp. 123–137. Springer, Heidelberg (2007)
78. van Breugel, F., Worrell, J.: A behavioural pseudometric for probabilistic transition systems. Theor. Comput. Sci. **331**(1), 115–142 (2005)
79. van Breugel, F., Hermida, C., Makkai, M., Worrell, J.: An accessible approach to behavioural pseudometrics. Proceedings of the 32nd International Colloquium on Automata, Languages and Programming. Lecture Notes in Computer Science, vol. 3580, pp. 1018–1030. Springer, Heidelberg (2005)
80. van Breugel, F., Hermida, C., Makkai, M., Worrell, J.: Recursively defined metric spaces without contraction. Theor. Comput. Sci. **380**(1–2), 143–163 (2007)
81. Desharnais, J., Jagadeesan, R., Gupta, V., Panangaden, P.: Metrics for labeled Markov systems. Proceedings of the 10th International Conference on Concurrency Theory. Lecture Notes in Computer Science, vol. 1664, pp. 258–273. Springer-Verlag, Heidelberg (1999)
82. Ferns, N., Panangaden, P., Precup, D.: Metrics for finite Markov decision processes. Proceedings of the 20th Conference in Uncertainty in Artificial Intelligence, pp. 162–169. AUAI Press (2004)
83. Ferns, N., Panangaden, P., Precup, D.: Metrics for Markov decision processes with infinite state spaces. Proceedings of the 21st Conference in Uncertainty in Artificial Intelligence, pp. 201–208. AUAI Press (2005)
84. Deng, Y., Chothia, T., Palamidessi, C., Pang, J.: Metrics for action-labelled quantitative transition systems. Electron. Notes Theor. Comput. Sci. **153**(2), 79–96 (2006)
85. Chatzikokolakis, K., Gebler, D., Palamidessi, C., Xu, L.: Generalized bisimulation metrics. Proceedings of the 25th International Conference on Concurrency Theory. Lecture Notes in Computer Science, vol. 8704, pp. 32–46. Springer, Heidelberg (2014)
86. Ying, M.: Bisimulation indexes and their applications. Theor. Comput. Sci. **275**, 1–68 (2002)
87. Zhang, L., Hermanns, H., Eisenbrand, F., Jansen, D.N.: Flow faster: Efficient decision algorithms for probabilistic simulations. Log. Methods Comput. Sci. **4**(4:6) (2008)
88. Gallo, G., Grigoriadis, M.D., Tarjan, R.E.: A fast parametric maximum flow algorithm and applications. SIAM J. Comput. **18**(1), 30–55 (1989)
89. Crafa, S., Ranzato, F.: Bisimulation and simulation algorithms on probabilistic transition systems by abstract interpretation. Form. Methods Syst. Des. **40**(3), 356–376 (2012)
90. Cousot, P., Cousot, R.: Abstract interpretation: a unified lattice model for static analysis of programs by construction or approximation of fixpoints. Proceedings of the 4th ACM Symposium on Principles of Programming Languages, pp. 238–252. ACM (1977)
91. Panangaden, P.: Labelled Markov Processes. Imperial College Press, London (2009)

Chapter 4
Probabilistic Testing Semantics

Abstract In this chapter we extend the traditional testing theory of De Nicola and Hennessy to the probabilistic setting. We first set up a general testing framework, and then introduce a vector-based testing approach that employs multiple success actions. It turns out that for finitary systems, i.e. finite-state and finitely branching systems, vector-based testing is equivalent to scalar testing that uses only one success action. Other variants, such as reward testing and extremal reward testing, are also discussed. They all coincide with vector-based testing as far as finitary systems are concerned.

Keywords Testing semantics · Vector-based testing · Scalar testing · Reward testing

4.1 A General Testing Framework

It is natural to view the semantics of processes as being determined by their ability to pass tests [1–4]; processes P_1 and P_2 are deemed to be semantically equivalent unless there is a test that can distinguish them. The actual tests used typically represent the ways in which users, or indeed other processes, can interact with P_i.

Let us first set up a general testing scenario within which this idea can be formulated. It assumes:

- a set of processes $\mathcal{P}roc$
- a set of tests $\mathcal{T}$, which can be applied to processes
- a set of outcomes $\mathcal{O}$, the possible results from applying a test to a process
- a function $\mathcal{A} : \mathcal{T} \times \mathcal{P}roc \rightarrow \mathcal{P}^+(\mathcal{O})$, representing the possible results of applying a specific test to a specific process.

Here $\mathcal{P}^+(\mathcal{O})$ denotes the collection of nonempty subsets of $\mathcal{O}$; so the result of applying a test T to a process P, $\mathcal{A}(T, P)$, is in general a *nonempty set* of outcomes, representing the fact that the behaviour of processes, and indeed tests, may be nondeterministic.

Moreover, some outcomes are considered better then others; for example, the application of a test may simply succeed, or it may fail, with success being better than failure. So we can assume that $\mathcal{O}$ is endowed with a partial order, in which $o_1 \leq o_2$ means that o_2 is a better outcome than o_1.

DOI 10.1007/978-3-662-45198-4_4

When comparing the results of applying tests to processes, we need to compare subsets of $\mathcal{O}$. There are two standard approaches to make this comparison, based on viewing these sets as elements of either the Hoare or Smyth powerdomain [5, 6] of $\mathcal{O}$. For $O_1, O_2 \in \mathcal{P}^+(\mathcal{O})$, we let:

(i) $O_1 \leq_{\text{Ho}} O_2$ if for every $o_1 \in O_1$ there exists some $o_2 \in O_2$ such that $o_1 \leq o_2$
(ii) $O_1 \leq_{\text{Sm}} O_2$ if for every $o_2 \in O_2$ there exists some $o_1 \in O_1$ such that $o_1 \leq o_2$.

Using these two comparison methods we obtain two different semantic preorders for processes:

(i) For $P, Q \in \mathcal{P}roc$, let $P \sqsubseteq_{\text{may}} Q$ if $\mathcal{A}(T, P) \leq_{\text{Ho}} \mathcal{A}(T, Q)$ for every test T
(ii) Similarly, let $P \sqsubseteq_{\text{must}} Q$ if $\mathcal{A}(T, P) \leq_{\text{Sm}} \mathcal{A}(T, Q)$ for every test T.

We use $P \simeq_{\text{may}} Q$ and $P \simeq_{\text{must}} Q$ to denote the associated equivalences.

The terminology *may* and *must* refers to the following reformulation of the same idea. Let $Pass \subseteq \mathcal{O}$ be an *upwards-closed* subset of $\mathcal{O}$, i.e. satisfying that $o' \geq o$ and $o \in Pass$ imply $o' \in Pass$. It is thought of as the set of outcomes that can be regarded as *passing* a test. Then we say that a process P *may* pass a test T with an outcome in *Pass*, notation "P **may** *Pass* T", if there is an outcome $o \in \mathcal{A}(P, T)$ with $o \in Pass$, and likewise P *must* pass a test T with an outcome in *Pass*, notation "P **must** *Pass* T", if for all $o \in \mathcal{A}(P, T)$ one has $o \in Pass$. Now

$$P \sqsubseteq_{\text{may}} Q \text{ iff } \forall T \in \mathcal{T}\, \forall Pass \in \mathcal{P}^{\uparrow}(\mathcal{O})\, (P \textbf{ may } Pass\ T \Rightarrow Q \textbf{ may } Pass\ T)$$
$$P \sqsubseteq_{\text{must}} Q \text{ iff } \forall T \in \mathcal{T}\, \forall Pass \in \mathcal{P}^{\uparrow}(\mathcal{O})\, (P \textbf{ must } Pass\ T \Rightarrow Q \textbf{ must } Pass\ T)$$

where $\mathcal{P}^{\uparrow}(\mathcal{O})$ is the set of upwards-closed subsets of $\mathcal{O}$.

Let us have a look at some typical instances of the set $\mathcal{O}$ and its associated partial order $\leq$.

1. The original theory of testing [1, 2] is obtained by using as the set of outcomes $\mathcal{O}$ the two-point lattice

with $\top$ representing the success of a test application, and $\bot$ failure.
2. However, for probabilistic processes we consider an application of a test to a process to succeed with a given probability. Thus we take as the set of outcomes the unit interval $[0, 1]$ with the standard mathematical ordering; if $0 \leq p < q \leq 1$, then succeeding with probability q is considered better than succeeding with probability p. This yields two preorders for probabilistic processes which, for convenience, we rename $\sqsubseteq_{\text{pmay}}$ and $\sqsubseteq_{\text{pmust}}$, with the associated equivalences $\simeq_{\text{pmay}}$ and $\simeq_{\text{pmust}}$, respectively. These preorders, and their associated equivalences, were first defined by Wang and Larsen [3]. We will refer to this approach as *scalar testing*.

3. Another approach of testing [4] employs a countable set of special actions $\Omega = \{\omega_1, \omega_2, \ldots\}$ to report success. When applied to probabilistic processes, this approach uses the function space $[0,1]^{\Omega}$ as the set of outcomes and the standard partial order for real functions: for any $o_1, o_2 \in \mathcal{O}$, we have $o_1 \leq o_2$ if and only if $o_1(\omega) \leq o_2(\omega)$ for every $\omega \in \Omega$. When Ω is fixed, an outcome $o \in \mathcal{O}$ can be considered as a vector $\langle o(\omega_1), o(\omega_2), \ldots \rangle$, with $o(\omega_i)$ representing the probability of success observed by action ω_i. Therefore, this approach is called *vector-based testing*.

For the second instance above, there is a useful simplification: the Hoare and Smyth preorders on *closed* subsets of $[0, 1]$, in particular on finite subsets of $[0, 1]$, are determined by their maximum and minimum elements, respectively.

Proposition 4.1 *Let* O_1, O_2 *be nonempty closed subsets of* $[0, 1]$, *we have*

(i) $O_1 \leq_{\mathrm{Ho}} O_2$ *if and only if* $max(O_1) \leq max(O_2)$
(ii) $O_1 \leq_{\mathrm{Sm}} O_2$ *if and only if* $min(O_1) \leq min(O_2)$.

Proof Straightforward calculations. □

Remark 4.1 Another formulation of the Hoare and Smyth preorders can be given as follows. Let $(X, \leq)$ be a partially ordered set. If $A \subseteq X$, then the *upper set* and the *lower set* of A are defined by $\uparrow A = \{a' \in X \mid a \leq a' \text{ for some } a \in A\}$ and dually $\downarrow A = \{a' \in X \mid a' \leq a \text{ for some } a \in A\}$, respectively. Then for any nonempty $A, B \subseteq X$, we have

1. $A \leq_{\mathrm{Ho}} B$ if and only if $\downarrow A \subseteq \downarrow B$;
2. $A \leq_{\mathrm{Sm}} B$ if and only if $\uparrow A \supseteq \uparrow B$.

As in the nonprobabilistic case [1], we could also define a testing preorder combining the may–must-preorders; we will not study this combination here.

4.2 Testing Probabilistic Processes

We start with some notation. For Θ, a distribution over R and function $f : R \to S$, $\mathrm{Img}_f(\Theta)$ is the distribution over S given by

$$\mathrm{Img}_f(\Theta)(s) = \sum\{\Theta(r) \mid r \in R \text{ and } f(r) = s\}$$

for any $s \in S$. For Δ, a distribution over S and function $f : S \to X$ into a vector space X, we sometimes write $\mathrm{Exp}_\Delta(f)$ or simply $f(\Delta)$ for $\sum_{s \in S} \Delta(s) \cdot f(s)$, the *expected value* of f. Our primary use of this notation is with X being the vector space of reals, but we will also use it with tuples of reals, or distributions over some set. In the latter case, this amounts to the notation $\sum_{i \in I} p_i \cdot \Delta_i$, where I is a finite index set and $\sum_{i \in I} p_i = 1$. When $p \in [0, 1]$, we also write $f_1\, {}_p\!\oplus f_2$ for $p \cdot f_1 + (1-p) \cdot f_2$. More generally, for function $F : S \to \mathcal{P}^+(X)$ with $\mathcal{P}^+(X)$ being the collection of nonempty subsets of X, we define $\mathrm{Exp}_\Delta F := \{\mathrm{Exp}_\Delta(f) \mid f \in F\}$; here $f \in F$ means that $f : S \to X$ is a *choice function*, that is, it satisfies the constraint that $f(s) \in F(s)$ for all $s \in S$. Recall that a set O is *convex* if $o_1, o_2 \in O$ and $p \in [0, 1]$, then the weighted average $o_1\,{}_p\!\oplus o_2$ is also in O.

Definition 4.1 A *(probabilistic) process* is defined by a tuple $\langle S, \Delta^\circ, \mathsf{Act}_\tau, \rightarrow\rangle$, where Act_τ is a set of actions Act augmented with a special action τ, $\langle S, \mathsf{Act}_\tau, \rightarrow\rangle$ is a pLTS and Δ° is a distribution over S, called the *initial distribution* of the pLTS. We sometimes identify a process with its initial distribution when the underlying pLTS is clear from the context.

We now define the parallel composition of two processes.

Definition 4.2 Let $P_1 = \langle S_1, \Delta_1^\circ, \mathsf{Act}_\tau, \rightarrow_1\rangle$ and $P_2 = \langle S_2, \Delta_2^\circ, \mathsf{Act}_\tau, \rightarrow_2\rangle$ be two processes, and A a set of visible actions. The *parallel composition* $P_1|_A P_2$ is the composite process $\langle S_1 \times S_2, \Delta_1^\circ \times \Delta_2^\circ, \mathsf{Act}_\tau, \rightarrow\rangle$ where $\rightarrow$ is the least relation satisfying the following rules:

$$\frac{s_1 \xrightarrow{\alpha}_1 \Delta \quad \alpha \notin A}{(s_1, s_2) \xrightarrow{\alpha} \Delta \times \overline{s_2}} \qquad \frac{s_2 \xrightarrow{\alpha}_2 \Delta \quad \alpha \notin A}{(s_1, s_2) \xrightarrow{\alpha} \overline{s_1} \times \Delta}$$

$$\frac{s_1 \xrightarrow{a}_1 \Delta_1, \quad s_2 \xrightarrow{a}_2 \Delta_2 \quad a \in A}{(s_1, s_2) \xrightarrow{\tau} \Delta_1 \times \Delta_2}.$$

Parallel composition is the basis of testing: it models the interaction of the observer with the process being tested; and models the observer himself—as a process. Let $\Omega := \{\omega_1, \omega_2, \cdots\}$ be a countable set of *success actions*, disjoint from Act_τ, define an Ω *test* to be a process $\langle S, \Delta^\circ, \mathsf{Act}_\tau \cup \Omega, \rightarrow\rangle$ with the constraint that $s \xrightarrow{\omega_i}$ and $s \xrightarrow{\omega_j}$ imply $i = j$. Let $\mathbb{T}$ be the class of all such tests, and write $\mathbb{T}_n$ for the subclass of $\mathbb{T}$ that uses at most n success actions $\omega_1, \ldots, \omega_n$; we write $\mathbb{T}_\mathbb{N}$ for $\bigcup_{n\in\mathbb{N}} \mathbb{T}_n$.

To *apply* test T to process P, we first form the composition $T|_{\mathsf{Act}}P$ and then resolve all nondeterministic choices into probabilistic choices. Thus, we obtain a set of resolutions as in Definition 4.3 below. For each resolution, any particular success action ω_i will have some probability of occurring; and those probabilities, taken together, give us a single *success tuple* for the whole resolution, so that if o is the tuple, then $o(\omega_i)$ is the recorded probability of ω_i's occurrence. The set of all those tuples, i.e. overall resolutions of $T|_{\mathsf{Act}}P$, is then the complete outcome of applying test T to process P: as such, it will be a subset of $[0, 1]^\Omega$.

Definition 4.3 A *resolution* of a distribution $\Delta \in \mathcal{D}(S)$ in a pLTS $\langle S, \Omega_\tau, \rightarrow\rangle$ is a triple $\langle R, \Theta, \rightarrow_R\rangle$ where $\langle R, \Omega_\tau, \rightarrow_R\rangle$ is a deterministic pLTS and $\Theta \in \mathcal{D}(R)$, such that there exists a *resolving function* $f \in R \to S$ satisfying

(i) $\mathrm{Img}_f(\Theta) = \Delta$
(ii) if $r \xrightarrow{\alpha}_R \Theta'$ for $\alpha \in \Omega_\tau$, then $f(r) \xrightarrow{\alpha} \mathrm{Img}_f(\Theta')$
(iii) if $f(r) \xrightarrow{\alpha}$ for $\alpha \in \Omega_\tau$, then $r \xrightarrow{\alpha}_R$.

A resolution $\langle R, \Omega_\tau, \rightarrow_R\rangle$ is said to be *static* if its resolving function f_R is injective.

A resolution has as its main effect the choice in any state of a single outgoing transition from all available ones with a same label; but f can be noninjective, so that the choice can vary between different departures from that state, depending, e.g., on the history of states and actions that led there. Further, since a single state of S can be split into a distribution over several states of R, all mapped to it by f, probabilistic interpolation between distinct choices is obtained automatically.

Static restrictions are particularly simple, in that they do not allow states to be resolved into distributions, or computation steps to be interpolated.

Example 4.1 Consider the pLTS in Fig. 4.1(a), with initial distribution $\overline{s_1}$. One of its resolutions is described in (b). The associated resolving function f is the following.

$$\begin{aligned} f(r_1) = f(r_2) = f(r_4) = f(r_5) &= s_1 \\ f(r_3) = f(r_6) &= s_2 \\ f(r_7) &= s_3 \\ f(r_8) &= s_4. \end{aligned}$$

From state s_1, there are three outgoing transitions all labelled with τ. According to the resolution in (b), the first time s_1 is visited, the transition $s_1 \xrightarrow{\tau} \overline{s_1}$ is chosen with probability $2/3$ and the transition $s_1 \xrightarrow{\tau} \overline{s_2}$ with probability $1/3$. The second time s_1 is visited, the transition $s_1 \xrightarrow{\tau} s_2$ is taken with probability $1/4$ and the transition $s_1 \xrightarrow{\tau} \overline{s_3}\,{}_{\frac{1}{2}}\!\oplus \overline{s_4}$ with probability $3/4$.

We now explain how to associate an outcome with a particular resolution which, in turn, will associate a set of outcomes with a distribution in a pLTS. Given a deterministic pLTS $\langle R, \Omega_\tau, \rightarrow \rangle$, consider the function $\mathcal{C} : (R \rightarrow [0,1]^\Omega) \rightarrow (R \rightarrow [0,1]^\Omega)$ defined by

$$\mathcal{C}(f)(r)(\omega) := \begin{cases} 1 & \text{if } r \xrightarrow{\omega} \\ 0 & \text{if } r \not\xrightarrow{\omega} \text{ and } r \not\xrightarrow{\tau} \\ \mathrm{Exp}_\Delta(f)(\omega) & \text{if } r \not\xrightarrow{\omega} \text{ and } r \xrightarrow{\tau} \Delta. \end{cases} \tag{4.1}$$

We view the unit interval $[0, 1]$ ordered in the standard manner as a complete lattice; this induces the structure of a complete lattice on the product $[0, 1]^\Omega$ and in turn on the set of functions $R \rightarrow [0,1]^\Omega$ via the partial order $\leq$ defined pointwise by letting $f \leq g$ iff $f(r) \leq f(g)$ for all $r \in R$. The function $\mathcal{C}$ is easily seen to be monotone and therefore has a least fixed point, which we denote by $\mathbb{V}_{\langle R,\Omega_\tau,\rightarrow\rangle}$; this is abbreviated to $\mathbb{V}$ when the resolution in question is understood.

Now let $\mathcal{A}(T, P)$ denote the set of vectors

$$\mathcal{A}(T,P) := \{\mathrm{Exp}_\Theta\left(\mathbb{V}_{\langle R,\Omega_\tau,\rightarrow\rangle}\right) \mid \langle R, \Theta, \rightarrow\rangle \text{ is a resolution of } [\![T|_{\mathsf{Act}}P]\!]\} \tag{4.2}$$

where $[\![T|_{\mathsf{Act}}P]\!]$ stands for the initial distribution of the process $T|_{\mathsf{Act}}P$.

We note that the result set $\mathcal{A}(T, P)$ is convex.

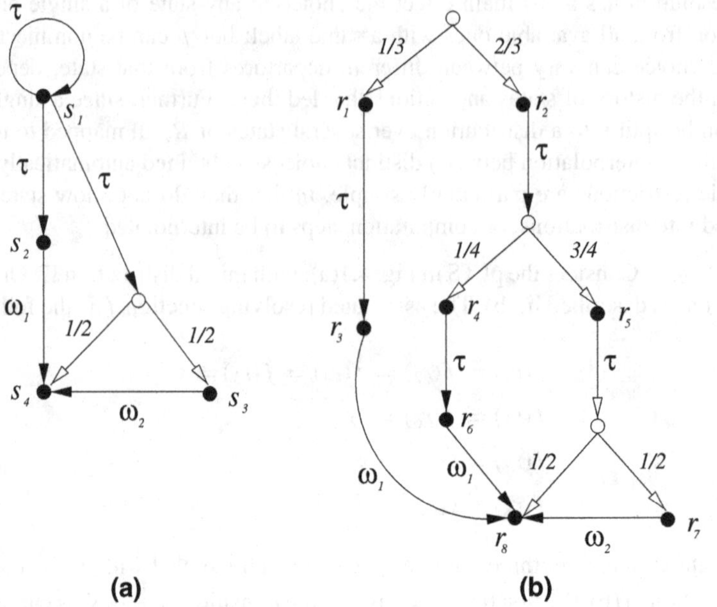

Fig. 4.1 A resolution

Lemma 4.1 *For any test T and process P, if $o_1, o_2 \in \mathcal{A}(T, P)$, then their weighted average $o_1\,{}_p\!\oplus o_2$ is also in $\mathcal{A}(T, P)$ for any $p \in [0, 1]$.*

Proof Let $\langle R_1, \Theta_1, \rightarrow_1\rangle, \langle R_2, \Theta_2, \rightarrow_2\rangle$ be the resolutions of $[\![T|_{\mathsf{Act}}P]\!]$ that give rise to o_1, o_2. We take their disjoint union, except initially we let $\Theta := \Theta_1\,{}_p\!\oplus\Theta_2$, to define the new resolution $\langle R_1 \cup R_2, \Theta, \rightarrow_1 \cup \rightarrow_2\rangle$. It is easy to see that the new resolution generates the interpolated tuple $o_1\,{}_p\!\oplus o_2$ as required. □

Definition 4.4 (Probabilistic Testing Preorders)

(i) $P \sqsubseteq^{\Omega}_{\mathrm{pmay}} Q$ if for every Ω test T, $\mathcal{A}(T, P) \leq_{\mathrm{Ho}} \mathcal{A}(T, Q)$.
(ii) $P \sqsubseteq^{\Omega}_{\mathrm{pmust}} Q$ if for every Ω test T, $\mathcal{A}(T, P) \leq_{\mathrm{Sm}} \mathcal{A}(T, Q)$.

These preorders are abbreviated to $P \sqsubseteq_{\mathrm{pmay}} Q$ and $P \sqsubseteq_{\mathrm{pmust}} Q$, when $|\Omega| = 1$, and their kernels are denoted by $\simeq_{\mathrm{pmay}}$ and $\simeq_{\mathrm{pmust}}$, respectively.

If $|\Omega| = 1$, then vector-based testing degenerates into scalar testing. In general, if $|\Omega| > 1$, then vector-based testing appears more discriminating than scalar testing. Surprisingly, if we restrict ourselves to finitary processes, then scalar testing is equally powerful as vector-based testing. A key ingredient of the proof is to exploit a method of testing called *reward testing*, which will be introduced in Sect. 4.4.

4.3 Bounded Continuity

In this section we introduce a notion of bounded continuity for real-valued functions. We will exploit it to show the continuity of the function C defined in (4.1), and then use it again in Chap. 6. We begin with a handy lemma below.

Lemma 4.2 (Exchange of Suprema) *Let function $g : \mathbb{N} \times \mathbb{N} \to \mathbb{R}$ be such that it is*

(i) monotone in both of its arguments, so that $i_1 \leq i_2$ implies $g(i_1, j) \leq g(i_2, j)$, and $j_1 \leq j_2$ implies $g(i, j_1) \leq g(i, j_2)$, for all $i, i_1, i_2, j, j_1, j_2 \in \mathbb{N}$, and
(ii) bounded above, so that there is a $c \in \mathbb{R}_{\geq 0}$ with $g(i, j) \leq c$ for all $i, j \in \mathbb{N}$.

Then

$$\lim_{i\to\infty} \lim_{j\to\infty} g(i, j) = \lim_{j\to\infty} \lim_{i\to\infty} g(i, j).$$

Proof Conditions (i) and (ii) guarantee the existence of all the limits. Moreover, for a nondecreasing sequence, its limit and supremum agree, and both sides equal the supremum of all $g(i, j)$ for $i, j \in \mathbb{N}$. In fact, $(\mathbb{R}, \leq)$ is a complete partially ordered set (CPO) and it is a basic result of CPOs [7] that

$$\bigsqcup_{i\in\mathbb{N}} \left(\bigsqcup_{j\in\mathbb{N}} g(i, j) \right) = \bigsqcup_{j\in\mathbb{N}} \left(\bigsqcup_{i\in\mathbb{N}} g(i, j) \right).$$ □

The following proposition states that some real functions satisfy the property of *bounded continuity*, which allows the exchange of limit and sum operations.

Proposition 4.2 (Bounded Continuity—Nonnegative Function) *Suppose a function $f : \mathbb{N} \times \mathbb{N} \to \mathbb{R}_{\geq 0}$ satisfies the following conditions:*

C1. *f is monotone in the second parameter, i.e. $j_1 \leq j_2$ implies $f(i, j_1) \leq f(i, j_2)$ for all $i, j_1, j_2 \in \mathbb{N}$;*
C2. *for any $i \in \mathbb{N}$, the limit $\lim_{j\to\infty} f(i, j)$ exists;*
C3. *the partial sums $S_n = \sum_{i=0}^{n} \lim_{j\to\infty} f(i, j)$ are bounded, i.e. there exists some $c \in \mathbb{R}_{\geq 0}$ such that $S_n \leq c$ for all $n \geq 0$.*

Then the following equality holds:

$$\sum_{i=0}^{\infty} \lim_{j\to\infty} f(i, j) = \lim_{j\to\infty} \sum_{i=0}^{\infty} f(i, j).$$

Proof Let $g : \mathbb{N} \times \mathbb{N} \to \mathbb{R}_{\geq 0}$ be the function defined by $g(n, j) = \sum_{i=0}^{n} f(i, j)$. It is easy to see that g is monotone in both arguments. By **C1** and **C2**, we have that $f(i, j) \leq \lim_{j\to\infty} f(i, j)$ for any $i, j \in \mathbb{N}$. So for any $j, n \in \mathbb{N}$, we have

$$g(n, j) = \sum_{i=0}^{n} f(i, j) \leq \sum_{i=0}^{n} \lim_{j\to\infty} f(i, j) \leq c$$

according to **C3**. In other words, g is bounded above. Therefore, we can apply Lemma 4.2 and obtain

$$\lim_{n\to\infty}\lim_{j\to\infty}\sum_{i=0}^{n} f(i,j) = \lim_{j\to\infty}\lim_{n\to\infty}\sum_{i=0}^{n} f(i,j). \tag{4.3}$$

For any $j \in \mathbb{N}$, the sequence $\{g(n,j)\}_{n\geq 0}$ is nondecreasing and bounded, so its limit $\sum_{i=0}^{\infty} f(i,j)$ exists. That is,

$$\lim_{n\to\infty}\sum_{i=0}^{n} f(i,j) = \sum_{i=0}^{\infty} f(i,j). \tag{4.4}$$

In view of **C2**, we have that for any given $n \in \mathbb{N}$, the limit $\lim_{j\to\infty}\sum_{i=0}^{n} f(i,j)$ exists and

$$\sum_{i=0}^{n}\lim_{j\to\infty} f(i,j) = \lim_{j\to\infty}\sum_{i=0}^{n} f(i,j). \tag{4.5}$$

By **C3**, the sequence $\{S_n\}_{n\geq 0}$ is bounded. Since it is also nondecreasing, it converges to $\sum_{i=0}^{\infty}\lim_{j\to\infty} f(i,j)$. That is,

$$\lim_{n\to\infty}\sum_{i=0}^{n}\lim_{j\to\infty} f(i,j) = \sum_{i=0}^{\infty}\lim_{j\to\infty} f(i,j). \tag{4.6}$$

Hence the left-hand side of the desired equality exists. By combining (4.3)–(4.6), we obtain the result, $\sum_{i=0}^{\infty}\lim_{j\to\infty} f(i,j) = \lim_{j\to\infty}\sum_{i=0}^{\infty} f(i,j)$. □

Proposition 4.3 (Bounded Continuity—General Function) *Let* $f : \mathbb{N}\times\mathbb{N} \to \mathbb{R}$ *be a function that satisfies the following conditions*

C1. $j_1 \leq j_2$ *implies* $|f(i,j_1)| \leq |f(i,j_2)|$ *for all* $i, j_1, j_2 \in \mathbb{N}$*;*
C2. *for any* $i \in \mathbb{N}$*, the limit* $\lim_{j\to\infty}|f(i,j)|$ *exists;*
C3. *the partial sums* $S_n = \sum_{i=0}^{n}\lim_{j\to\infty}|f(i,j)|$ *are bounded, i.e. there exists some* $c \in \mathbb{R}_{\geq 0}$ *such that* $S_n \leq c$*, for all* $n \geq 0$*;*
C4. *for all* $i, j_1, j_2 \in \mathbb{N}$*, if* $j_1 \leq j_2$ *and* $f(i,j_1) > 0$*, then* $f(i,j_2) > 0$*.*

Then the following equality holds:

$$\sum_{i=0}^{\infty}\lim_{j\to\infty} f(i,j) = \lim_{j\to\infty}\sum_{i=0}^{\infty} f(i,j).$$

Proof By Proposition 4.2 and conditions **C1**, **C2** and **C3**, we infer that

$$\lim_{j\to\infty}\sum_{i=0}^{\infty}|f(i,j)| = \sum_{i=0}^{\infty}\lim_{j\to\infty}|f(i,j)|. \tag{4.7}$$

C4 implies that the function $g : \mathbb{N} \times \mathbb{N} \to \mathbb{R}_{\geq 0}$ given by $g(i, j) := f(i, j) + |f(i, j)|$ satisfies conditions **C1**, **C2** and **C3** of Proposition 4.2. In particular, the limit $\lim_{j\to\infty} g(i, j) = 0$ if $f(i, j) \leq 0$ for all $j \in \mathbb{N}$, and $\lim_{j\to\infty} g(i, j) = 2\lim_{j\to\infty} |f(i, j)|$ otherwise. Hence

$$\lim_{j\to\infty} \sum_{i=0}^{\infty} (f(i, j) + |f(i, j)|) = \sum_{i=0}^{\infty} \lim_{j\to\infty} (f(i, j) + |f(i, j)|). \quad (4.8)$$

Since $\sum_{i=0}^{\infty} f(i, j) = \sum_{i=0}^{\infty} (f(i, j) + |f(i, j)|) - \sum_{i=0}^{\infty} |f(i, j)|$, we have

$$\begin{aligned}
\lim_{j\to\infty} \sum_{i=0}^{\infty} f(i, j) &= \lim_{j\to\infty} \left(\sum_{i=0}^{\infty} (f(i, j) + |f(i, j)|) - \sum_{i=0}^{\infty} |f(i, j)|\right) \\
&\quad \text{[existence of the two limits by (4.7) and (4.8)]} \\
&= \lim_{j\to\infty} \sum_{i=0}^{\infty} (f(i, j) + |f(i, j)|) - \lim_{j\to\infty} \sum_{i=0}^{\infty} |f(i, j)| \\
&\quad \text{[by (4.7) and (4.8)]} \\
&= \sum_{i=0}^{\infty} \lim_{j\to\infty} (f(i, j) + |f(i, j)|) - \sum_{i=0}^{\infty} \lim_{j\to\infty} |f(i, j)| \\
&= \sum_{i=0}^{\infty} (\lim_{j\to\infty} (f(i, j) + |f(i, j)|) - \lim_{j\to\infty} |f(i, j)|) \\
&= \sum_{i=0}^{\infty} \lim_{j\to\infty} (f(i, j) + |f(i, j)| - |f(i, j)|) \\
&= \sum_{i=0}^{\infty} \lim_{j\to\infty} f(i, j).
\end{aligned}$$

□

Lemma 4.3 *The function $\mathcal{C}$ defined in* (4.1) *is continuous.*

Proof Let $f_0 \leq f_1 \leq \ldots$ be an increasing chain in $R \to [0, 1]^{\Omega}$. We need to show that

$$\mathcal{C}\left(\bigsqcup_{n\geq 0} f_n\right) = \bigsqcup_{n\geq 0} \mathcal{C}(f_n). \quad (4.9)$$

For any $r \in R$, we are in one of the following three cases:

1. $r \xrightarrow{\omega}$ for some $\omega \in \Omega$. We have

$$\begin{aligned}
\mathcal{C}\left(\bigsqcup_{n\geq 0} f_n\right)(r)(\omega) &= 1 && \text{by (4.1)} \\
&= \bigsqcup_{n\geq 0} 1 \\
&= \bigsqcup_{n\geq 0} \mathcal{C}(f_n)(r)(\omega) \\
&= \left(\bigsqcup_{n\geq 0} \mathcal{C}(f_n)\right)(r)(\omega)
\end{aligned}$$

and

$$\mathcal{C}\left(\bigsqcup_{n\geq 0} f_n\right)(r)(\omega') = 0 = \left(\bigsqcup_{n\geq 0}\mathcal{C}(f_n)\right)(r)(\omega')$$

for all $\omega' \neq \omega$.

2. $r \not\rightarrow$. Similar to the last case; we have

$$\mathcal{C}\left(\bigsqcup_{n\geq 0} f_n\right)(r)(\omega) = 0 = \left(\bigsqcup_{n\geq 0}\mathcal{C}(f_n)\right)(r)(\omega)$$

for all $\omega \in \Omega$.

3. Otherwise, $r \xrightarrow{\tau} \Delta$ for some $\Delta \in \mathcal{D}(R)$. Then we infer that for any $\omega \in \Omega$,

$$\begin{aligned}
\mathcal{C}\left(\bigsqcup\nolimits_{n\geq 0} f_n\right)(r)(\omega) &= \left(\bigsqcup\nolimits_{n\geq 0} f_n\right)(\Delta)(\omega) && \text{by (4.1)}\\
&= \sum\nolimits_{r\in\lceil\Delta\rceil} \Delta(r)\cdot\left(\bigsqcup\nolimits_{n\geq 0} f_n\right)(r)(\omega)\\
&= \sum\nolimits_{r\in\lceil\Delta\rceil} \Delta(r)\cdot\left(\bigsqcup\nolimits_{n\geq 0} f_n(r)\right)(\omega)\\
&= \sum\nolimits_{r\in\lceil\Delta\rceil} \bigsqcup\nolimits_{n\geq 0} \Delta(r)\cdot f_n(r)(\omega)\\
&= \sum\nolimits_{r\in\lceil\Delta\rceil} \lim_{n\to\infty} \Delta(r)\cdot f_n(r)(\omega)\\
&= \lim_{n\to\infty} \sum\nolimits_{r\in\lceil\Delta\rceil} \Delta(r)\cdot f_n(r)(\omega) && \text{by Proposition 4.2}\\
&= \bigsqcup\nolimits_{n\geq 0} \sum\nolimits_{r\in\lceil\Delta\rceil} \Delta(r)\cdot f_n(r)(\omega)\\
&= \bigsqcup\nolimits_{n\geq 0} f_n(\Delta)(\omega)\\
&= \bigsqcup\nolimits_{n\geq 0} \mathcal{C}(f_n)(r)(\omega)\\
&= \left(\bigsqcup\nolimits_{n\geq 0} \mathcal{C}(f_n)\right)(r)(\omega).
\end{aligned}$$

In the above reasoning, Proposition 4.2 is applicable because we can define the function $f : R \times \mathbb{N} \to \mathbb{R}_{\geq 0}$ by letting $f(r,n) = \Delta(r)\cdot f_n(r)(\omega)$ and check that f satisfies the three conditions in Proposition 4.2. If R is finite, we can extend it to a countable set $R' \supseteq R$ and require $f(r',n) = 0$ for all $r' \in R'\backslash R$ and $n \in \mathbb{N}$.

a) f satisfies condition **C1**. For any $r \in R$ and $n_1, n_2 \in \mathbb{N}$, if $n_1 \leq n_2$, then $f_{n_1} \leq f_{n_2}$. It follows that

$$f(r,n_1) \;=\; \Delta(r)\cdot f_{n_1}(r)(\omega) \;\leq\; \Delta(r)\cdot f_{n_2}(r)(\omega) \;=\; f(r,n_2).$$

b) f satisfies condition **C2**. For any $r \in R$, the sequence $\{\Delta(r)\cdot f_n(r)(\omega)\}_{n=0}^{\infty}$ is nondecreasing and bounded by $\Delta(r)$. Therefore, the limit $\lim_{n\to\infty} f(r,n)$ exists.

c) f satisfies condition **C3**. For any $R'' \subseteq R$, we can see that the partial sum $\sum_{r \in R''} \lim_{n\to\infty} f(r,n)$ is bounded because

$$\sum_{r \in R''} \lim_{n\to\infty} f(r,n) = \sum_{r \in R''} \lim_{n\to\infty} \Delta(r) \cdot f_n(r) \leq \sum_{r \in R''} \Delta(r) \leq \sum_{r \in R} \Delta(r) = 1. \quad \square$$

Because of Lemma 4.3 and Proposition 2.1, the least fixed point of C can be written as $\mathbb{V} = \bigsqcup_{n \in \mathbb{N}} C^n(\bot)$, where $\bot(r) = 0$ for all $r \in R$.

4.4 Reward Testing

In this section we introduce an alternative testing approach based on the probabilistic testing discussed in Sect. 4.2. The idea is to associate each success action $\omega \in \Omega$ a reward, and performing a success action means accumulating some reward. The outcomes of this reward testing are expected rewards.

4.4.1 A Geometric Property

We have seen in Proposition 4.1 that the comparison of two sets with respect to the Hoare and Smyth preorders can be simplified if they are closed subsets of $[0,1]$, as it suffices to consider their maximum and minimum elements. This simplification does not apply if we want to compare two subsets of $[0,1]^n$, even if they are closed, because maximum and minimum elements might not exist for sets of vectors. However, we can convert a set of vectors into a set of scalars by taking the expected reward entailed by each vector with respect to a reward vector. Interestingly, the comparison of two sets of vectors is related to the comparison of suprema and infima of two sets of scalars, provided some closure conditions are imposed. Therefore, to some extent we generalise Proposition 4.1 from the comparison of sets of scalars to the comparison of sets of vectors. Mathematically, the result can be viewed as an analytic property in geometry, which could be of independent interest.

Suppose that in vector-based testing we use at most n success actions taken from the set $\Omega = \{\omega_1, ..., \omega_n\}$ with $n > 1$. A testing outcome $o \in [0,1]^\Omega$ can be viewed as the n-dimensional vector $\langle o(\omega_1), ..., o(\omega_n)\rangle$ where $o(\omega_i)$ is the probability of successfully reaching success action ω_i. Similarly, a *reward vector* $h \in [0,1]^\Omega$ can be regarded as the vector $\langle h(\omega_1), ..., h(\omega_n)\rangle$ where $h(\omega_i)$ is the reward given to ω_i. We sometimes take the dot product of a particular vector $h \in [0,1]^n$ and a set of vectors $O \subseteq [0,1]^n$, resulting in a set of scalars given by $h \cdot O := \{h \cdot o \mid o \in O\}$.

Definition 4.5 A subset O of the n-dimensional Euclidean space is *p-closed* (for *probabilistically* closed) iff

- It is *convex*, and
- It is *Cauchy closed*, that is, it contains all its limit points in the usual Euclidean metric, and is *bounded*.[1]

We are now ready to generalise Proposition 4.1 in order to compare two sets of vectors. Here we require one set to be *p*-closed, which allows us to appeal to the separation theorem from discrete geometry (cf. Theorem 2.7); Cauchy closure alone is not enough.

Theorem 4.1 *Let* A, B *be subsets of* $[0,1]^n$*; then we have*

$$A \le_{\mathrm{Ho}} B \text{ iff } \forall h \in [0,1]^n : \bigsqcup h \cdot A \le \bigsqcup h \cdot B \quad \text{if } B \text{ is p-closed, and}$$
$$A \le_{\mathrm{Sm}} B \text{ iff } \forall h \in [0,1]^n : \bigsqcap h \cdot A \le \bigsqcap h \cdot B \quad \text{if } A \text{ is p-closed.}$$

Proof We consider first the *only-if* direction for the Smyth case:

$$\begin{array}{lll} & A \le_{\mathrm{Sm}} B & \\ \textit{iff} & \forall b \in B : \exists a \in A : a \le b & \text{Definition of } \le_{\mathrm{Ho}} \\ \text{implies} & \forall h \in [0,1]^n : \forall b \in B : \exists a \in A : h \cdot a \le h \cdot b & h \ge 0 \\ \text{implies} & \forall h \in [0,1]^n : \forall b \in B : \bigsqcap h \cdot A \le h \cdot b & \bigsqcap h \cdot A \le h \cdot a \\ \text{implies} & \forall h \in [0,1]^n : \bigsqcap h \cdot A \le \bigsqcap h \cdot B & \text{Definition of infimum.} \end{array}$$

For the *if*-direction, we use separating hyperplanes, proving the contrapositive:

$$\begin{array}{lll} & A \not\le_{\mathrm{Sm}} B & \\ \textit{iff} & \forall a \in A : \neg(a \le b) & \text{Definition of } \le_{\mathrm{Sm}}\text{; for some } b \in B \\ \textit{iff} & A \cap B' = \emptyset & \text{define } B' := \{b' \in \mathbb{R}^n \mid b' \le b\} \\ \textit{iff} & \exists h \in \mathbb{R}^n, c \in \mathbb{R} : & \\ & \quad \forall a \in A, b' \in B' : & \\ & \qquad h \cdot b' < c < h \cdot a. & \end{array}$$

In the last step of the reasoning above, we have used Theorem 2.7 as A is *p*-closed and B' is convex and Cauchy closed by construction; see Fig. 4.2. Moreover, without loss of generality, the inequality can be in the direction shown, else we simply multiply h, c by -1.

We now argue that h is nonnegative, whence by scaling of h, c we obtain without loss of generality that $h \in [0,1]^n$. Assume for a contradiction that $h_i < 0$. Choose

[1] Cauchy closure and boundedness together amount to *compactness*.

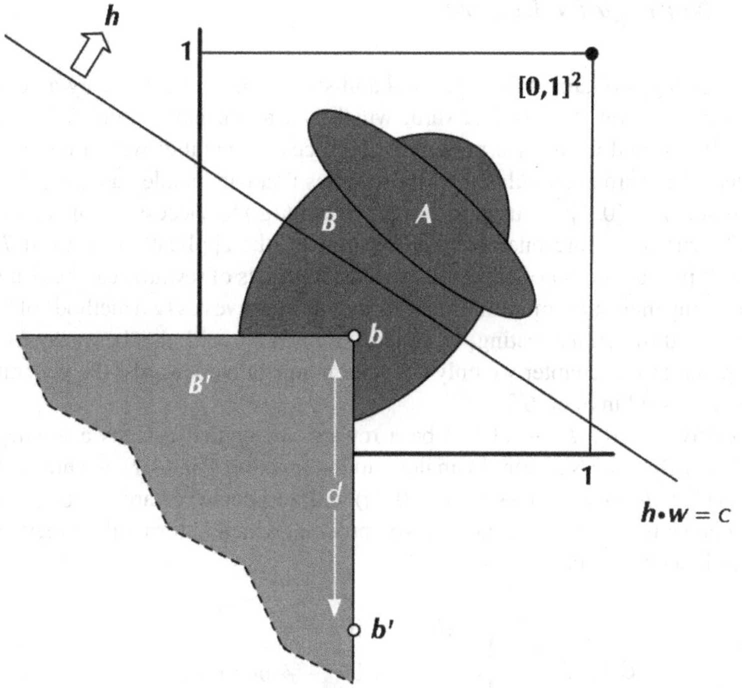

Fig. 4.2 Separation by a hyperplane. (Reprinted from [8], with kind permission from Springer Science+Business Media)

scalar $d \geq 0$ large enough so that the point $b' := (b_1, \cdots, b_i - d, \cdots, b_n)$ falsifies $h \cdot b' < c$; since b' is still in B', it contradicts the separation. Thus we continue

	$\exists h \in [0,1]^n, c \in \mathbb{R}:$	
iff	$\forall a \in A, b' \in B':$	above comments concerning d
	$h \cdot b' < c < h \cdot a$	
iff	$\exists h \in [0,1]^n, c \in \mathbb{R}: \forall a \in A: h \cdot b < c < h \cdot a$	set b' to b; note $b \in B'$
implies	$\exists h \in [0,1]^n, c \in \mathbb{R}: h \cdot b < c \leq \bigsqcap h \cdot A$	property of infimum
implies	$\exists h \in [0,1]^n, c \in \mathbb{R}: \bigsqcap h \cdot B < c \leq \bigsqcap h \cdot A$	$b \in B$, hence $\bigsqcap h \cdot B \leq h \cdot b$
implies	$\neg(\forall h \in [0,1]^n: \bigsqcap h \cdot A \leq \bigsqcap h \cdot B)$	

The proof for the Hoare case is analogous. □

4.4.2 Nonnegative Rewards

Reward testing is obtained from the probabilistic testing in Sect. 4.2 by associating each success action $\omega \in \Omega$ a reward, which is a nonnegative number in the unit interval $[0, 1]$, and a run of a probabilistic process in parallel with a test yielding an expected reward accumulated by those states that can enable success actions. A reward tuple $h \in [0, 1]^\Omega$ is used to assign reward $h(\omega)$ to success action ω, for each $\omega \in \Omega$. Due to the presence of nondeterminism, the application of a test T to a process P produces a set of expected rewards. Two sets of rewards can be compared by examining their supremum/infimum elements; this gives us two methods of testing called reward may/must testing. In general, a reward could also be a negative real number. But in this chapter we only consider nonnegative rewards; the general case will be discussed in Sect. 6.9.

Formally, let $h : \Omega \rightarrow [0, 1]$ be a reward vector that assigns a nonnegative reward to each success action. In analogy to the function $\mathcal{C}$ in (4.1), we now define a function $\mathcal{C}^h : (R \rightarrow [0, 1]) \rightarrow (R \rightarrow [0, 1])$ with respect to reward vector h in order to associate a reward with a deterministic process, which in turn will associate a set of rewards with a process.

$$\mathcal{C}^h(f)(r) := \begin{cases} h(\omega) & \text{if } r \xrightarrow{\omega} \\ 0 & \text{if } r \stackrel{\omega}{\not\rightarrow} \text{ and } r \stackrel{\tau}{\not\rightarrow} \\ \mathrm{Exp}_\Delta(f) & \text{if } r \stackrel{\omega}{\not\rightarrow} \text{ and } r \xrightarrow{\tau} \Delta. \end{cases} \tag{4.10}$$

The function $\mathcal{C}^h$ is also continuous, thus has a least fixed point $\mathbb{V}^h$. Let $\mathcal{A}^h(T, P)$ denote the set of rewards

$$\{\mathrm{Exp}_\Theta(\mathbb{V}^h) \mid \langle R, \Theta, \rightarrow \rangle \text{ is a resolution of } [\![T|_{\mathsf{Act}}P]\!]\} .$$

Definition 4.6 Let P and Q be two probabilistic processes. We define two reward testing preorders:

(i) $P \sqsubseteq^\Omega_{\mathbf{nrmay}} Q$ if for every Ω test T and nonnegative reward tuple $h \in [0, 1]^\Omega$, $\bigsqcup \mathcal{A}^h(T, P) \leq \bigsqcup \mathcal{A}^h(T, Q)$.
(ii) $P \sqsubseteq^\Omega_{\mathbf{nrmust}} Q$ if for every Ω test T and nonnegative reward tuple $h \in [0, 1]^\Omega$, $\bigsqcap \mathcal{A}^h(T, P) \leq \bigsqcap \mathcal{A}^h(T, Q)$.

These preorders are abbreviated to $P \sqsubseteq_{\mathbf{nrmay}} Q$ and $P \sqsubseteq_{\mathbf{nrmust}} Q$, when $|\Omega| = 1$.

Example 4.2 Let us use the notation $a.\omega$ from the calculus of communicating systems to mean the test that can perform action a followed by ω before reaching a deadlock state. By applying this test to the process Q_1 in Fig. 4.3a, we obtain the pLTS in Fig. 4.3b that is already deterministic, hence has only one resolution itself. Moreover, the outcome $\mathbb{V}^h$ associated with it is determined by its value at the state s_0. This in turn is the least solution of the equation

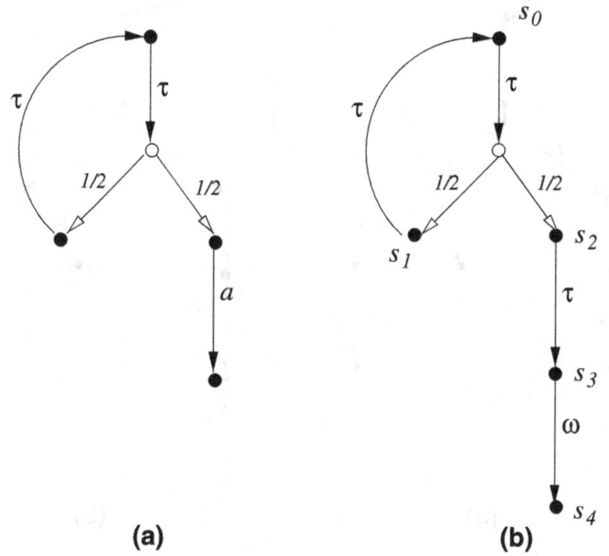

Fig. 4.3 Testing the process Q_1

$$\mathbb{V}^h(s_0) = \frac{1}{2} \cdot \mathbb{V}^h(s_0) + \frac{1}{2}h(\omega).$$

In fact this equation has a unique solution in $[0, 1]$, namely $h(\omega)$. Therefore, $\mathcal{A}^h(a.\omega, Q_1) = \{h(\omega)\}$.

Example 4.3 Consider the process Q_2 depicted in Fig. 4.4a and the application of the test $T = a.\omega$ to it; this is outlined in Fig. 4.4b. In the pLTS of $T\,|_{\mathsf{Act}}Q_2$, for each $k \geq 1$ there is a resolution R_k such that $\mathbb{V}^h(R_k) = (1 - \frac{1}{2^k})$; intuitively it goes around the loop $(k-1)$ times before, at last, taking the right-hand τ action. Thus $\mathcal{A}^h(T, Q_2)$ contains $(1 - \frac{1}{2^k})h(\omega)$ for every $k \geq 1$. But it also contains $h(\omega)$, because of the resolution that takes the left-hand τ move every time. Therefore, $\mathcal{A}^h(T, Q_2)$ includes the set

$$\left\{\left(1 - \frac{1}{2}\right) h\,(\omega),\ \left(1 - \frac{1}{2^2}\right) h(\omega), \ldots,\ \left(1 - \frac{1}{2^k}\right) h(\omega), \ldots\ h(\omega)\right\}.$$

From later results it will follow that $\mathcal{A}^h(T, Q_2)$ is actually the convex closure of this set, namely $[\frac{1}{2}h(\omega), h(\omega)]$.

4.5 Extremal Reward Testing

In the previous section, our approach to testing consisted of two parts:

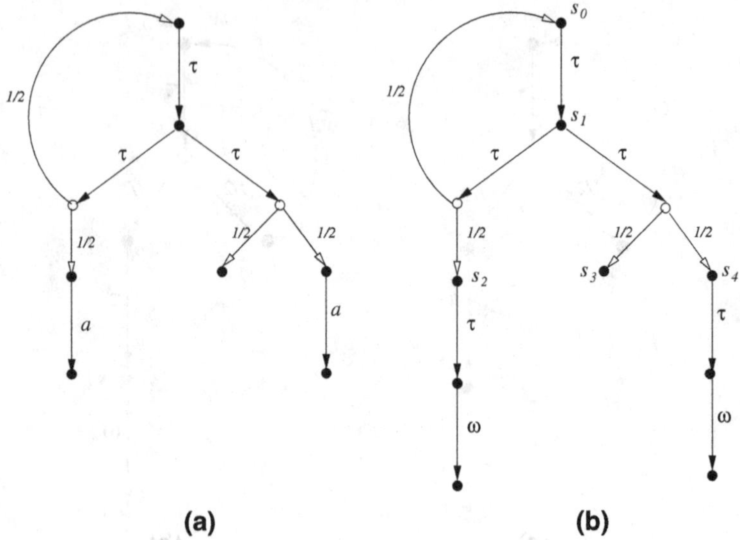

Fig. 4.4 Testing the process Q_2

(1) For each test T, process P and reward vector h, calculate a set of outcomes $\mathcal{A}^h(T, P)$, which is a subset of $[0, 1]$.
(2) For each pair of processes P, Q, compare the corresponding sets of outcomes, $\mathcal{A}^h(T, P)$ and $\mathcal{A}^h(T, Q)$ for every test T and reward vector h, in terms of their suprema and infima.

But our methods for comparing sets of outcomes do not necessarily require us to calculate the entire set of outcomes. For example, Proposition 4.1 says that for closed sets, it suffices to compare extremal outcomes. Here we propose an alternative approach to testing based on calculating directly the extremal values of possible outcomes.

Note that the results in Sects. 4.5 and 4.6 are not used in the rest of this book, except for a notion of extreme policy (Definition 4.8) that is referred to in Sect. 6.5.1. So the reader may skip these two sections in the first reading, and return back when necessary.

The function $\mathcal{C}^h$ used to associate a reward with a resolution is only defined in (4.10) above for deterministic pLTSs. Here we consider generalisations to an arbitrary finitely branching pLTS $\langle S, \Omega_\tau, \rightarrow \rangle$.

Now consider the function $\mathcal{C}^h_{\min} : (S \rightarrow [0,1]) \rightarrow (S \rightarrow [0,1])$ defined by:

$$\mathcal{C}^h_{\min}(f)(s) = \begin{cases} h(\omega) & \text{if } s \xrightarrow{\omega} \\ 0 & \text{if } s \stackrel{\omega}{\not\rightarrow} \text{ and } s \stackrel{\tau}{\not\rightarrow} \\ min\{\operatorname{Exp}_\Delta(f) \mid s \xrightarrow{\tau} \Delta\} & \text{if s } \stackrel{\omega}{\not\rightarrow} \text{ and } s \xrightarrow{\tau}. \end{cases}$$

In a similar fashion, we can define the function $\mathcal{C}^h_{\max} : (S \rightarrow [0,1]) \rightarrow (S \rightarrow [0,1])$ that uses the *max* function in place of *min*. Both these functions are monotone and therefore have least fixed points, which we abbreviate to $\mathbb{V}^h_{\min}$ and $\mathbb{V}^h_{\max}$, respectively.

Lemma 4.4 *For any finitely branching pLTS and reward vector h,*

(a) both functions $\mathcal{C}^h_{min}$ and $\mathcal{C}^h_{max}$ are continuous;
(b) both results functions $\mathbb{V}^h_{min}$ and $\mathbb{V}^h_{max}$ are continuous.

Proof Again the proof of part (a) is nontrivial. See Lemma 4.3 for the continuity of $\mathcal{C}$. The continuity of $\mathcal{C}^h_{\min}$ and $\mathcal{C}^h_{\min}$ can be similarly shown. However part (b) is an immediate consequence.
So in analogy with the evaluation function $\mathbb{V}^h$, these results functions can be captured by a chain of approximants:

$$\mathbb{V}^h_{\min} = \bigsqcup_{n\in\mathbb{N}} \mathbb{V}^{h,n}_{\min} \qquad \text{and} \qquad \mathbb{V}^h_{\max} = \bigsqcup_{n\in\mathbb{N}} \mathbb{V}^{h,n}_{\max} \tag{4.11}$$

where $\mathbb{V}^{h,0}_{\min}(s) = \mathbb{V}^{h,0}_{\max}(s) = 0$ for every state $s \in S$, and

- $\mathbb{V}^{h,(k+1)}_{\min} = \mathcal{C}_{\min}(\mathbb{V}^{h,k}_{\min})$
- $\mathbb{V}^{h,(k+1)}_{\max} = \mathcal{C}_{\max}(\mathbb{V}^{h,k}_{\max})$.

Now for a test T, a process P, and a reward vector h, we have two ways of defining the outcome of the application of T to P:

$$\mathcal{A}^h_{\min}(T,P) = \mathbb{V}^h_{\min}([\![T\,|_{\mathsf{Act}}\,P]\!])$$
$$\mathcal{A}^h_{\max}(T,P) = \mathbb{V}^h_{\max}([\![T\,|_{\mathsf{Act}}\,P]\!]).$$

Here $\mathcal{A}_{\min}(T,P)$ returns a single probability p, estimating the minimal probability of success; it is a pessimistic estimate. On the other hand, $\mathcal{A}_{\max}(T,P)$ is optimistic, in that it gives the maximal probability of success.

Definition 4.7 The extremal reward testing preorders $\sqsubseteq^{\Omega}_{\text{ermay}}$ and $\sqsubseteq^{\Omega}_{\text{ermust}}$ are defined as follows:

(i) $P \sqsubseteq^{\Omega}_{\text{ermay}} Q$ if for every Ω test T and nonnegative reward tuple $h \in [0,1]^{\Omega}$, $\mathcal{A}^h_{\max}(T,P) \le \mathcal{A}^h_{\max}(T,Q)$.
(ii) $P \sqsubseteq^{\Omega}_{\text{ermust}} Q$ if for every Ω test T and nonnegative reward tuple $h \in [0,1]^{\Omega}$, $\mathcal{A}^h_{\min}(T,P) \le \mathcal{A}^h_{\min}(T,Q)$.

These preorders are abbreviated to $P \sqsubseteq_{\text{ermay}} Q$ and $P \sqsubseteq_{\text{ermust}} Q$ when $|\Omega| = 1$. We use the obvious notation for the kernels of these preorders.

Example 4.4 Let T be the test $a.\omega$. By applying it to the process P in Fig. 4.3a, we obtain the pLTS in (b) that is deterministic and therefore all three functions $\mathbb{V}^h_{\max}$, $\mathbb{V}^h_{\min}$ and $\mathbb{V}^h$ coincide, giving $\mathcal{A}^h_{\max}(a.\omega, P) = \mathcal{A}^h_{\min}(T,P) = h(\omega)$.
Again it is straightforward to establish

$$P \simeq_{\text{ermay}} a.\mathbf{0} \qquad\qquad P \simeq_{\text{ermust}} a.\mathbf{0}$$

where $a.\mathbf{0}$ is the process with the only behaviour of performing action a before halting.

Example 4.5 Consider the pLTS from Fig. 4.4 resulting from the application of the test $T = a.\omega$ to the process Q. It is easy to see that the function $\mathbb{V}^h_{\max}$ satisfies

$$\mathbb{V}^h_{\max}(s_0) = max\left\{\frac{1}{2}h(\omega),\ x\right\} \tag{4.12}$$

$$x = \frac{1}{2}h(\omega) + \frac{1}{2}\cdot \mathbb{V}^h_{\max}(s_0).$$

It is not difficult to show that this has a unique solution, namely $\mathbb{V}^h_{\max}(s_0) = h(\omega)$. With further analysis, one can conclude that

$$Q \simeq_{\text{ermay}} a.\mathbf{0}.$$

If *max* is replaced by *min* in (4.12) above, then the resulting equation also has a unique solution, giving $\mathbb{V}^h_{\min}(s_0) = \frac{1}{2}h(\omega)$. It follows that

$$a.\mathbf{0} \not\sqsubseteq_{\text{ermust}} Q$$

because $\mathbb{V}^h_{\min}([\![T\,|_{\mathsf{Act}}a.\mathbf{0}]\!]) = h(\omega)$. Again further analysis will show

$$Q \sqsubseteq_{\text{ermust}} a.\mathbf{0}$$

4.6 Extremal Reward Testing Versus Resolution-Based Reward Testing

In this section we compare two approaches of testing introduced in the previous two subsections. Our first result is that in the most general setting, they lead to different testing preorders.

Example 4.6 Consider the infinite-state pLTS in Fig. 4.5.

We compare the state s_1 with the process $a.\mathbf{0}$. With the test $a.\omega$ using resolutions, we get:

$$\begin{aligned} \mathcal{A}^h(a.\omega, s_1) &= \updownarrow\left\{0, \left(1-\tfrac{1}{2}\right)h(\omega), \ldots, \left(1-\tfrac{1}{2^k}\right)h(\omega), \ldots\right\} \\ \mathcal{A}^h(a.\omega, a.\mathbf{0}) &= \{h(\omega)\}, \end{aligned} \tag{4.13}$$

which means that $a.\mathbf{0} \not\sqsubseteq_{\text{pmay}} s_1$.

However when we use extremal testing, the test $a.\omega$ cannot distinguish these processes. It is straightforward to see that $\mathbb{V}^h_{\max}(a.\omega\,|_{\mathsf{Act}}\,a.\mathbf{0}) = h(\omega)$. To see that

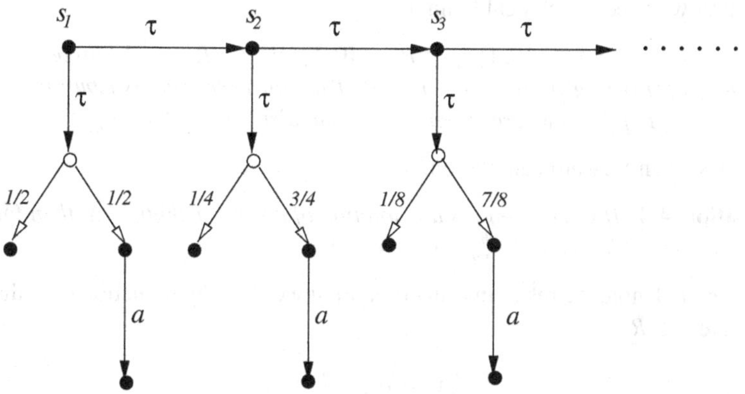

Fig. 4.5 An infinite-state pLTS

$\mathbb{V}^h_{\max}(a.\omega \,|_{\mathsf{Act}}\, s_1)$ also evaluates to $h(\omega)$, we let $x_k = \mathbb{V}^h_{\max}(a.\omega \,|_{\mathsf{Act}}\, s_k)$, for all $k \geq 1$, and we have the following infinite equation system.

$$\begin{aligned} x_1 &= max\left\{\tfrac{1}{2}h(\omega), x_2\right\} \\ x_2 &= max\left\{(1-\tfrac{1}{4})h(\omega), x_3\right\} \\ &\vdots \\ x_k &= max\left\{\left(1-\tfrac{1}{2^k}\right)h(\omega), x_{k+1}\right\} \\ &\vdots \end{aligned}$$

We have $x_k = h(\omega)$ for all $k \geq 1$ as the least solution of the above equation system.

With some more work one can go on to show that no test can distinguish between these processes using optimistic extremal testing, meaning that $a.\mathbf{0} \sqsubseteq_{\mathrm{ermay}} s_1$.

In the remainder of this section, we show that provided some finitary constraints are imposed on the pLTS, extremal reward testing and resolution-based reward testing coincide. First we examine *must* testing, which is easier than the *may* case; this in turn is treated in the following section.

4.6.1 Must Testing

Here we show that provided we restrict our attention to finitely branching processes, there is no difference between extremal *must* testing and resolution-based *must* testing.

Let us consider a pLTS $\langle S, \Omega_\tau, \rightarrow\rangle$, obtained perhaps from applying a test T to a process P in $(T \,|_{\mathsf{Act}}\, P)$. We have two ways of obtaining a result for a distribution of states from S, by applying the function $\mathbb{V}^h_{\min}$, or by using resolutions of the pLTS to realise $\mathbb{V}^h$. Our first result says that regardless of the actual resolution used, the value obtained from the latter will always dominate the former.

But first we need a technical lemma.

Lemma 4.5 *Let $g : S \to [0,1]$, $g' : R \to [0,1]$ and $f : R \to S$ be three functions satisfying $g(f(r)) \leq g'(r)$ for every $r \in R$. Then for every subdistribution Θ over R, $Exp_\Delta(g) \leq Exp_\Theta(g')$ where Δ denotes the subdistribution $Img_\Theta(f)$.*

Proof A straightforward calculation. □

Proposition 4.4 *If $\langle R, \Theta, \to_R \rangle$ is a resolution of a subdistribution Δ, then for any reward vector h it holds that $\mathbb{V}^h_{min}(\Delta) \leq \mathbb{V}^h(\Theta)$.*

Proof Let f denote the resolving function. First we show by induction on n that for every state $r \in R$

$$\mathbb{V}^{h,n}_{\min}(f(r)) \leq \mathbb{V}^{h,n}(r). \tag{4.14}$$

For $n = 0$, this is trivial. We consider the inductive step; note that by the previous lemma, the inductive hypothesis implies that

$$\mathrm{Exp}_\Gamma\left(\mathbb{V}^{h,n}_{\min}\right) \leq \mathrm{Exp}_\Theta\left(\mathbb{V}^{h,n}\right) \tag{4.15}$$

for any pair of subdistributions satisfying $\Gamma = \mathrm{Img}_\Theta(f)$.

If $r \xrightarrow{\omega}_R \Theta$, then $f(r) \xrightarrow{\omega}$, and thus $\mathbb{V}^{h,n+1}_{\min}(f(r)) = h(\omega) = \mathbb{V}^{h,n+1}(r)$. A similar argument applies if $r \not\rightarrow$, that is $r \stackrel{\tau}{\not\rightarrow}$ and $s \stackrel{\omega}{\not\rightarrow}$. So the remaining possibility is that $r \xrightarrow{\tau}_R \Theta$ for some Θ, and $r \stackrel{\omega}{\not\rightarrow}$, where we know $f(r) \xrightarrow{\tau} \mathrm{Img}_\Theta(f)$.

$$\begin{aligned}
\mathbb{V}^{h,n+1}_{\min}(f(r)) &= min\{\mathrm{Exp}_\Delta(\mathbb{V}^{h,n}_{\min}) \mid f(r) \xrightarrow{\tau} \Delta\} \\
&\leq \mathrm{Exp}_\Gamma(\mathbb{V}^{h,n}_{\min}) \qquad \text{where } \Gamma \text{ denotes } \mathrm{Img}_\Theta(f) \\
&\leq \mathbb{V}^{h,n}(\Theta) \qquad \text{by induction and (4.15) above} \\
&= \mathbb{V}^{h,n+1}(r)
\end{aligned}$$

Now by continuity we have from (4.14) that

$$\mathbb{V}^h_{\min}(f(r)) \leq \mathbb{V}^h(r). \tag{4.16}$$

The result now follows by the previous lemma, since if $\langle R, \Theta, \to_R \rangle$ is a resolution of a subdistribution Δ with resolving function f then, by definition, $\Delta = \mathrm{Img}_\Theta(f)$. □

Our next result says that in any finitely branching computation structure, we can find a resolution that realises the function $\mathbb{V}_{\min}$. Moreover this resolution will be static.

Definition 4.8 A (static) *extreme policy* for a pLTS $\langle S, \Omega_\tau, \to \rangle$ is a partial function $\mathsf{ep} : S \rightharpoonup \mathcal{D}(S)$ satisfying:

(a) $s \xrightarrow{\omega}$ implies $s \xrightarrow{\omega} \mathsf{ep}(s)$
(b) otherwise, if $s \xrightarrow{\tau}$, then $s \xrightarrow{\tau} \mathsf{ep}(s)$.

Intuitively an extreme policy ep determines a computation through the pLTS. But this set of possible computations, unlike resolutions as defined in Definition 4.3, is very restrictive. Policy ep decides at each state, once and for all, which of the available τ choices to take; it does not interpolate, and since it is a function of the state, it makes the same choice on every visit. But there are two constraints:

(i) Condition (a) ensures an in-built preference for reporting success; if the state is successful, the policy must also report success;
(ii) Condition (b), together with (a), means that $\mathsf{ep}(s)$ is defined whenever $s \longrightarrow$. This ensures that the policy cannot decide to stop at a state s if there is a possibility of proceeding from s; the computation must proceed, if it is possible to proceed.

An extreme policy ep determines a deterministic pLTS $\langle S, \Omega_\tau, \rightarrow_{\mathsf{ep}}\rangle$, where $\rightarrow_{\mathsf{ep}}$ is determined by $s \rightarrow_{\mathsf{ep}} \mathsf{ep}(s)$. Moreover for any subdistribution Δ over S, it determines the obvious resolution $\langle S, \Delta, \rightarrow_{\mathsf{ep}}\rangle$, with the identity as the associated resolving function. Indeed it is possible to show that every static resolution is determined in this manner by some extreme policy.

Proposition 4.5 *Let Δ be a subdistribution in a finitely branching pLTS $\langle S, \Omega_\tau, \rightarrow\rangle$. Then there exists a static resolution of Δ, say $\langle R, \Theta, \rightarrow_R\rangle$ such that*

$$Exp_\Theta\left(\mathbb{V}^h\right) = Exp_\Delta\left(\mathbb{V}^h_{min}\right)$$

for any reward vector h.

Proof We exhibit the required resolution by defining an extreme policy over S; in other words, the resolution will take the form $\langle S, \Theta, \rightarrow_{\mathsf{ep}}\rangle$ for some extreme policy $\mathsf{ep}(-)$.

We say the extreme policy $\mathsf{ep}(-)$ is *min-seeking* if its domain is $\{ s \in S \mid s \longrightarrow \}$ and it satisfies:

$$\text{if } s \stackrel{\omega}{\not\longrightarrow} \text{ but } s \stackrel{\tau}{\longrightarrow} \text{ then } \mathbb{V}^h_{\min}(\mathsf{ep}(s)) \leq \mathbb{V}^h_{\min}(\Delta) \text{ whenever } s \stackrel{\tau}{\longrightarrow} \Delta.$$

Note that by design, a min-seeking policy satisfies:

$$\text{if } s \stackrel{\omega}{\not\longrightarrow} \text{ but } s \stackrel{\tau}{\longrightarrow} \text{ then } \mathbb{V}^h_{\min}(s) = \mathbb{V}^h_{\min}(\mathsf{ep}(s)). \qquad (4.17)$$

In a finitely branching pLTS, it is straightforward to define a min-seeking extreme policy:

(i) If $s \stackrel{\omega}{\longrightarrow}$, then let $\mathsf{ep}(s)$ be any Δ such that $s \stackrel{\omega}{\longrightarrow} \Delta$.
(ii) Otherwise, if $s \stackrel{\tau}{\longrightarrow}$, let $\{\Delta_1, \dots \Delta_n\}$ be the finite nonempty set $\{ \Delta \mid s \stackrel{\tau}{\longrightarrow} \Delta\}$. Now let $\mathsf{ep}(s)$ be any Δ_k satisfying the property $\mathbb{V}^h_{\min}(\Delta_k) \leq \mathbb{V}^h_{\min}(\Delta_j)$ for every $1 \leq j \leq n$; at least one such Δ_k must exist.

We now show that the static resolution determined by such a policy, $\langle S, \Theta, \rightarrow_{\mathsf{ep}}\rangle$ satisfies the requirements of the proposition. For the sake of clarity, let us write $\mathbb{V}^h_{\mathsf{ep}}(\Delta)$ for the value realised for Δ in this resolution.

We already know, from Proposition 4.4, that $\mathbb{V}^h_{\min}(\Delta) \leq \mathbb{V}^h_{\mathsf{ep}}(\Delta)$ and so we concentrate on the converse, $\mathbb{V}^h_{\mathsf{ep}}(\Delta) \leq \mathbb{V}^h_{\min}(\Delta)$. Recall that the function $\mathbb{V}^h_{\mathsf{ep}}$ is the least fixed point of the function $\mathcal{C}^h$ defined in (4.10) above, and interpreted in the above resolution. So the result follows if we can show that the function $\mathbb{V}^h_{\min}$ is also a fixed point. This amounts to proving

$$\mathbb{V}^h_{\min}(s) = \begin{cases} h(\omega) & \text{if } s \xrightarrow{\omega} \\ 0 & \text{if } s \not\rightarrow \\ \mathbb{V}^h_{\min}(\mathsf{ep}(s)) & \text{otherwise.} \end{cases}$$

However this is a straightforward consequence of (4.17) above. □

Theorem 4.2 *Let P and Q be any finitely branching processes. Then $P \sqsubseteq^{\Omega}_{\mathbf{ermust}} Q$ if and only if $P \sqsubseteq^{\Omega}_{\mathbf{nrmust}} Q$*

Proof It follows from two previous propositions that

$$\mathbb{V}^h_{\min}([\![T\,|_{\mathsf{Act}}\,P]\!]) = \min\{\mathbb{V}^h(\Theta) \mid \langle R, \Theta, \rightarrow\rangle \text{ is a resolution of } [\![T\,|_{\mathsf{Act}}\,P]\!]\}$$

for any test T, process P and reward vector h. Thus, it is immediate that $\sqsubseteq_{\mathbf{ermust}}$ coincides with $\sqsubseteq_{\mathbf{nrmust}}$. □

4.6.2 May Testing

Here we can try to apply the same proof strategy as in the previous section. The analogue to Proposition 4.4 goes through:

Proposition 4.6 *If $\langle R, \Theta, \rightarrow_R\rangle$ is a resolution of Δ, then for any reward vector h, we have $\mathbb{V}^h(\Delta) \leq \mathbb{V}^h_{max}(\Theta)$.*

Proof Similar to the proof of Proposition 4.4.

However the proof strategy used in Proposition 4.5 cannot be used to show that $\mathbb{V}^h_{\max}$ can be realised by some static resolution, as the following example shows.

Example 4.7 In analogy with the definition used in the proof of Proposition 4.5, we say that an extreme policy $\mathsf{ep}(-)$ is *max-seeking* if its domain is precisely the set $\{s \in S \mid s \longrightarrow\}$, and

$$\text{if } s \stackrel{\omega}{\not\longrightarrow} \text{ but } s \xrightarrow{\tau} \text{ then } \mathbb{V}^h_{\max}(\Delta) \leq \mathbb{V}^h_{\max}(\mathsf{ep}(s)) \text{ whenever } s \xrightarrow{\tau} \Delta.$$

This ensures that $\mathbb{V}^h_{\max}(s) = \mathbb{V}^h_{\max}(\mathsf{ep}(s))$, whenever $s \xrightarrow{\tau}$ and $s \stackrel{\omega}{\not\longrightarrow}$, and again it is straightforward to define a max-seeking extreme policy in a finitely branching pLTS. However the resulting resolution does not in general realise the function $\mathbb{V}^h_{\max}$.

To see this, let us consider the (finitely branching) pLTS used in Example 4.6. Here in addition to the two states $\omega.\mathbf{0}$ and $\mathbf{0}$, there is the infinite set $\{s_1, \ldots s_k, \ldots\}$ and the transitions

- $s_k \xrightarrow{\tau} \overline{s_{k+1}}$
- $s_k \xrightarrow{\tau} [\![\mathbf{0}\ _{\frac{1}{2^k}}\!\oplus \omega.\mathbf{0}]\!]$.

We have calculated $\mathbb{V}^h_{\max}(s_k)$ to be $h(\omega)$ for every k, and a max-seeking extreme policy is determined by $\mathsf{ep}(s_k) = s_{k+1}$; indeed this is essentially the only such policy. However the resolution associated with this policy does not realise $\mathbb{V}^h_{\max}$, as $\mathbb{V}^h_{\mathsf{ep}}(s_k) = 0$.

Nevertheless, we will show that if we restrict attention to finitary pLTSs, then there will always exist some static resolution that realises $\mathbb{V}^h_{\max}$. The proof relies on techniques used in Markov process theory [9], and unlike that of Proposition 4.5 is nonconstructive; we simply prove that some such resolution exists, without actually showing how to construct it. Although such techniques are relatively standard in the theory of Markov decision processes, see [9] for example, they are virtually unknown in concurrency theory. So we give a detailed exposition that cumulates in Theorem 4.3.

Consider a set of all functions from a finite set R to $[0, 1]$, denoted by $[0, 1]^R$, and the distance function d over $[0, 1]^R$ defined by $d(f, g) = max|f(r) - g(r)|_{r\in R}$. We can check that $([0, 1]^R, d)$ constitutes a complete metric space. Let $\delta \in (0, 1)$ be a discount factor. The discounted version of the function $\mathcal{C}^h$ in (4.10), $\mathcal{C}^{\delta,h} : (R \to [0, 1]) \to (R \to [0, 1])$ defined by

$$\mathcal{C}^{\delta,h}(f)(r) = \begin{cases} h(\omega) & \text{if } r \xrightarrow{\omega} \\ 0 & \text{if } r \stackrel{\omega}{\not\longrightarrow} \text{ and } r \stackrel{\tau}{\not\longrightarrow} \\ \delta \cdot \mathrm{Exp}_{\Delta}(f) & \text{if } r \stackrel{\omega}{\not\longrightarrow} \text{ and } r \xrightarrow{\tau} \Delta \end{cases} \tag{4.18}$$

is a contraction mapping with constant δ. It follows from the Banach fixed-point theorem (cf. Theorem 2.8) that $\mathcal{C}^{\delta,h}$ has a unique fixed point when $\delta < 1$, which we denote by $\mathbb{V}^{\delta,h}$. On the other hand, it can be shown that $\mathcal{C}^{\delta,h}$ is a continuous function over the complete lattice $[0, 1]^R$. So $\mathbb{V}^{\delta,h}$, as the least fixed point of $\mathcal{C}^{\delta,h}$, has the characterisation $\mathbb{V}^{\delta,h} = \bigsqcup_{n\in\mathbb{N}} \mathbb{V}^{\delta,h,n}$, where $\mathbb{V}^{\delta,h,n}$ is the nth iteration of $\mathcal{C}^{\delta,h}$ over $\bot$. Note that if there is no discount, i.e. $\delta = 1$, we see that $\mathcal{C}^{\delta,h}$ and $\mathbb{V}^{\delta,h}$ coincide with $\mathcal{C}^h$and$\mathbb{V}^h$, respectively. Similarly, we can define $\mathbb{V}^{\delta,h}_{\min}$ and $\mathbb{V}^{\delta,h}_{\max}$.

The functions $\mathcal{C}^{\delta,h}$ and $\mathbb{V}^{\delta,h}_{\max}$ have the following properties.

Lemma 4.6 *Let $h \in [0, 1]^{\Omega}$ be any reward vector.*

1. *For any $\delta \in (0, 1]$, the functions $\mathcal{C}^{\delta,h}$ and $\mathcal{C}^{\delta,h}_{max}$ are continuous;*
2. *If $\delta_1, \delta_2 \in (0, 1]$ and $\delta_1 \leq \delta_2$, then we have $\mathcal{C}^{\delta_1,h} \leq \mathcal{C}^{\delta_2,h}$ and $\mathcal{C}^{\delta_1,h}_{max} \leq \mathcal{C}^{\delta_2,h}_{max}$;*
3. *Let $\{\delta_n\}_{n\geq 1}$ be a nondecreasing sequence of discount factors converging to 1. It holds that $\bigsqcup_{n\in\mathbb{N}} \mathcal{C}^{\delta_n,h} = \mathcal{C}^h$ and $\bigsqcup_{n\in\mathbb{N}} \mathcal{C}^{\delta_n,h}_{max} = \mathcal{C}^h_{max}$.*

Proof We only consider $\mathcal{C}$; the case for $\mathcal{C}_{\max}$ is similar.

1. Similar to the proof of Lemma 4.3.
2. Straightforward by the definition of $\mathcal{C}$.
3. For any $f \in S \to [0, 1]$ and $s \in S$, we show that

$$\mathcal{C}^h(f)(s) = \left(\bigsqcup_{n\in\mathbb{N}} \mathcal{C}^{\delta_n,h}\right)(f)(s). \tag{4.19}$$

We focus on the nontrivial case that $s \xrightarrow{\alpha} \Delta$ for some action α and distribution $\Delta \in \mathcal{D}(S)$.

$$\begin{aligned}\left(\bigsqcup_{n\in\mathbb{N}} \mathcal{C}^{\delta_n,h}\right)(f)(s) &= \bigsqcup_{n\in\mathbb{N}} \mathcal{C}^{\delta_n,h}(f)(s)\\ &= \bigsqcup_{n\in\mathbb{N}} \delta_n \cdot f(\Delta)\\ &= f(\Delta)\cdot\left(\bigsqcup_{n\in\mathbb{N}} \delta_n\right)\\ &= f(\Delta)\cdot 1\\ &= \mathcal{C}^h(f)(s).\end{aligned}$$

□

Lemma 4.7 *Let $h \in [0,1]^\Omega$ be a reward vector and $\{\delta_n\}_{n\geq 1}$ be a nondecreasing sequence of discount factors converging to* 1.

- $\mathbb{V}^h = \bigsqcup_{n\in\mathbb{N}} \mathbb{V}^{\delta_n,h}$
- $\mathbb{V}^h_{max} = \bigsqcup_{n\in\mathbb{N}} \mathbb{V}^{\delta_n,h}_{max}$.

Proof We only consider $\mathbb{V}^h$; the case for $\mathbb{V}^h_{\max}$ is similar. We use the notation $\mathit{lfp}(f)$ for the least fixed point of the function f over a complete lattice. Recall that $\mathbb{V}^h$ and $\mathbb{V}^{\delta_n,h}$ are the least fixed points of $\mathcal{C}^h$ and $\mathcal{C}^{\delta_n,h}$, respectively, so we need to prove that

$$\mathit{lfp}(\mathcal{C}^h) = \bigsqcup_{n\in\mathbb{N}} \mathit{lfp}(\mathcal{C}^{\delta_n,h}). \tag{4.20}$$

We now show two inequations.

For any $n \in \mathbb{N}$, we have $\delta_n \leq 1$, so Lemma 4.6 (2) yields $\mathcal{C}^{\delta_n,h} \leq \mathcal{C}^h$. It follows that $\mathit{lfp}(\mathcal{C}^{\delta_n,h}) \leq \mathit{lfp}(\mathcal{C}^h)$, thus $\bigsqcup_{n\in\mathbb{N}} \mathit{lfp}(\mathcal{C}^{\delta_n,h}) \leq \mathit{lfp}(\mathcal{C}^h)$.

For the other direction, that is $\mathit{lfp}(\mathcal{C}^h) \leq \bigsqcup_{n\in\mathbb{N}} \mathit{lfp}(\mathcal{C}^{\delta_n,h})$, it suffices to show that $\bigsqcup_{n\in\mathbb{N}} \mathit{lfp}(\mathcal{C}^{\delta_n,h})$ is a prefixed point of $\mathcal{C}^h$, i.e.

$$\mathcal{C}^h\left(\bigsqcup_{n\in\mathbb{N}} \mathit{lfp}(\mathcal{C}^{\delta_n,h})\right) \leq \bigsqcup_{n\in\mathbb{N}} \mathit{lfp}(\mathcal{C}^{\delta_n,h}),$$

which we derive as follows. Let $\{\delta_n\}_{n\geq 1}$ be a nondecreasing sequence of discount factors converging to 1.

$$\begin{aligned}&\mathcal{C}^h\left(\bigsqcup_{n\in\mathbb{N}} \mathit{lfp}(\mathcal{C}^{\delta_n,h})\right)\\ &= \left(\bigsqcup_{m\in\mathbb{N}} \mathcal{C}^{\delta_m,h}\right)\left(\bigsqcup_{n\in\mathbb{N}} \mathit{lfp}(\mathcal{C}^{\delta_n,h})\right) && \text{by Lemma 4.6 (3)}\\ &= \bigsqcup_{m\in\mathbb{N}} \mathcal{C}^{\delta_m,h}\left(\bigsqcup_{n\in\mathbb{N}} \mathit{lfp}(\mathcal{C}^{\delta_n,h})\right)\\ &= \bigsqcup_{m\in\mathbb{N}} \bigsqcup_{n\in\mathbb{N}} \mathcal{C}^{\delta_m,h}(\mathit{lfp}(\mathcal{C}^{\delta_n,h})) && \text{by Lemma 4.6 (1)}\\ &= \bigsqcup_{m\in\mathbb{N}} \bigsqcup_{n\geq m} \mathcal{C}^{\delta_m,h}(\mathit{lfp}(\mathcal{C}^{\delta_n,h}))\end{aligned}$$

$$\leq \bigsqcup_{m\in\mathbb{N}} \bigsqcup_{n\geq m} \mathcal{C}^{\delta_n,h}(lfp(\mathcal{C}^{\delta_n,h})) \qquad \text{by Lemma 4.6 (2)}$$
$$= \bigsqcup_{n\in\mathbb{N}} \mathcal{C}^{\delta_n,h}(lfp(\mathcal{C}^{\delta_n,h}))$$
$$= \bigsqcup_{n\in\mathbb{N}} lfp(\mathcal{C}^{\delta_n,h}).$$

This completes the proof of (4.20). □

Lemma 4.8 *Suppose* $\delta \in (0,1]$ *and* Δ° *is a subdistribution in a pLTS* $\langle S, \Omega_\tau, \rightarrow\rangle$. *If* $\langle T, \Theta^\circ, \rightarrow\rangle$ *is a resolution of* Δ°, *then we have* $\mathbb{V}^{\delta,h}(\Theta^\circ) \leq \mathbb{V}^{\delta,h}_{max}(\Delta^\circ)$ *for any reward vector* h.

Proof Let $f : T \rightarrow S$ be the resolving function associated with the resolution $\langle T, \Theta^\circ, \rightarrow\rangle$, we show by induction on n that

$$\mathbb{V}^{\delta,h,n}_{\max}(f(t)) \geq \mathbb{V}^{\delta,h,n}(t) \text{ for any } t \in T. \tag{4.21}$$

The base case $n = 0$ is trivial. We consider the inductive step. If $t \xrightarrow{\omega} \Theta$, then $f(t) \xrightarrow{\omega} f(\Theta)$, thus $\mathbb{V}^{\delta,h,n}_{\max}(f(t)) = h(\omega) = \mathbb{V}^{\delta,h,n}(t)$. Now suppose that $t \stackrel{\omega}{\not\longrightarrow}$ and $t \xrightarrow{\tau} \Theta$. Then, $f(t) \stackrel{\omega}{\not\longrightarrow}$ and $f(t) \xrightarrow{\tau} f(\Theta)$. We can infer that

$$\begin{aligned}
\mathbb{V}^{\delta,h,(n+1)}_{\max}(f(t)) &= \delta \cdot max\{\mathbb{V}^{\delta,h,n}_{\max}(\Delta) \mid f(t) \xrightarrow{\tau} \Delta\} \\
&\geq \delta \cdot \mathbb{V}^{\delta,h,n}_{\max}(f(\Theta)) \\
&= \delta \cdot \textstyle\sum_{s\in S} f(\Theta)(s) \cdot \mathbb{V}^{\delta,h,n}_{\max}(s) \\
&= \delta \cdot \textstyle\sum_{t'\in T} \Theta(t') \cdot \mathbb{V}^{\delta,h,n}_{\max}(f(t')) \\
&\geq \delta \cdot \textstyle\sum_{t'\in T} \Theta(t') \cdot \mathbb{V}^{\delta,h,n}(t') \qquad \text{by induction} \\
&= \delta \cdot \mathbb{V}^{\delta,h,n}(\Theta) \\
&= \mathbb{V}^{\delta,h,(n+1)}(t).
\end{aligned}$$

So we have proved (4.21), from which it follows that

$$\mathbb{V}^{\delta,h}_{\max}(f(t)) \geq \mathbb{V}^{\delta,h}(t) \text{ for any } t \in T. \tag{4.22}$$

Therefore, we have that

$$\begin{aligned}
\mathbb{V}^{\delta,h}_{\max}(\Delta^\circ) &= \mathbb{V}^{\delta,h}_{\max}(f(\Theta^\circ)) \\
&= \sum_{s\in S} f(\Theta^\circ)(s) \cdot \mathbb{V}^{\delta,h}_{\max}(s) \\
&= \sum_{t\in T} \Theta^\circ(t) \cdot \mathbb{V}^{\delta,h}_{\max}(f(t)) \\
&\geq \sum_{t\in T} \Theta^\circ(t) \cdot \mathbb{V}^{\delta,h}(t) \qquad \text{by (4.22)} \\
&= \mathbb{V}^{\delta,h}(\Theta^\circ).
\end{aligned}$$

□

Lemma 4.9 *Suppose $\delta < 1$ and Δ° is a subdistribution in a finitary pLTS given by $\langle S, \Omega_\tau, \rightarrow\rangle$. There exists a static resolution $\langle T, \Theta^\circ, \rightarrow\rangle$ such that*

$$\mathbb{V}^{\delta,h}_{max}(\Delta^\circ) \;=\; \mathbb{V}^{\delta,h}(\Theta^\circ)$$

for any reward vector h.

Proof Let $\langle T, \Theta^\circ, \rightarrow\rangle$ be a resolution with an injective resolving function f such that

$$\text{if } t \xrightarrow{\tau} \Theta \text{ then } \mathbb{V}^{\delta,h}_{\max}(f(\Theta)) = max\{\mathbb{V}^{\delta,h}_{\max}(\Delta) \mid f(t) \xrightarrow{\tau} \Delta\}.$$

A finitary pLTS is finitely branching, which ensures the existence of such resolving function f.

Let $g : T \rightarrow [0,1]$ be the function defined by $g(t) = \mathbb{V}^{\delta,h}_{\max}(f(t))$ for all $t \in T$. Below we show that g is a fixed point of $\mathcal{C}^{\delta,h}$. If $t \xrightarrow{\omega}$, then $f(t) \xrightarrow{\omega}$. Therefore, $\mathcal{C}^{\delta}(g)(t) = h(\omega) = \mathbb{V}^{\delta,h}_{\max}(f(t)) = g(t)$. Now suppose $t \not\xrightarrow{\omega}$ and $t \xrightarrow{\tau} \Theta$. By the definition of f, we have $f(t) \not\xrightarrow{\omega}$, $f(t) \xrightarrow{\tau} f(\Theta)$ such that the condition, $\mathbb{V}^{\delta,h}_{\max}(f(\Theta)) = max\{\mathbb{V}^{\delta,h}_{\max}(\Delta) \mid f(t) \xrightarrow{\tau} \Delta\}$, holds. Therefore,

$$\begin{aligned}
\mathcal{C}^{\delta,h}(g)(t) &= \delta \cdot g(\Theta)\\
&= \delta \cdot \textstyle\sum_{t\in T} \Theta(t) \cdot g(t)\\
&= \delta \cdot \textstyle\sum_{t\in T} \Theta(t) \cdot \mathbb{V}^{\delta,h}_{\max}(f(t))\\
&= \delta \cdot \textstyle\sum_{s\in S} f(\Theta)(s) \cdot \mathbb{V}^{\delta,h}_{\max}(s)\\
&= \delta \cdot \mathbb{V}^{\delta,h}_{\max}(f(\Theta))\\
&= \delta \cdot max\{\mathbb{V}^{\delta,h}_{\max}(\Delta) | f(t) \xrightarrow{\tau} \Delta\}\\
&= \mathbb{V}^{\delta,h}_{\max}(f(t))\\
&= g(t).
\end{aligned}$$

Since $\mathcal{C}^{\delta}$ has a unique fixed point $\mathbb{V}^{\delta,h}$, we derive that g coincides with $\mathbb{V}^{\delta,h}$, i.e. $\mathbb{V}^{\delta,h}(t) = g(t) = \mathbb{V}^{\delta,h}_{\max}(f(t))$ for all $t \in T$, from which we can obtain the required result $\mathbb{V}^{\delta,h}(\Theta^\circ) = \mathbb{V}^{\delta,h}_{\max}(\Delta^\circ)$. □

Theorem 4.3 *Let Δ° be a subdistribution in a finitary pLTS $\langle S, \Omega_\tau, \rightarrow\rangle$. There exists a static resolution $\langle T, \Theta^\circ, \rightarrow\rangle$ such that $Exp_{\Theta^\circ}(\mathbb{V}^h) = Exp_{\Delta^\circ}(\mathbb{V}^h_{max})$.*

Proof By Lemma 4.9, for every discount factor $d \in (0,1)$, there exists a static resolution that achieves the maximum expected reward. Since the pLTS is finitary, there are finitely many different static resolutions. There must exist a static resolution that achieves the maximum expected reward for infinitely many discount factors. In other words, for every nondecreasing sequence $\{\delta_n\}_{n\geq 1}$ converging to 1, there exists a subsequence $\{\delta_{n_k}\}_{k\geq 1}$ and a static resolution $\langle T, \Theta^\circ, \longrightarrow\rangle$ with resolving function

f_0 such that $\mathbb{V}^{\delta_{n_k},h}(t) = \mathbb{V}_{\max}^{\delta_{n_k},h}(f_0(t))$ for all $t \in T$ and $k = 1, 2, \ldots$. By Lemma 4.7, we have that, for every $t \in T$,

$$\begin{aligned}\mathbb{V}^h(t) &= \bigsqcup_{k\in\mathbb{N}} \mathbb{V}^{\delta_{n_k},h}(t)\\ &= \bigsqcup_{k\in\mathbb{N}} \mathbb{V}_{\max}^{\delta_{n_k},h}(f_0(t))\\ &= \mathbb{V}_{\max}^h(f_0(t)).\end{aligned}$$

It follows that $\mathbb{V}^h(\Theta^\circ) = \mathbb{V}_{\max}^h(\Delta^\circ)$. □

Theorem 4.4 *For finite-state processes, $P \sqsubseteq^{\Omega}_{\mathbf{er}may} Q$, if and only if $P \sqsubseteq^{\Omega}_{\mathbf{nr}may} Q$.*

Proof Similar to that of Theorem 4.2 but employing Theorem 4.3 in place of Proposition 4.5. □

4.7 Vector-Based Testing Versus Scalar Testing

In this section we show that for finitary processes, scalar testing is as powerful as vector-based testing. As a stepping stone, we use resolution-based reward testing, which is shown to be equivalent to vector-based testing.

Theorem 4.5 *For any Ω and finitary processes P, Q, we have*

$$\begin{aligned} P \sqsubseteq^{\Omega}_{pmay} Q &\quad \text{iff} \quad P \sqsubseteq^{\Omega}_{\mathbf{nr}may} Q\\ P \sqsubseteq^{\Omega}_{pmust} Q &\quad \text{iff} \quad P \sqsubseteq^{\Omega}_{\mathbf{nr}must} Q.\end{aligned}$$

Proof Given test T, process P and reward vector h, we introduce the following notation:

$$\begin{aligned}\mathcal{A}_f(T,P) &:= \updownarrow\{\mathrm{Exp}_\Theta(\mathbb{V}) \mid \langle R,\Theta,\rightarrow\rangle \text{ is a static resolution of } [\![T|_{\mathsf{Act}}P]\!]\}\\ \mathcal{A}^h_f(T,P) &:= \updownarrow\{\mathrm{Exp}_\Theta(\mathbb{V}^h) \mid \langle R,\Theta,\rightarrow\rangle \text{ is a static resolution of } [\![T|_{\mathsf{Act}}P]\!]\}.\end{aligned}$$

We have the following two claims:

Claim 1 For any test T, process P, and reward vector h, we always have that $\mathcal{A}_f(T,P) \subseteq \mathcal{A}(T,P)$. Moreover, if P and T are finitary, then $\mathcal{A}_f(T,P)$ is p-closed.

Claim 2 Let $h \in [0,1]^m$ be a reward tuple, $T \in \mathbb{T}_m$ and P are finitary test and process, respectively.

$$\begin{aligned}\bigsqcup \mathcal{A}^h_f(T,P) &= \bigsqcup \mathcal{A}^h(T,P)\\ \bigsqcap \mathcal{A}^h_f(T,P) &= \bigsqcap \mathcal{A}^h(T,P).\end{aligned}$$

For the first claim, we observe that static resolutions are still resolutions and by Lemma 4.1, the set $\mathcal{A}(T, P)$ is convex. If P and T are finitary, their composition $P|_{\mathsf{Act}}T$ is finitary too. The set $\mathcal{A}_f(T, P)$ is the convex closure of a finite number of points, so it is clearly Cauchy closed.

For the second claim, we observe that Proposition 4.6 and Theorem 4.3 imply that

$$\bigsqcup \mathcal{A}^h_f(T, P) \;=\; \mathbb{V}^h_{\max}([\![T|_{\mathsf{Act}}P]\!]) \;=\; \bigsqcup \mathcal{A}^h(T, P). \tag{4.23}$$

Similarly, Propositions 4.4 and 4.5 imply that

$$\bigsqcap \mathcal{A}^h_f(T, P) \;=\; \mathbb{V}^h_{\min}([\![T|_{\mathsf{Act}}P]\!]) \;=\; \bigsqcap \mathcal{A}^h(T, P). \tag{4.24}$$

We also note that for any deterministic process $\langle R, \Theta, \rightarrow\rangle$, it holds that

$$h \cdot \mathbb{V}(r) = \mathbb{V}^h(r) \tag{4.25}$$

for any $r \in R$, since an easy inductive proof establishes that

$$h \cdot \mathbb{V}^n(r) = \mathbb{V}^{h,n}(r)$$

for all $n \in \mathbb{N}$.

For the *only-if* direction, we apply Theorem 4.1; this direction does not require p-closure.

For the *if* direction, we prove the must-case in the contrapositive; the may-case is similar.

	$P \not\sqsubseteq^{\Omega}_{\mathbf{pmust}} Q$	
iff	$\mathcal{A}(T, P) \not\leq_{\mathrm{Sm}} \mathcal{A}(T, Q)$	for some Ω test T
implies	$\mathcal{A}_f(T, P) \not\leq_{\mathrm{Sm}} \mathcal{A}(T, Q)$	by **Claim 1**
iff	$\bigsqcap h \cdot \mathcal{A}_f(T, P) \not\leq \bigsqcap h \cdot \mathcal{A}(T, Q)$	for some $h \in [0, 1]^{\Omega}$; **Claim 1**; Theorem 4.1
iff	$\bigsqcap \mathcal{A}^h_f(T, P) \not\leq \bigsqcap \mathcal{A}^h(T, Q)$	by (4.25)
iff	$\bigsqcap \mathcal{A}^h(T, P) \not\leq \bigsqcap \mathcal{A}^h(T, Q)$	**Claim 2**
implies	$P \not\sqsubseteq^{\Omega}_{\mathbf{nrmust}} Q$.	

□

We now show that for finitary processes, scalar testing is equally powerful as finite-dimensional reward testing.

Theorem 4.6 *For any $n \in \mathbb{N}$ and finitary processes P, Q, we have*

$$P \sqsubseteq^{\Omega}_{\mathbf{nrmay}} Q \;\; \textit{iff} \;\; P \sqsubseteq_{\mathbf{nrmay}} Q$$
$$P \sqsubseteq^{\Omega}_{\mathbf{nrmust}} Q \;\; \textit{iff} \;\; P \sqsubseteq_{\mathbf{nrmust}} Q.$$

Proof The *only-if* direction is trivial in both cases. For *if* we prove the must-case in the contrapositive; the may-case is similar.

Suppose thus that $P \not\sqsubseteq^{\Omega}_{\mathbf{nrmust}} Q$, then P, Q are distinguished by some Ω-test $T \in \mathbb{T}_n$ and reward $h \in [0,1]^{\Omega}$, so that

$$\bigsqcap \mathcal{A}^h(T,P) \not\leq \bigsqcap \mathcal{A}^h(T,Q). \tag{4.26}$$

Without loss of generality, assume that the success actions in T are $\omega_1, \ldots, \omega_n$. We construct a new test T' with only one success action ω as follows. For each transition $s \xrightarrow{\alpha} \Delta$ in T, if no state in $\lceil \Delta \rceil$ can perform a success action, we keep this transition; otherwise, we form a distribution Δ' to substitute for Δ in this transition.

1. First we partition $\lceil \Delta \rceil$ into two disjoint sets $\{s_i\}_{i\in I}$ and $\{s_j\}_{j\in J}$. For each $i \in I$, we have $s_i \xrightarrow{\omega_{i'}} \Delta_i$ for some distribution Δ_i and success action $\omega_{i'}$; for each $j \in J$, no success action is possible from s_j.
2. Next, we introduce a new state s' to replace the states in the first state set, together with a deadlock state u. So we set $\lceil \Delta' \rceil := \{s_j\}_{j\in J} \cup \{s', u\}$ and

$$\begin{aligned} \Delta'(s') &:= \sum_{i\in I} h(\omega_{i'}) \cdot \Delta(s_i) \\ \Delta'(s_j) &:= \Delta(s_j) \qquad \text{for each } j \in J \\ \Delta'(u) &:= 1 - \Delta'(s') - \sum_{j\in J} \Delta'(s_j). \end{aligned}$$

3. Finally, a new transition $s' \xrightarrow{\omega} \overline{u}$ is added.

We do similar modifications for other transitions in T. Since there are finitely many transitions in total, the above procedure will terminate and result in a new test T'.

The effect of changing T into T' is to replace each occurrence of $\omega_{i'}$ by ω with discount factor $h(\omega_{i'})$ (the other part $1 - h(\omega_{i'})$ is consumed by a deadlock). For any process P, the overall probability of ω's occurrence, in any resolution of $[\![T' |_{\mathsf{Act}} P]\!]$, is therefore the h-weighted reward $h \cdot o$ for the tuple o in the corresponding resolution of $[\![T |_{\mathsf{Act}} P]\!]$.

Thus from (4.26), we have that P, Q can be distinguished using the scalar test T' with its single success action ω; that is, we achieve $P \not\sqsubseteq_{\mathbf{nrmust}} Q$ as required. □

We are now in a position to prove that scalar testing is as powerful as finite-dimensional vector-based testing.

Theorem 4.7 *For any $n \in \mathbb{N}$ and finitary processes P, Q, we have*

$$\begin{aligned} P \sqsubseteq^{\Omega}_{pmay} Q \quad &\textit{iff} \quad P \sqsubseteq_{pmay} Q \\ P \sqsubseteq^{\Omega}_{pmust} Q \quad &\textit{iff} \quad P \sqsubseteq_{pmust} Q. \end{aligned}$$

Proof Combining Theorems 4.5 and 4.6 yields the coincidence of $\sqsubseteq^{\Omega}_{\mathrm{pmay}}$ with $\sqsubseteq_{\mathbf{nrmay}}$, no matter what is the size of Ω, as long as it is finite. It follows that $\sqsubseteq^{\Omega}_{\mathrm{pmay}}$ is the same as $\sqsubseteq_{\mathrm{pmay}}$. The must case is similar. □

4.8 Bibliographic Notes

Probabilistic extensions of testing equivalences [1] have been widely studied. There are two different proposals on how to include probabilistic choice: (i) a test should be nonprobabilistic, i.e. there is no occurrence of probabilistic choice in a test [10–14]; or (ii) a test can be probabilistic, i.e. probabilistic choice may occur in tests as well as processes [3, 4, 15–19]. This book adopts the second approach.

Some work [10, 11, 15, 16] does not consider nondeterminism but deals exclusively with *fully probabilistic* processes. In this setting, a process passes a test with a unique probability instead of a set of probabilities, and testing preorders in the style of [1] have been characterised in terms of *probabilistic traces* [15] and *probabilistic acceptance trees* [16].

Generalisations of the testing theory of [1] to probabilistic systems first appear in [11] and [20], for generative processes without nondeterministic choice. The application of testing to the probabilistic processes we consider here stems from [3]. The idea of vector-based testing originally comes from [4].

In [21], reactive processes are tested against three classes of tests: reactive probabilistic tests, fully nondeterministic tests, and nondeterministic and probabilistic tests. Thus three testing equivalences are obtained. It is shown that the one based on the third class of tests is strictly finer than the first two, which are incomparable with each other.

In [12], a testing theory is proposed that associates a *reward*, a nonnegative real number, to every success state in a test process; in calculating the set of results of applying a test to a process, the probabilities of reaching a success state are multiplied by the reward associated to that state. [12] allows nonprobabilistic tests only, but applies these to arbitrary nondeterministic probabilistic processes, and provides a trace-like denotational characterisation of the resulting may-testing preorder. Denotational characterisations of the variant of our testing preorders in which only τ-free processes are allowed as test processes appear in [17, 22]. These characterisations are improved in [18].

In [19], a testing theory for nondeterministic probabilistic processes is developed in which, as in [23], all probabilistic choices are resolved first. A consequence of this is that the idempotence of internal choice must be sacrificed. The work [19] extends the results of [16] with nondeterminism, but suffers from the same problems as [23]. Similarly, in [24], probabilistic choices are resolved first and the resulting resolutions are compared under the traditional testing semantics of [1]. Some papers distill preorders for probabilistic processes by means of *testing scenarios* in which the premise that a test is itself a process is given up. These include [10, 13] and [25].

In our testing framework given in Sect. 4.1, applying a test to a process yields a composite process from which success probabilities are extracted cumulatively on all resolutions. In [26], a different testing framework is presented where success probabilities are considered in a trace-by-trace fashion. Different variants of testing equivalences are compared and placed in spectrum of trace and testing equivalences.

In Definition 4.3, we use resolving functions to model the behaviour of schedulers or environments to resolve nondeterminism. The schedulers are powerful enough to examine the structure of a probabilistic process. For certain applications one might like to restrict the power of the schedulers [27–31], so as to obtain coarser testing preorders or equivalences than those in Definition 4.4. Then a natural question is: to what extent should a scheduler be restricted? There are proposals for making probabilistic or/and internal choice unobservable to the scheduler. But so far we have not seen widely accepted criteria.

References

1. De Nicola, R., Hennessy, M.: Testing equivalences for processes. Theor. Comput. Sci. **34**, 83–133 (1984)
2. Hennessy, M.: An Algebraic Theory of Processes. The MIT Press, Cambridge (1988)
3. Yi, W., Larsen, K.G.: Testing probabilistic and nondeterministic processes. Proceedings of the IFIP TC6/WG6.1 12th International Symposium on Protocol Specification, Testing and Verification, IFIP Transactions, vol. C-8, pp. 47–61. North-Holland (1992)
4. Segala, R.: Testing probabilistic automata. Proceedings of the 7th International Conference on Concurrency Theory. Lecture Notes in Computer Science, vol. 1119, pp. 299–314. Springer, Heidelberg (1996)
5. Hennessy, M.: Powerdomains and nondeterministic recursive definitions. Proceedings of the 5th International Symposium on Programming. Lecture Notes in Computer Science, vol. 137, pp. 178–193. Springer, Heidelberg (1982)
6. Abransky, S., Jung, A.: Domain Theory. Handbook of Logic and Computer Science, vol. 3, pp. 1–168. Clarendon Press, Oxford (1994)
7. Winskel, G.: The Formal Semantics of Programming Languages: An Introduction. The MIT Press, Cambridge (1993)
8. Deng, Y., van Glabbeek, R., Morgan, C.C., Zhang, C.: Scalar outcomes suffice for finitary probabilistic testing. Proceedings of the 16th European Symposium on Programming. Lecture Notes in Computer Science, vol. 4421, pp. 363–378. Springer, Heidelberg (2007)
9. Puterman, M.L.: Markov Decision Processes. Wiley, New York (1994)
10. Larsen, K.G., Skou, A.: Bisimulation through probabilistic testing. Inf. Comput. **94**(1), 1–28 (1991)
11. Christoff, I.: Testing equivalences and fully abstract models for probabilistic processes. Proceedings of the 1st International Conference on Concurrency Theory. Lecture Notes in Computer Science, vol. 458, pp. 126–140. Springer, Heidelberg (1990)
12. Jonsson, B., Ho-Stuart, C., Yi, W.: Testing and refinement for nondeterministic and probabilistic processes. Proceedings of the 3rd International Symposium on Formal Techniques in Real-Time and Fault-Tolerant Systems. Lecture Notes in Computer Science, vol. 863, pp. 418–430. Springer, Heidelberg (1994)
13. Kwiatkowska, M.Z., Norman, G.: A testing equivalence for reactive probabilistic processes. Electr. Notes Theor. Comput. Sci. **16**(2), 1–19 (1998)
14. Gregorio-Rodríguez, C., Núñez, M.: Denotational semantics for probabilistic refusal testing. Electron. Notes Theor. Comput. Sci. **22**, 111–137 (1999)
15. Cleaveland, R., Dayar, Z., Smolka, S.A., Yuen, S.: Testing preorders for probabilistic processes. Inf. Comput. **154**(2), 93–148 (1999)
16. Núñez, M.: Algebraic theory of probabilistic processes. J. Log. Algebr. Program. **56**, 117–177 (2003)
17. Jonsson, B., Yi, W.: Compositional testing preorders for probabilistic processes. Proceedings of the 10th Annual IEEE Symposium on Logic in Computer Science, pp. 431–441. Computer Society Press (1995)

18. Jonsson, B., Yi, W.: Testing preorders for probabilistic processes can be characterized by simulations. Theor. Comput. Sci. **282**(1), 33–51 (2002)
19. Cazorla, D., Cuartero, F., Ruiz, V., Pelayo, F., Pardo, J.: Algebraic theory of probabilistic and nondeterministic processes. J. Log. Algebr. Program. **55**(1–2), 57–103 (2003)
20. Cleaveland, R., Smolka, S.A., Zwarico, A.: Testing preorders for probabilistic processes. Proceedings of the 19th International Colloquium on Automata, Languages and Programming. Lecture Notes in Computer Science, vol. 623, pp. 708–719. Springer, Heidelberg (1992)
21. Bernardo, M., Sangiorgi, D., Vignudelli, V.: On the discriminating power of testing equivalences for reactive probabilistic systems: Results and open problems. Proceedings of the 11th International Conference on Quantitative Evaluation of Systems. Lecture Notes in Computer Scienc, vol. 8567, pp. 281–296. Springer, Heidelberg (2014)
22. Jonsson, B., Yi, W.: Fully abstract characterization of probabilistic may testing. Proceedings of the 5th International AMAST Workshop on Formal Methods for Real-Time and Probabilistic Systems. Lecture Notes in Computer Science, vol. 1601, pp. 1–18. Springer, Heidelberg (1999)
23. Morgan, C.C., McIver, A.K., Seidel, K., Sanders, J.: Refinement-oriented probability for CSP. Form. Asp. Comput. **8**(6), 617–647 (1996)
24. Acciai, L., Boreale, M., De Nicola, R.: Linear-time and may-testing semantics in a probabilistic reactive setting. Proceedings of FMOODS-FORTE'11. Lecture Notes in Computer Science, vol. 6722, pp. 29–43. Springer, Heidelberg (2011)
25. Stoelinga, M., Vaandrager, F.W.: A testing scenario for probabilistic automata. Proceedings of the 30th International Colloquium on Automata, Languages and Programming. Lecture Notes in Computer Science, vol. 2719, pp. 407–18. Springer, Heidelberg (2003)
26. Bernardo, M., De Nicola, R., Loreti, M.: Revisiting trace and testing equivalences for nondeterministic and probabilistic processes. Log. Methods Comput. Sci. **10**(1:16), 1–42 (2014)
27. de Alfaro, L., Henzinger, T., Jhala, R.: Compositional methods for probabilistic systems. Proceedings of the 12th International Conference on Concurrency Theory. Lecture Notes in Computer Science, vol. 2154, pp. 351–365. Springer, Heidelberg (2001)
28. Cheung, L., Lynch, N., Segala, R., Vaandrager, F.: Switched PIOA: Parallel composition via distributed scheduling. Theor. Comput. Sci. **365**(1–2), 83–108 (2006)
29. Chatzikokolakis, K., Palamidessi, C.: Making random choices invisible to the scheduler. Proceedings of the 18th International Conference on Concurrency Theory. Lecture Notes in Computer Science, vol. 4703, pp. 42–58. Springer, Heidelberg (2007)
30. Giro, S., D'Argenio, P.: On the expressive power of schedulers in distributed probabilistic systems. Electron. Notes Theor. Comput. Sci. **253**(3), 45–71 (2009)
31. Georgievska, S., Andova, S.: Probabilistic may/must testing: Retaining probabilities by restricted schedulers. Form. Asp. Comput. **24**, 727–748 (2012)

Chapter 5
Testing Finite Probabilistic Processes

Abstract In this chapter we focus on finite processes and understand testing semantics from three different aspects. First, we coinductively define simulation relations. Unlike the nonprobabilistic setting, where there is a clear gap between testing and simulation semantics, here testing semantics is as strong as simulation semantics. Second, a probabilistic logic is presented to completely determine testing preorders. Therefore, both positive and negative results can be established. For example, if two finite processes P and Q are related by may preorder then we can construct a simulation relation to witness this, otherwise a modal formula can be constructed that is satisfiable by P but not by Q. Moreover, the distinguishing formula can be turned into a test that P can pass but Q cannot. Finally, for finite processes, both may and must testing preorders can be completely axiomatised.

Keywords Finite processes · Probabilistic simulation · Testing preorders · Modal logic · Axiomatisation

5.1 Introduction

To succinctly represent the behaviour of probabilistic processes, we introduce a simple process calculus that is a probabilistic extension of Hoare's communicating sequential processes (CSP) [1], called pCSP. It has *three* choice operators, external $P \Box Q$, internal $P \sqcap Q$ and a probabilistic choice $P\,_{p}\!\oplus Q$. So a semantic theory for pCSP will have to provide a coherent account of the precise relationships between these operators.

We aim to give alternative characterisations of the testing preorders $\sqsubseteq_{\text{pmay}}$ and $\sqsubseteq_{\text{pmust}}$ (cf. Definition 4.4). This problem was addressed previously by Segala in [2], but using testing preorders ($\sqsubseteq^{\Omega}_{\text{pmay}}$ and $\sqsubseteq^{\Omega}_{\text{pmust}}$) that differ in two ways from the ones in [3–6] and ours. First of all, in [2] the success of a test is achieved by the *actual execution* of a predefined *success action*, rather than reaching of a *success state*. We call this an *action-based* approach, as opposed to the *state-based* approach used in this book. Second, [2] employs a countable number of success actions instead of a single one, so this is *vector-based*, as opposed to *scalar* testing. Segala's results in [2] depend crucially on this form of testing. To achieve our current results, we need vector-based testing preorders as a stepping stone. We relate them to ours by using

DOI 10.1007/978-3-662-45198-4_5

Theorem 4.7 from the last chapter saying that for finitary processes the preorders $\sqsubseteq^{\Omega}_{\text{pmay}}$ and $\sqsubseteq^{\Omega}_{\text{pmust}}$ coincide with $\sqsubseteq_{\text{pmay}}$ and $\sqsubseteq_{\text{pmust}}$. We will proceed in two steps: finite processes are considered in this chapter and finitary processes, which may have loops, are dealt with in Chap. 6.

We will first introduce the syntax and operational semantics of the language pCSP and then instantiate the general testing framework in Sect. 4.1 to pCSP processes. In Sect. 5.5 we use the transitions $s \xrightarrow{\alpha} \Delta$ to define two coinductive preorders, the *forward simulation* (or simply called *simulation*) preorder $\sqsubseteq_S$ [7, 8], and the *failure simulation* preorder $\sqsubseteq_{FS}$ over pCSP processes. The latter extends the failure simulation preorder of [9] to probabilistic processes. Both preorders use a natural generalisation of the transitions, first to take the form $\Delta \xrightarrow{\alpha} \Delta'$, and then to *weak* versions $\Delta \xRightarrow{\alpha} \Delta'$. The second preorder differs from the first one in the use of a *failure* predicate $s \not\xrightarrow{X}$, indicating that in the state s none of the actions in X can be performed.

Both preorders are preserved by all the operators in pCSP, and are *sound* with respect to the testing preorders; that is $P \sqsubseteq_S Q$ implies $P \sqsubseteq_{\text{pmay}} Q$ and $P \sqsubseteq_{FS} Q$ implies $P \sqsubseteq_{\text{pmust}} Q$. The soundness is proved in a very standard method. But *completeness*, that the testing preorders imply the respective simulation preorders, requires some ingenuity. We prove it indirectly, involving a characterisation of the testing and simulation preorders in terms of a modal logic.

Our modal logic, defined in Sect. 5.6, uses conjunction $\varphi_1 \wedge \varphi_2$, the modality $\langle a\rangle\varphi$ from the Hennessy–Milner Logic, and a probabilistic construct $\varphi_1\, {}_p\!\oplus \varphi_2$. A satisfaction relation between processes and formulae then gives, in a natural manner, a *logical preorder* between processes: $P \sqsubseteq^{\mathcal{L}} Q$ means that every $\mathcal{L}$ formula satisfied by P is also satisfied by Q. We establish that $\sqsubseteq^{\mathcal{L}}$coincides with $\sqsubseteq_S$ (and hence $\sqsubseteq_{\text{pmay}}$ also).

To capture failures, we add, for every set of actions X, a formula **ref**(X) to our logic, satisfied by any process that, after it can do no further internal actions, can perform none of the actions in X either. The constructs $\bigwedge$, $\langle a\rangle$ and **ref** stem from the modal characterisation of the nonprobabilistic failure simulation preorder, given in [9]. We show that $\sqsubseteq_{\text{pmust}}$, as well as $\sqsubseteq_{FS}$, can be characterised in a similar manner with this extended modal logic.

We prove these characterisation results through two cycles of inclusions:

$$\begin{array}{ccccccccc}
\sqsubseteq^{\mathcal{L}} & \subseteq & \sqsubseteq_S & \subseteq & \sqsubseteq_{\text{pmay}} & = & \sqsubseteq^{\Omega}_{\text{pmay}} & \subseteq & \sqsubseteq^{\mathcal{L}} \\
\sqsubseteq^{\mathcal{F}} & \subseteq & \sqsubseteq_{FS} & \subseteq & \sqsubseteq_{\text{pmust}} & = & \sqsubseteq^{\Omega}_{\text{pmust}} & \subseteq & \sqsubseteq^{\mathcal{F}} \\
\multicolumn{2}{c}{\underbrace{\qquad}_{\text{Sect. 5.6}}} & \multicolumn{2}{c}{\underbrace{\qquad}_{\text{Sect. 5.5}}} & \underbrace{\qquad}_{\text{Sect. 4.1}} & \multicolumn{2}{c}{\underbrace{\qquad}_{\text{Sect. 4.7}}} & \multicolumn{2}{c}{\underbrace{\qquad}_{\text{Sect. 5.7}}}
\end{array}$$

In Sect. 5.6 we show that $P \sqsubseteq^{\mathcal{L}} Q$ implies $P \sqsubseteq_S Q$ (and hence $P \sqsubseteq_{\text{pmay}} Q$), and likewise for $\sqsubseteq^{\mathcal{F}}$ and $\sqsubseteq_{FS}$; the proof involves constructing, for each pCSP process P, a *characteristic formula* φ_P. To obtain the other direction, in Sect. 5.7 we show how every modal formula φ can be captured, in some sense, by a test T_φ; essentially the ability of a pCSP process to satisfy φ is determined by its ability to pass the test T_φ.

We capture the conjunction of two formulae by a probabilistic choice between the corresponding tests; in order to prevent the results from these tests getting mixed up, we employ the vector-based tests of [2], so that we can use different success actions in the separate probabilistic branches. Therefore, we complete our proof by recalling Theorem 4.7 that the scalar testing preorders imply the vector-based ones.

It is well-known that may and must testing for standard CSP can be captured equationally [3, 4, 10]. In Sect. 5.3 we show that most of the standard equations are no longer valid in the probabilistic setting of pCSP. However, we show in Sect. 5.9 that both $P \sqsubseteq_{\text{pmay}} Q$ and $P \sqsubseteq_{\text{pmust}} Q$ can still be captured equationally over full pCSP. In the may case the essential (in)equation required is

$$a.(P\,{}_{p}{\oplus}\,Q) \quad \sqsubseteq \quad a.P\,{}_{p}{\oplus}\,a.Q.$$

The must case is more involved: in the absence of the distributivity of the external and internal choices over each other, to obtain completeness we require a complicated inequational schema.

5.2 The Language pCSP

We first define the language and its operational semantics. Then we show how the general probabilistic testing theory outlined in Sect. 4.1 can be applied to processes from this language.

5.2.1 The Syntax

Let Act be a set of *visible* (or *external*) actions, ranged over by $a, b, c, \ldots$, which processes can perform. Then the finite probabilistic CSP processes are given by the following two sorted syntax:

$$P ::= S \mid P\,{}_{p}{\oplus}\,P$$
$$S ::= \mathbf{0} \mid a.P \mid P \sqcap P \mid S \Box S \mid S \mid_A S$$

We write pCSP, ranged over by P, Q, for the set of process terms defined by this grammar, and sCSP, ranged over by s, t, for the subset comprising only the *state-based* process terms (the subsort S above).

The process $P\,{}_{p}{\oplus}\,Q$, for $0 \leq p \leq 1$, represents a *probabilistic choice* between P and Q: with probability p it will act like P and with probability $1 - p$ it will act like Q. Any process is a probabilistic combination of state-based processes built by repeated application of the operator ${}_{p}{\oplus}$. The state-based processes have a CSP-like syntax, involving the stopped process $\mathbf{0}$, action prefixing $a.\_$ for $a \in \mathsf{Act}$, *internal-* and *external choices* $\sqcap$ and $\Box$, and a *parallel composition* $\mid_A$ for $A \subseteq \mathsf{Act}$.

The process $P \sqcap Q$ will first do a so-called *internal action* $\tau \notin \mathsf{Act}$, choosing *nondeterministically* between P and Q. Therefore $\sqcap$, like $a.\_$, acts as a *guard*, in the sense that it converts any process arguments into a state-based process.

The process $s \,\Box\, t$ on the other hand does not perform actions itself but rather allows its arguments to proceed, disabling one argument as soon as the other has done a visible action. In order for this process to start from a state rather than a probability distribution of states, we require its arguments to be state-based as well; the same requirement applies to $|_A$.

Finally, the expression $s\,|_A\,t$, where $A \subseteq \mathsf{Act}$, represents processes s and t running in parallel. They may synchronise by performing the same action from A simultaneously; such a synchronisation results in τ. In addition s and t may independently do any action from $(\mathsf{Act}_\tau \backslash A)$, where $\mathsf{Act}_\tau := \mathsf{Act} \cup \{\tau\}$.

Although, formally the operators $\Box$ and $|_A$ can only be applied to state-based processes, informally we use expressions of the form $P \,\Box\, Q$ and $P\,|_A\,Q$, where P and Q are *not* state-based, as syntactic sugar for expressions in the above syntax obtained by distributing $\Box$ and $|_A$ over ${}_p\oplus$. Thus, for example $s \,\Box\, (t_1 \,{}_p\oplus\, t_2)$ abbreviates the term $(s \,\Box\, t_1)\,{}_p\oplus\,(s \,\Box\, t_2)$.

The full language of CSP [1, 10, 11] has many more operators; we have simply chosen a representative selection, and have added probabilistic choice. Our parallel operator is not a CSP primitive, but it can easily be expressed in terms of them—in particular $P\,|_A\,Q = (P \|_A Q)\backslash A$, where $\|_A$ and $\backslash A$ are the parallel composition and hiding operators of [11]. It can also be expressed in terms of the parallel composition, renaming and restriction operators of CCS. We have chosen this (nonassociative) operator for convenience in defining the application of tests to processes.

As usual we may elide $\mathbf{0}$; the prefixing operator $a.\_$ binds stronger than any binary operator; and precedence between binary operators is indicated via brackets or spacing. We will also sometimes use indexed binary operators, such as $\bigoplus_{i \in I} p_i \cdot P_i$ with $\sum_{i \in I} p_i = 1$ and all $p_i > 0$, and $\Box_{i \in I} P_i$, for some finite index set I.

5.2.2 *The Operational Semantics*

The above intuitive reading of the various operators can be formalised by an *operational semantics* that associates with each process term a graph-like structure representing the manner in which it may react to users' requests. Let us briefly recall this procedure for nonprobabilistic CSP.

The operational semantics of CSP is obtained by endowing the set of terms with the structure of an LTS. Specifically

(i) the set of states S is taken to be all terms from the language CSP
(ii) the action relations $P \xrightarrow{\alpha} Q$ are defined inductively on the syntax of terms.

A precise definition may be found in [11].

$$a.P \xrightarrow{a} [\![P]\!]$$

$$P \sqcap Q \xrightarrow{\tau} [\![P]\!] \qquad\qquad P \sqcap Q \xrightarrow{\tau} [\![Q]\!]$$

$$\frac{s_1 \xrightarrow{a} \Delta}{s_1 \,\Box\, s_2 \xrightarrow{a} \Delta} \qquad\qquad \frac{s_2 \xrightarrow{a} \Delta}{s_1 \,\Box\, s_2 \xrightarrow{a} \Delta}$$

$$\frac{s_1 \xrightarrow{\tau} \Delta}{s_1 \,\Box\, s_2 \xrightarrow{\tau} \Delta \,\Box\, s_2} \qquad\qquad \frac{s_2 \xrightarrow{\tau} \Delta}{s_1 \,\Box\, s_2 \xrightarrow{\tau} s_1 \,\Box\, \Delta}$$

$$\frac{s_1 \xrightarrow{\alpha} \Delta \quad \alpha \notin A}{s_1 \,|_A\, s_2 \xrightarrow{\alpha} \Delta \,|_A\, s_2} \qquad\qquad \frac{s_2 \xrightarrow{\alpha} \Delta \quad \alpha \notin A}{s_1 \,|_A\, s_2 \xrightarrow{\alpha} s_1 \,|_A\, \Delta}$$

$$\frac{s_1 \xrightarrow{a} \Delta_1,\ s_2 \xrightarrow{a} \Delta_2 \ a \in A}{s_1 \,|_A\, s_2 \xrightarrow{\tau} \Delta_1 \,|_A\, \Delta_2}$$

Fig. 5.1 Operational semantics of pCSP

In order to interpret the full pCSP operationally we need to use pLTSs, the probabilistic generalisation of LTSs (see Sect. 3.2). We mimic the operational interpretation of CSP as an LTS by associating with pCSP a particular pLTS $\langle \mathsf{sCSP}, \mathsf{Act}_\tau, \rightarrow \rangle$ in which sCSP is the set of states and Act_τ is the set of transition labels. However there are two major differences:

(i) only a subset of terms in pCSP will be used as the set of states in the pLTS
(ii) terms in pCSP will be interpreted as distributions over sCSP, rather than as elements of sCSP.

We interpret pCSP processes P as distributions $[\![P]\!] \in \mathcal{D}(\mathsf{sCSP})$ via the function $[\![-]\!] : \mathsf{rpCSP} \rightarrow \mathcal{D}(\mathsf{sCSP})$ defined by

$$[\![s]\!] := \overline{s} \quad \text{for } s \in \mathsf{sCSP}, \quad \text{and} \quad [\![P\,_p\!\oplus Q]\!] := [\![P]\!]\,_p\!\oplus [\![Q]\!]\,.$$

The definition of the relations $\xrightarrow{\alpha}$ is given in Fig. 5.1, where a ranges over Act and α over Act_τ.

These rules are very similar to the standard ones used to interpret CSP as a labelled transition system [11], but are modified so that the result of an action is a distribution. The rules for external choice and parallel composition use an obvious notation for distributing an operator over a distribution; for example $\Delta \,\Box\, s$ represents the distribution, given by

$$(\Delta \,\Box\, s)(t) = \begin{cases} \Delta(s') & \text{if } t = s' \,\Box\, s \\ 0 & \text{otherwise.} \end{cases}$$

We sometimes write $\tau.P$ for $P \sqcap P$, thus giving $\tau.P \xrightarrow{\tau} [\![P]\!]$.

5.2.3 *The Precedence of Probabilistic Choice*

Our operational semantics entails that $\Box$ and $|_A$ distribute over probabilistic choice:

$$\begin{aligned}\llbracket P \Box (Q\,{}_{p}\!\oplus R)\rrbracket &= \llbracket (P \Box Q)\,{}_{p}\!\oplus (P \Box R)\rrbracket \\ \llbracket P\,|_A\,(Q\,{}_{p}\!\oplus R)\rrbracket &= \llbracket (P\,|_A\,Q)\,{}_{p}\!\oplus (P\,|_A\,R)\rrbracket\end{aligned}$$

These identities are not a consequence of our testing methodology: they are hardwired in our interpretation $\llbracket\_\rrbracket$ of pCSP expressions as distributions.

A consequence of our operational semantics is that, for example, in the process $a.(b\,{}_{\frac{1}{2}}\!\oplus c)|_{\emptyset}d$ the action d can be scheduled either before a, or after the probabilistic choice between b and c—but it *cannot* be scheduled after a and before this probabilistic choice. We justify this by thinking of $P\,{}_{p}\!\oplus Q$ not as a process that starts with making a probabilistic choice, but rather as one that *has* just made such a choice, and with probability p *is* no more and no less than the process P. Thus $a.(P\,{}_{p}\!\oplus Q)$ is a process that in doing the a-step makes a probabilistic choice between the alternative targets P and Q.

This design decision is in full agreement with previous work featuring nondeterminism, probabilistic choice and parallel composition [5, 7, 12]. Moreover, a probabilistic choice between processes P and Q that does not take precedence over actions scheduled in parallel can simply be written as $\tau.(P\,{}_{p}\!\oplus Q)$. Here $\tau.P$ is an abbreviation for $P \sqcap P$. Using the operational semantics of $\sqcap$ in Fig. 5.1, $\tau.P$ is a process whose sole initial transition is $\tau.P \xrightarrow{\tau} P$, hence $\tau.(P\,{}_{p}\!\oplus Q)$ is a process that starts with making a probabilistic choice, modelled as an internal action, and with probability p proceeds as P. Any activity scheduled in parallel with $\tau.(P\,{}_{p}\!\oplus Q)$ can now be scheduled before or after this internal action, hence before or after the making of the choice. In particular, $a.\tau.(b\,{}_{\frac{1}{2}}\!\oplus c)|_{\emptyset}d$ allows d to happen between a and the probabilistic choice.

5.2.4 *Graphical Representation of pCSP Processes*

The set of states *reachable* from a subdistribution Δ is the smallest set that contains $\lceil\Delta\rceil$ and is closed under transitions, meaning that if some state s is reachable and $s \xrightarrow{\alpha} \Theta$ then every state in $\lceil\Theta\rceil$ is reachable as well. We graphically depict the operational semantics of a pCSP expression P by drawing the part of the pLTS defined above that is reachable from $\llbracket P\rrbracket$ as a finite acyclic directed graph in the way described in Sect. 3.2.

Two examples are described in Fig. 5.2. The interpretation of the simple process $(b \sqcap c) \Box (d\,{}_{\frac{1}{2}}\!\oplus a)$ is the distribution $\overline{(b \sqcap c)} \Box \Delta$, where Δ is the distribution resulting from the interpretation of $(d\,{}_{\frac{1}{2}}\!\oplus a)$, itself a two-point distribution mapping both the states $d.\mathbf{0}$ and $a.\mathbf{0}$ to the probability $\frac{1}{2}$. The result is again a two-point distribution, this time mapping the two states $(b \sqcap c) \Box d$ and $(b \sqcap c) \Box a$ to $\frac{1}{2}$. The

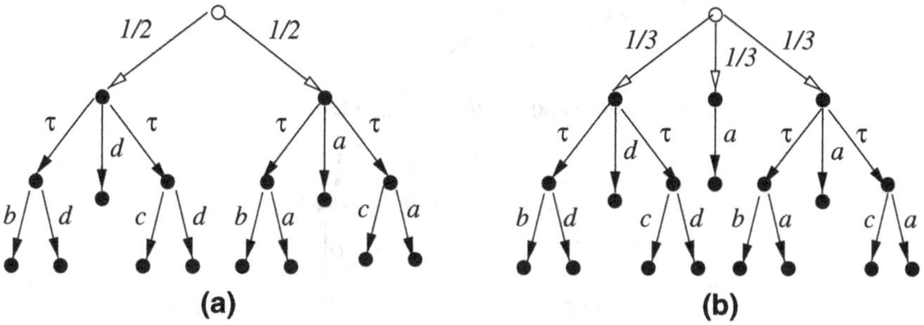

Fig. 5.2 The pLTSs of example processes

end result in (a) is obtained by further interpreting these two states using the rules in Fig. 5.1. In (b) we show the graphical representation that results when this term is combined probabilistically with the simple process $a.\mathbf{0}$.

To sum up, the operational semantics endows pCSP with the structure of a pLTS, and the function $[\![-]\!]$ interprets process terms in pCSP as distributions in this pLTS, which can be represented by finite acyclic directed graphs (typically drawn as trees), with edges labelled either by probabilities or actions, so that edges labelled by actions and probabilities alternate (although in pictures we may suppress 1-labelled edges and point distributions).

5.2.5 *Testing pCSP Processes*

Let us now turn to applying the testing theory from Sect. 4.1 to pCSP. As with the standard theory [3, 4], we use as tests any process from pCSP itself, which in addition can use a special symbol ω to report success. For example, the term $a.\omega\,_{\frac{1}{4}}\!\oplus (b \;\Box\; c.\omega)$ is a probabilistic test, which 25 % of the time requests an a action, and 75 % requests that c can be performed. If it is used as *must* test, the 75 % that requests a c action additionally requires that b is not possible. As in [3, 4], it is not the execution of ω that constitutes success, but the arrival in a state where ω is possible. The introduction of the ω action is simply a method of defining a success predicate on states without having to enrich the language of processes explicitly with such predicates.

Formally, let $\omega \notin \mathsf{Act}_\tau$ and write Act^ω for $\mathsf{Act} \cup \{\omega\}$ and Act^ω_τ for $\mathsf{Act} \cup \{\tau, \omega\}$. In Fig. 5.1 we now let a range over Act^ω and α over Act^ω_τ. Tests may have subterms $\omega.P$, but other processes may not. We write pCSP^ω for the set of all tests. To apply the test T to the process P we run them in parallel, tightly synchronised; that is, we run the combined process $T\,|_{\mathsf{Act}}\,P$. Here P can only synchronise with T, and in turn the test T can only perform the success action ω, in addition to synchronising with the process being tested; of course both tester and testee can also perform internal actions. An example is provided in Fig. 5.3, where the test $a.\omega\,_{\frac{1}{4}}\!\oplus (b \;\Box\; c.\omega)$ is applied

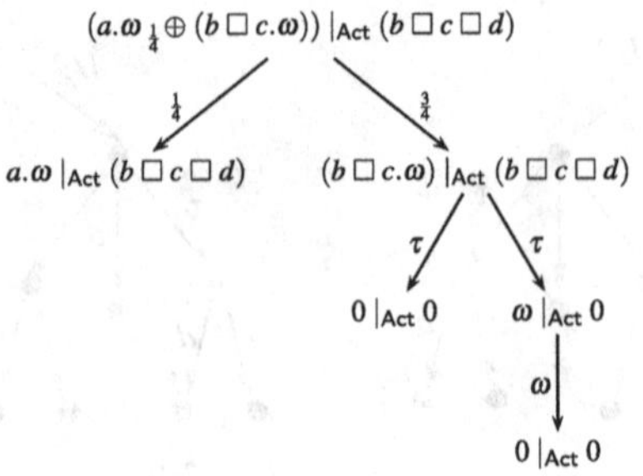

$$\mathcal{A}((a.\omega \;{}_{\frac{1}{4}}\!\oplus (b \,\Box\, c.\omega)), (b \,\Box\, c \,\Box\, d)) = \tfrac{1}{4}\cdot\{0\} + \tfrac{3}{4}\cdot\{0,1\} = \{0, \tfrac{3}{4}\}$$

Fig. 5.3 Example of testing

to the process $b \,\Box\, c \,\Box\, d$. We see that 25 % of the time the test is unsuccessful, in that it does not reach a state where ω is possible, and 75 % of the time it *may* be successful, depending on how the now internal choice between the actions b and c is resolved, but it is not the case that it *must* be successful.

$[\![T\,|_{\mathsf{Act}}\,P]\!]$ is representable as a finite graph that encodes all possible interactions of the test T with the process P. It only contains the actions τ and ω. Each occurrence of τ represents a nondeterministic choice, either in T or P themselves, or as a nondeterministic response by P to a request from T, while the distributions represent the resolution of underlying probabilities in T and P. We use the structure $[\![T\,|_{\mathsf{Act}}\,P]\!]$ to define $\mathcal{A}(T, P)$, a nonempty subset of $[0, 1]$ representing the set of probabilities that applying T to P will be a success.

Definition 5.1 We define a function $\mathbb{V}_f : \mathsf{sCSP} \to \mathcal{P}^+([0, 1])$, which when applied to any state in sCSP returns a finite subset of $[0, 1]$; it is extended to the type $\mathcal{D}(\mathsf{sCSP}) \to \mathcal{P}^+([0, 1])$ via the convention $\mathbb{V}_f(\Delta) := \mathrm{Exp}_\Delta \mathbb{V}_f$ (cf. Sect. 4.2).

$$\mathbb{V}_f(s) = \begin{cases} \{1\} & \text{if } s \xrightarrow{\omega}, \\ \bigcup\{\mathbb{V}_f(\Delta) \mid s \xrightarrow{\tau} \Delta\} & \text{if } s \not\xrightarrow{\omega},\ s \xrightarrow{\tau}, \\ \{0\} & \text{otherwise} \end{cases}$$

We will tend to write the expected value of $\mathbb{V}_f$ explicitly and use the convenient notation

$$\mathbb{V}_f(\Delta) \;=\; \Delta(s_1)\cdot\mathbb{V}_f(s_1) + \ldots + \Delta(s_n)\cdot\mathbb{V}_f(s_n)$$

where $\lceil\Delta\rceil = \{s_1, \ldots s_n\}$. Note that $\mathbb{V}_f(\_)$ is indeed a well-defined function, because the pLTS $\langle \mathsf{sCSP}, \mathsf{Act}_\tau, \to\rangle$ is finitely branching and well-founded.

For example consider the transition systems in Fig. 5.4, where for reference we have labelled the nodes. Then $\mathbb{V}_f(s_1) = \{1, 0\}$ while $\mathbb{V}_f(s_2) = \{1\}$, and therefore

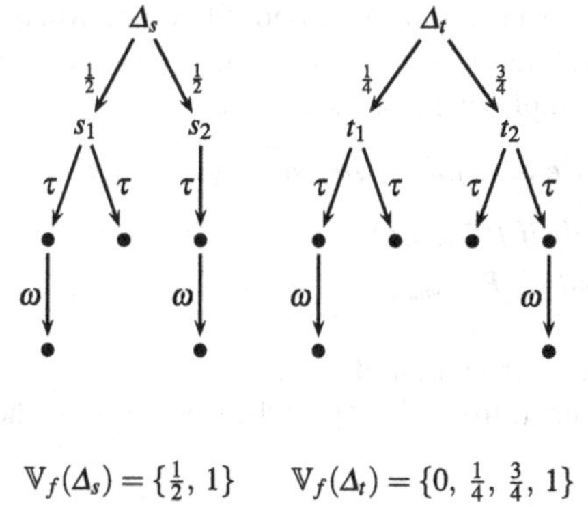

Fig. 5.4 Collecting results

$\mathbb{V}_f(\Delta_s) = \frac{1}{2} \cdot \{1, 0\} + \frac{1}{2} \cdot \{1\}$ that, since there are only two possible choices, evaluates further to $\{\frac{1}{2}, 1\}$. Similarly $\mathbb{V}_f(t_1) = \mathbb{V}_f(t_2) = \{0, 1\}$ and we calculate that

$$\mathbb{V}_f(\Delta_t) = \frac{1}{4} \cdot \{0, 1\} + \frac{3}{4} \cdot \{0, 1\} = \{0, \frac{1}{4}, \frac{3}{4}, 1\}.$$

Definition 5.2 For any process $P \in$ pCSP and test T, let $\mathcal{A}(T, P) = \mathbb{V}_f[\![T \mid_{\mathsf{Act}} P]\!]$. With this definition we now have two testing preorders for pCSP, one based on *may testing*, $P \sqsubseteq_{\text{pmay}} Q$, and the other on *must testing*, $P \sqsubseteq_{\text{pmust}} Q$.

Comparing the results-gathering function $\mathbb{V}_f$ in definition 5.1 with the one given in Sect. 4.2, we notice that the former only records those testing outcomes obtained by using static resolutions of a pLTS while the latter records the outcomes of all resolutions. Here, we prefer to use the former that is simpler because we are dealing with scalar testing and as a matter of fact applying convex closure to subsets of the one-dimensional interval $[0, 1]$ (such as those arisen from applying scalar tests to processes) has no effect on the Hoare and Smyth orders between these subsets.

Lemma 5.1 *Suppose* $X, Y \subseteq [0, 1]$. *Then*

1. $X \leq_{Ho} Y$, *if and only if* $\updownarrow X \leq_{Ho} \updownarrow Y$.
2. $X \leq_{Sm} Y$, *if and only if* $\updownarrow X \leq_{Sm} \updownarrow Y$.

Proof We restrict attention to the first clause; the proof of the second one goes likewise. It suffices to show that (i) $X \leq_{\text{Ho}} \updownarrow X$ and (ii) $\updownarrow X \leq_{\text{Ho}} X$. We only prove (ii) since (i) is obvious. Suppose $x \in \updownarrow X$, then $x = \sum_{i \in I} p_i x_i$ for a finite set I with $\sum_{i \in I} p_i = 1$ and $x_i \in X$. Let $x^* = max\{x_i \mid i \in I\}$, then

$$x = \sum_{i \in I} p_i x_i \leq \sum_{i \in I} p_i x^* = x^* \in X.$$

□

It follows that, for scalar testing it makes no difference whether convex closure is employed or not. Therefore, vector-based testing is also a conservative extension of scalar testing without employing convex closure.

Corollary 5.1 *Suppose Ω is the singleton set $\{\omega\}$. Then*

1. $P \sqsubseteq^{\Omega}_{pmay} Q$ *if and only if* $P \sqsubseteq_{pmay} Q$.
2. $P \sqsubseteq^{\Omega}_{pmust} Q$ *if and only if* $P \sqsubseteq_{pmust} Q$.

Proof The result follows from Lemma 5.1. □

Lemma 5.1 does not generalise to $[0,1]^{\Omega}$, when $|\Omega| > 1$, as the following example demonstrates:

Example 5.1 Let X, Y denote $\{(0.5, 0.5)\}$ and $\{(1,0), (0,1)\}$ respectively. Then it is easy to show that $\updownarrow X \leq_{\mathrm{Ho}} \updownarrow Y$, although obviously $X \not\leq_{\mathrm{Ho}} Y$.

This example can be exploited to show that for vector-based testing it *does* make a difference whether convex closure is employed.

Example 5.2 Consider the two processes

$$P := a\,{}_{\frac{1}{2}}\!\oplus b \qquad \text{and} \qquad Q := a \sqcap b\,.$$

Take $\Omega = \{\omega_1, \omega_2\}$. Employing a results-gathering function without convex closure, with the test $T := a.\omega_1 \,\Box\, b.\omega_2$ we would obtain

$$\begin{aligned} \mathcal{A}(T, P) &= \{(0.5, 0.5)\} \\ \mathcal{A}(T, Q) &= \{(1,0), (0,1)\}\,. \end{aligned}$$

As pointed out in Example 5.1, this entails $\mathcal{A}(T,P) \not\leq_{\mathrm{Ho}} \mathcal{A}(T,Q)$, although their convex closures *are* related under the Hoare preorder.

Convex closure is a uniform way of ensuring that internal choice can simulate an arbitrary probabilistic choice. For the processes P and Q of Example 5.2 we will see later on that $P \sqsubseteq_S Q$ and $P \sqsubseteq_{\mathrm{pmay}} Q$. This fits with the intuition that a probabilistic choice is an acceptable implementation of a nondeterministic choice occurring in a specification. Considering that we use $\sqsubseteq^{\Omega}_{\mathrm{pmay}}$ as a stepping stone in showing the coincidence of $\sqsubseteq_S$ and $\sqsubseteq_{\mathrm{pmay}}$, we must have $P \sqsubseteq^{\Omega}_{\mathrm{pmay}} Q$. For this reason we used convex closure in Definition 4.3.

5.3 Counterexamples

We will see in this section that many of the standard testing axioms are not valid in the presence of probabilistic choice. We also provide counterexamples for a few distributive laws involving probabilistic choice that might appear plausible at first sight. In all cases we establish a statement $P \not\sqsubseteq_{\mathrm{pmay}} Q$ by exhibiting a test T such that $max(\mathcal{A}(T,P)) \neq max(\mathcal{A}(T,Q))$ and a statement $P \not\sqsubseteq_{\mathrm{pmust}} Q$ by exhibiting a test T

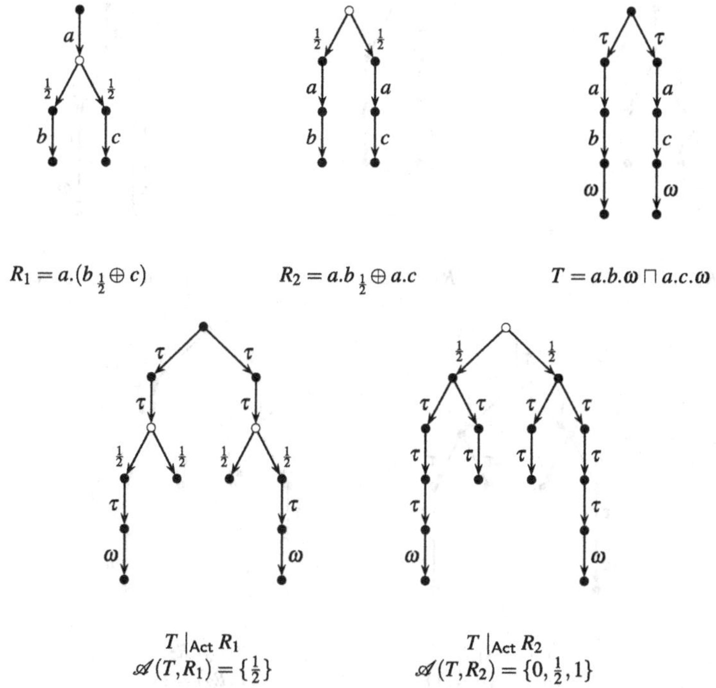

Fig. 5.5 Action prefix does not distribute over probabilistic choice

such that $min(\mathcal{A}(T,P)) \neq min(\mathcal{A}(T,Q))$. In case $max(\mathcal{A}(T,P)) > max(\mathcal{A}(T,Q))$ we find in particular that $P \not\sqsubseteq_{\text{pmay}} Q$, and in case $min(\mathcal{A}(T,P)) > min(\mathcal{A}(T,Q))$ we obtain $P \not\sqsubseteq_{\text{pmust}} Q$.

Example 5.3 The axiom $a.(P\,{}_p{\oplus}\,Q) = a.P\,{}_p{\oplus}\,a.Q$ is unsound.

Consider the example in Fig. 5.5. In R_1 the probabilistic choice between b and c is taken after the action a, while in R_2 the choice is made before the action has happened. These processes can be distinguished by the nondeterministic test $T = a.b.\omega \sqcap a.c.\omega$. First consider running this test on R_1. There is an immediate choice made by the test, effectively running either the test $a.b.\omega$ on R_1 or the test $a.c.\omega$; in fact the effect of running either test is exactly the same. Consider $a.b.\omega$. When run on R_1 the a action immediately happens, and there is a probabilistic choice between running $b.\omega$ on either b or c, giving as possible results $\{1\}$ or $\{0\}$; combining these according to the definition of the function $\mathbb{V}_f(\_)$ we get $\frac{1}{2}\cdot\{0\} + \frac{1}{2}\cdot\{1\} = \frac{1}{2}$. Since running the test $a.c.\omega$ has the same effect, $\mathcal{A}(T,R_1)$ turns out to be the same set $\frac{1}{2}$.

Now consider running the test T on R_2. Because R_2, and hence also $T\,|_{\mathsf{Act}} R_2$, starts with a probabilistic choice, due to the definition of the function $\mathbb{V}_f(\_)$, the test must first be applied to the probabilistic components, $a.b$ and $a.c$, respectively, and the results subsequently combined probabilistically. When the test is run on $a.b$, immediately a nondeterministic choice is made in the test, to run either $a.b.\omega$

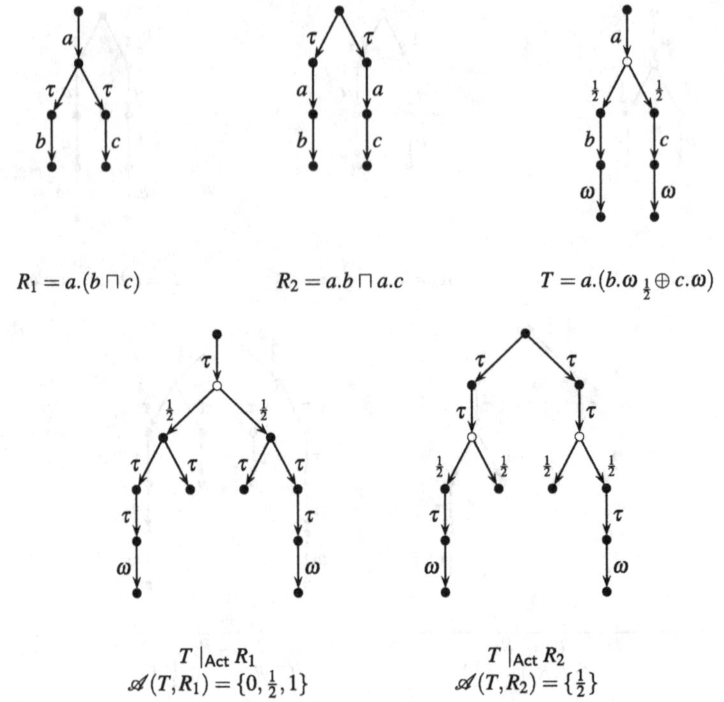

Fig. 5.6 Action prefix does not distribute over internal choice

or $a.c.\omega$. With the former we get the result $\{1\}$, with the latter $\{0\}$, so overall, for running T on a, b, we get the possible results $\{0, 1\}$. The same is true when we run it on $a.c$, and therefore $\mathcal{A}(T, R_2) = \frac{1}{2} \cdot \{0, 1\} + \frac{1}{2} \cdot \{0, 1\} = \{0, \frac{1}{2}, 1\}$.

So we have $R_2 \not\sqsubseteq_{\text{pmay}} R_1$ and $R_1 \not\sqsubseteq_{\text{pmust}} R_2$.

Example 5.4 The axiom $a.(P \sqcap Q) = a.P \sqcap a.Q$ is unsound.

It is well-known that this axiom is valid in the standard theory of testing, for non-probabilistic processes. However, consider the processes R_1 and R_2 in Fig. 5.6, and note that these processes do not contain any probabilistic choice. But they can be differentiated by the probabilistic test $T = a.(b.\omega \, {}_{\frac{1}{2}}\!\oplus c.\omega)$; the details are in Fig. 5.6. There is only one possible outcome from applying T to R_2, the probability $\frac{1}{2}$, because the nondeterministic choice is made before the probabilistic choice. On the other hand when T is applied to R_1 there are three possible outcomes, 0, $\frac{1}{2}$ and 1, because effectively the probabilistic choice takes precedence over the nondeterministic choice. So we have $R_1 \not\sqsubseteq_{\text{pmay}} R_2$ and $R_2 \not\sqsubseteq_{\text{pmust}} R_1$.

Example 5.5 The axiom $a.(P \,\Box\, Q) = a.P \,\Box\, a.Q$ is unsound.

This axiom is valid in the standard may-testing semantics. However, let us consider the two processes $R_1 = a.(b \,\Box\, c)$ and $R_2 = a.b \,\Box\, a.c$. By applying the probabilistic test $T = a.(b.\omega\,{}_{\frac{1}{2}}\!\oplus c.\omega)$, we see that $\mathcal{A}(T, R_1) = \{1\}$ and $\mathcal{A}(T, R_2) = \{\frac{1}{2}\}$. Therefore $R_1 \not\sqsubseteq_{\text{pmay}} R_2$ and $R_2 \not\sqsubseteq_{\text{pmust}} R_1$.

Example 5.6 The axiom $P = P \;\Box\; P$ is unsound.

Let R_1 and R_2 denote $(a\,{}_{\frac{1}{2}}\!\oplus b)$ and $(a\,{}_{\frac{1}{2}}\!\oplus b) \;\Box\; (a\,{}_{\frac{1}{2}}\!\oplus b)$, respectively. It is easy to calculate that $\mathcal{A}(a.\omega, R_1) = \{\frac{1}{2}\}$ but, because of the way we interpret external choice as an operator over distributions of states in a pLTS, it turns out that $[\![R_2]\!] = [\![((a \;\Box\; a)\,{}_{\frac{1}{2}}\!\oplus (a \;\Box\; b))\,{}_{\frac{1}{2}}\!\oplus ((b \;\Box\; a)\,{}_{\frac{1}{2}}\!\oplus (b \;\Box\; b))]\!]$ and so $\mathcal{A}(a.\omega, R_2) = \{\frac{3}{4}\}$. Therefore, $R_2 \not\sqsubseteq_{\text{pmay}} R_1$ and $R_2 \not\sqsubseteq_{\text{pmust}} R_1$.

Example 5.7 The axiom $P\,{}_p\!\oplus (Q \sqcap R) = (P\,{}_p\!\oplus Q) \sqcap (P\,{}_p\!\oplus R)$ is unsound.

Consider the processes $R_1 = a\,{}_{\frac{1}{2}}\!\oplus (b \sqcap c)$ and $R_2 = (a\,{}_{\frac{1}{2}}\!\oplus b) \sqcap (a\,{}_{\frac{1}{2}}\!\oplus c)$, and the test $T_1 = a.\omega \sqcap (b.\omega\,{}_{\frac{1}{2}}\!\oplus c.\omega)$. In the best of possible worlds, when we apply T_1 to R_1 we obtain probability 1, that is $max(\mathcal{A}(T_1, R_1)) = 1$. Informally, this is because half the time when it is applied to the subprocess a of R_1, optimistically the subtest $a.\omega$ is actually run. The other half of the time, when it is applied to the subprocess $(b \sqcap c)$, optimistically the subtest $T_r = (b.\omega\,{}_{\frac{1}{2}}\!\oplus c.\omega)$ is actually used. And here again, optimistically, we obtain probability 1: whenever the test $b.\omega$ is used it might be applied to the subprocess b, while when $c.\omega$ is used it might be applied to c. Formally we have

$$\begin{aligned}
\mathcal{A}(T_1, R_1) &= \tfrac{1}{2} \cdot \mathcal{A}(T_1, a) + \tfrac{1}{2} \cdot \mathcal{A}(T_1, b \sqcap c)\\
&= \tfrac{1}{2} \cdot (\mathcal{A}(a.\omega, a) \cup \mathcal{A}(T_r, a)) \; +\\
&\tfrac{1}{2} \cdot (\mathcal{A}(T_1, b) \cup \mathcal{A}(T_1, c) \cup \mathcal{A}(a.\omega, b \sqcap c) \cup \mathcal{A}(T_r, b \sqcap c))\\
&= \tfrac{1}{2} \cdot (\{1\} \cup \{0\}) + \tfrac{1}{2} \cdot (\{0, \tfrac{1}{2}\} \cup \{0, \tfrac{1}{2}\} \cup \{0\} \cup \{0, \tfrac{1}{2}, 1\})\\
&= \{0, \tfrac{1}{4}, \tfrac{1}{2}, \tfrac{3}{4}, 1\}
\end{aligned}$$

However, no matter how optimistic we are when applying T_1 to R_2 we can never get probability 1; the most we can hope for is $\frac{3}{4}$, which might occur when T_1 is applied to the subprocess $(a\,{}_{\frac{1}{2}}\!\oplus b)$. Specifically, when the subprocess a is being tested the subtest $a.\omega$ might be used, giving probability 1, and when the subprocess b is being tested the subtest $(b.\omega\,{}_{\frac{1}{2}}\!\oplus c.\omega)$ might be used, giving probability $\frac{1}{2}$. We leave the reader to check that formally

$$\mathcal{A}(T_1, R_2) = \{0, \tfrac{1}{4}, \tfrac{1}{2}, \tfrac{3}{4}\}$$

from which we can conclude $R_1 \not\sqsubseteq_{\text{pmay}} R_2$.

We can also show that $R_2 \not\sqsubseteq_{\text{pmust}} R_1$, using the test

$$T_2 = (b.\omega \;\Box\; c.\omega) \sqcap (a.\omega\,{}_{\frac{1}{3}}\!\oplus (b.\omega\,{}_{\frac{1}{2}}\!\oplus c.\omega)).$$

Reasoning pessimistically, the worst that can happen when applying T_2 to R_1 is that we get probability 0. Each time the subprocess a is tested the worst probability will occur when the subtest $(b.\omega \;\Box\; c.\omega)$ is used; this results in probability 0. Similarly, when the subprocess $(b \sqcap c)$ is being tested the subtest $(a.\omega\,{}_{\frac{1}{3}}\!\oplus (b.\omega\,{}_{\frac{1}{2}}\!\oplus c.\omega))$ will give probability 0. In other words $min(\mathcal{A}(T_2, R_1)) = 0$. When applying T_2 to R_2, things can never be as bad. The worst probability will occur when T_2 is applied to the subprocess $(a\,{}_{\frac{1}{2}}\!\oplus b)$, namely probability $\frac{1}{6}$. We leave the reader to check that formally $\mathcal{A}(T_2, R_1) = \{0, \frac{1}{6}, \frac{1}{3}, \frac{1}{2}, \frac{2}{3}\}$ and $\mathcal{A}(T_2, R_2) = \{\frac{1}{6}, \frac{1}{3}, \frac{1}{2}, \frac{2}{3}\}$.

Example 5.8 The axiom $P \sqcap (Q\,{}_p\!\oplus R) = (P \sqcap Q)\,{}_p\!\oplus (P \sqcap R)$ is unsound.

Let $R_1 = a \sqcap (b\,{}_{\frac{1}{2}}\!\oplus c)$, $R_2 = (a \sqcap b)\,{}_{\frac{1}{2}}\!\oplus (a \sqcap c)$ and T be the test $a.(\omega\,{}_{\frac{1}{2}}\!\oplus \mathbf{0}) \,\Box\, b.\omega$. One can check that $\mathcal{A}(T, R_1) = \{\frac{1}{2}\}$ and $\mathcal{A}(T, R_2) = \frac{1}{2}\{\frac{1}{2}, 1\} + \frac{1}{2}\{\frac{1}{2}, 0\} = \{\frac{1}{4}, \frac{1}{2}, \frac{3}{4}\}$. Therefore, we have $R_2 \not\sqsubseteq_{\text{pmay}} R_1$ and $R_1 \not\sqsubseteq_{\text{pmust}} R_2$.

Example 5.9 The axiom $P \,\Box\, (Q \sqcap R) = (P \,\Box\, Q) \sqcap (P \,\Box\, R)$ is unsound.

Let $R_1 = (a\,{}_{\frac{1}{2}}\!\oplus b) \,\Box\, (c \sqcap d)$, $R_2 = ((a\,{}_{\frac{1}{2}}\!\oplus b) \,\Box\, c) \sqcap ((a\,{}_{\frac{1}{2}}\!\oplus b) \,\Box\, d)$ and T be the test $(a.\omega\,{}_{\frac{1}{2}}\!\oplus c.\omega) \sqcap (b.\omega\,{}_{\frac{1}{2}}\!\oplus d.\omega)$. This time we get $\mathcal{A}(T, R_1) = \{0, \frac{1}{4}, \frac{1}{2}, \frac{3}{4}, 1\}$ and $\mathcal{A}(T, R_2) = \{\frac{1}{4}, \frac{3}{4}\}$. So $R_1 \not\sqsubseteq_{\text{pmay}} R_2$ and $R_2 \not\sqsubseteq_{\text{pmust}} R_1$.

Example 5.10 The axiom $P \sqcap (Q \,\Box\, R) = (P \sqcap Q) \,\Box\, (P \sqcap R)$ is unsound.

Let $R_1 = (a\,{}_{\frac{1}{2}}\!\oplus b) \sqcap ((a\,{}_{\frac{1}{2}}\!\oplus b) \,\Box\, \mathbf{0})$ and $R_2 = ((a\,{}_{\frac{1}{2}}\!\oplus b) \sqcap (a\,{}_{\frac{1}{2}}\!\oplus b)) \Box ((a\,{}_{\frac{1}{2}}\!\oplus b) \sqcap \mathbf{0})$. One obtains $\mathcal{A}(a.\omega, R_1) = \{\frac{1}{2}\}$ and $\mathcal{A}(a.\omega, R_2) = \{\frac{1}{2}, \frac{3}{4}\}$. So $R_2 \not\sqsubseteq_{\text{pmay}} R_1$. Let R_3 and R_4 result from substituting $a\,{}_{\frac{1}{2}}\!\oplus b$ for each of P, Q and R in the axiom above. Now $\mathcal{A}(a.\omega, R_3) = \{\frac{1}{2}, \frac{3}{4}\}$ and $\mathcal{A}(a.\omega, R_4) = \{\frac{3}{4}\}$. So $R_4 \not\sqsubseteq_{\text{pmust}} R_3$.

Example 5.11 The axiom $P\,{}_p\!\oplus (Q \,\Box\, R) = (P\,{}_p\!\oplus Q) \,\Box\, (P\,{}_p\!\oplus R)$ is unsound.

Let $R_1 = a\,{}_{\frac{1}{2}}\!\oplus (b \,\Box\, c)$, $R_2 = (a\,{}_{\frac{1}{2}}\!\oplus b) \,\Box\, (a\,{}_{\frac{1}{2}}\!\oplus c)$ and $R_3 = (a \,\Box\, b)\,{}_{\frac{1}{2}}\!\oplus (a \,\Box\, c)$. R_1 is an instance of the left-hand side of the axiom, and R_2 an instance of the right-hand side. Here we use R_3 as a tool to reason about R_2, but in Sect. 5.11.2 we will need R_3 in its own right. Note that $[\![R_2]\!] = \frac{1}{2} \cdot [\![R_1]\!] + \frac{1}{2} \cdot [\![R_3]\!]$. Let $T = a.\omega$. It is easy to see that $\mathcal{A}(T, R_1) = \{\frac{1}{2}\}$ and $\mathcal{A}(T, R_3) = \{1\}$. Therefore, $\mathcal{A}(T, R_2) = \{\frac{3}{4}\}$. So we have $R_2 \not\sqsubseteq_{\text{pmay}} R_1$ and $R_2 \not\sqsubseteq_{\text{pmust}} R_1$.

Of all the examples in this section, this is the only one for which we can show that $\sqsubseteq_{\text{pmay}}$ and $\sqsupseteq_{\text{pmay}}$ both fail, i.e. both inequations that can be associated with the axiom are unsound for *may* testing. Let $T = a.(\omega\,{}_{\frac{1}{2}}\!\oplus \mathbf{0}) \sqcap (b.\omega\,{}_{\frac{1}{2}}\!\oplus c.\omega)$. It is not hard to check that $\mathcal{A}(T, R_1) = \{0, \frac{1}{4}, \frac{1}{2}, \frac{3}{4}\}$ and $\mathcal{A}(T, R_3) = \{\frac{1}{2}\}$. It follows that $\mathcal{A}(T, R_2) = \{\frac{1}{4}, \frac{3}{8}, \frac{1}{2}, \frac{5}{8}\}$. Therefore, we have $R_1 \not\sqsubseteq_{\text{pmay}} R_2$.

For future reference, we also observe that $R_1 \not\sqsubseteq_{\text{pmay}} R_3$ and $R_3 \not\sqsubseteq_{\text{pmay}} R_1$.

5.4 Must Versus May Testing

On pCSP there are two differences between the preorders $\sqsubseteq_{\text{pmay}}$ and $\sqsubseteq_{\text{pmust}}$:

- Must testing is more discriminating
- The preorders $\sqsubseteq_{\text{pmay}}$ and $\sqsubseteq_{\text{pmust}}$ are oriented in opposite directions.

In this section we substantiate these claims by proving that $P \sqsubseteq_{\text{pmust}} Q$ implies $Q \sqsubseteq_{\text{pmay}} P$, and by providing a counterexample that shows the implication is strict. We are only able to obtain the implication since our language is for finite processes and does not feature *divergence*, infinite sequences of τ actions. It is well-known from the nonprobabilistic theory of testing [3, 4] that in the presence of divergence $\simeq_{\text{may}}$ and $\simeq_{\text{must}}$ are incomparable.

To establish a relationship between must testing and may testing, we define the context $C[\_] = \_\,|_{\{\omega\}}(\omega \;\Box\; (\nu \sqcap \nu))$ so that for every test T we obtain a new test $C[T]$, by considering ν instead of ω as success action.

Lemma 5.2 *For any process P and test T, it holds that*

1. *if* $p \in \mathcal{A}(T, P)$*, then* $(1 - p) \in \mathcal{A}(C[T], P)$
2. *if* $p \in \mathcal{A}(C[T], P)$*, then there exists some* $q \in \mathcal{A}(T, P)$ *such that* $1 - q \leq p$.

Proof A state of the form $C[s]\,|_{\text{Act}}\, t$ can always do a τ move, and never directly a success action ν. The τ steps that $C[s]\,|_{\text{Act}}\, t$ can do fall into three classes: the resulting distribution is either

- a point distribution $\overline{u}$ with $u \xrightarrow{\nu}$; we call this a *successful* τ step, because it contributes 1 to the set $\mathbb{V}_f(C[s]\,|_{\text{Act}}\, t)$
- a point distribution $\overline{u}$ with u, a state from which the success action ν is unreachable; we call this an *unsuccessful* τ step, because it contributes 0 to the set $\mathbb{V}_f(C[s]\,|_{\text{Act}}\, t)$
- or a distribution of form $C[\Theta]\,|_{\text{Act}}\, \Delta$.

Note that

- $C[s]\,|_{\text{Act}}\, t$ can always do a successful τ step
- $C[s]\,|_{\text{Act}}\, t$ can do an unsuccessful τ step, iff $s\,|_{\text{Act}} t$ can do a ω step
- and $C[s]\,|_{\text{Act}}\, t \xrightarrow{\tau} C[\Theta]\,|_{\text{Act}}\, \Delta$, iff $s\,|_{\text{Act}}\, t \xrightarrow{\tau} \Theta\,|_{\text{Act}}\, \Delta$.

Using this, both claims follow by a straightforward induction on T and P. □

Proposition 5.1 *If* $P \sqsubseteq_{pmust} Q$ *then* $Q \sqsubseteq_{pmay} P$.

Proof Suppose $P \sqsubseteq_{\text{pmust}} Q$. We must show that, for any test T, if $p \in \mathcal{A}(T, Q)$ then there exists a $q \in \mathcal{A}(T, P)$ such that $p \leq q$. So suppose $p \in \mathcal{A}(T, Q)$. By the first clause of Lemma 5.2, we have $(1 - p) \in \mathcal{A}(C[T], Q)$. Given that $P \sqsubseteq_{\text{pmust}} Q$, there must be an $x \in \mathcal{A}(C[T], P)$ such that $x \leq 1 - p$. By the second clause of Lemma 5.2, there exists a $q \in \mathcal{A}(T, P)$ such that $1 - q \leq x$. It follows that $p \leq q$. Therefore $Q \sqsubseteq_{\text{pmay}} P$. □

Example 5.12 To show that must testing is strictly more discriminating than may testing, consider the processes $a \;\Box\; b$ and $a \sqcap b$, and expose them to test $a.\omega$. It is not hard to see that $\mathcal{A}(a.\omega, a \;\Box\; b) = \{1\}$, whereas $\mathcal{A}(a.\omega, a \sqcap b) = \{0, 1\}$. Since $min(\mathcal{A}(a.\omega, a \;\Box\; b)) = 1$ and $min(\mathcal{A}(a.\omega, a \sqcap b)) = 0$, using Proposition 4.1 we obtain that $(a \;\Box\; b) \not\sqsubseteq_{\text{pmust}} (a \sqcap b)$.

Since $max(\mathcal{A}(a.\omega, a \Box b)) = max(\mathcal{A}(a.\omega, a \sqcap b)) = 1$, as a *may* test, the test $a.\omega$ does not differentiate between the two processes $a \Box b$ and $a \sqcap b$. In fact, we have $(a \sqcap b) \sqsubseteq_{\text{pmay}} (a \Box b)$, and even $(a \Box b) \simeq_{\text{pmay}} (a \sqcap b)$, but this cannot be shown so easily, as we would have to consider all possible tests. In Sect. 5.5 we will develop a tool to prove statements $P \sqsubseteq_{\text{pmay}} Q$, and apply it to derive the equality above (axiom (**May0**) in Fig. 5.8).

5.5 Forward and Failure Simulation

The examples of Sect. 5.3 have been all negative, because one can easily demonstrate an inequivalence between two processes by exhibiting a test that distinguishes them in the appropriate manner. A direct application of the definition of the testing preorders is usually unsuitable for establishing positive results, as this involves a universal quantification over all possible tests that can be applied. To give positive results of the form $P \sqsubseteq_{\text{pmay}} Q$ (and similarly for $P \sqsubseteq_{\text{pmust}} Q$) we need to come up with a preorder $\sqsubseteq_{\text{finer}}$ such that $(P \sqsubseteq_{\text{finer}} Q) \Rightarrow (P \sqsubseteq_{\text{pmay}} Q)$ and statements $P \sqsubseteq_{\text{finer}} Q$ can be obtained by exhibiting a single witness.

In this section we introduce coinductive relations: *forward simulations* and *failure simulations*. For *may* testing we use forward simulations as our witnesses, and for *must* testing we use failure simulations as witnesses. The definitions are somewhat complicated by the fact that in a pLTS transitions go from states to distributions; consequently, if we are to use sequences of transitions, or *weak transitions* $\stackrel{a}{\Longrightarrow}$ that abstract from sequences of internal actions that might precede or follow the a transition, then we need to generalise transitions so that they go from distributions to distributions. We first recall the mathematical machinery developed in Sect. 3.3, where we have discussed various ways of lifting a relation $\mathcal{R} \subseteq S \times S$ to a relation $\mathcal{R}^\dagger \subseteq \mathcal{D}(S) \times \mathcal{D}(S)$. Exactly the same idea can be used to lift a relation $\mathcal{R} \subseteq S \times \mathcal{D}(S)$ to a relation $\mathcal{R}^\dagger \subseteq \mathcal{D}(S) \times \mathcal{D}(S)$. This justifies our slight abuse of notation here of keeping writing $\mathcal{R}^\dagger$ for the lifted relation.

Definition 5.3 Let $\mathcal{R} \subseteq S \times \mathcal{D}(S)$ be a relation from states to subdistributions. Then $\mathcal{R}^\dagger \subseteq \mathcal{D}(S) \times \mathcal{D}(S)$ is the smallest relation that satisfies:

(1) $s \mathcal{R} \Theta$ implies $\overline{s}\ \mathcal{R}^\dagger\ \Theta$, and
(2) (Linearity) $\Delta_i\ \mathcal{R}^\dagger\ \Theta_i$ for all $i \in I$ implies $(\sum_{i \in I} p_i \cdot \Delta_i)\ \mathcal{R}^\dagger\ (\sum_{i \in I} p_i \cdot \Theta_i)$, where I is a finite index set and $\sum_{i \in I} p_i = 1$.

From the above definition we immediately get the following property.

Lemma 5.3 *$\Delta\ \mathcal{R}^\dagger\ \Theta$ if and only if there is a collection of states $\{s_i\}_{i \in I}$, a collection of distributions $\{\Theta_i\}_{i \in I}$ and a collection of probabilities $\{p_i\}_{i \in I}$, for some finite index set I, such that $\sum_{i \in I} p_i = 1$ and Δ, Θ can be decomposed as follows:*

1. $\Delta = \sum_{i \in I} p_i \cdot \overline{s_i}$
2. $\Theta = \sum_{i \in I} p_i \cdot \Theta_i$
3. For each $i \in I$ *we have* $s_i \mathcal{R} \Theta_i$.

Proof Similar to the proof of Proposition 3.1. □

We apply this definition to the action relations $\xrightarrow{\alpha} \subseteq \mathsf{sCSP} \times \mathcal{D}(\mathsf{sCSP})$ in the operational semantics of pCSP, and obtain lifted relations between $\mathcal{D}(\mathsf{sCSP})$ and $\mathcal{D}(\mathsf{sCSP})$, which to ease the notation we write as $\Delta \xrightarrow{\alpha} \Theta$; then, using pCSP terms to represent distributions, a simple instance of a transition between distributions is given by

$$(a.b \,\Box\, a.c)\,_{\frac{1}{2}}\!\oplus a.d \xrightarrow{a} b\,_{\frac{1}{2}}\!\oplus d.$$

Note that we also have

$$(a.b \,\Box\, a.c)\,_{\frac{1}{2}}\!\oplus a.d \xrightarrow{a} (b\,_{\frac{1}{2}}\!\oplus c)\,_{\frac{1}{2}}\!\oplus d \tag{5.1}$$

because, viewed as a distribution, the term $(a.b \,\Box\, a.c)\,_{\frac{1}{2}}\!\oplus a.d$ may be rewritten as $((a.b \,\Box\, a.c)\,_{\frac{1}{2}}\!\oplus (a.b \,\Box\, a.c))\,_{\frac{1}{2}}\!\oplus\, a.d$ representing the sum of point distributions

$$\tfrac{1}{4} \cdot \overline{(a.b \,\Box\, a.c)} + \tfrac{1}{4} \cdot \overline{(a.b \,\Box\, a.c)} + \tfrac{1}{2} \cdot \overline{a.d}$$

from which the move (5.1) can easily be derived using the three moves from states

$$a.b \,\Box\, a.c \xrightarrow{a} \overline{b} \qquad a.b \,\Box\, a.c \xrightarrow{a} \overline{c} \qquad a.d \xrightarrow{a} \overline{d}.$$

The lifting construction can also be used to define the concept of a *partial* internal move between distributions, one where part of the distribution does an internal move and the remainder remains unchanged. Write $s \xrightarrow{\hat{\tau}} \Delta$ if either $s \xrightarrow{\tau} \Delta$ or $\Delta = \overline{s}$. This relation between states and distributions can be lifted to one between distributions and distributions, and again for notational convenience we use $\Delta_1 \xrightarrow{\hat{\tau}} \Delta_2$ to denote the lifted relation. As an example, again using process terms to denote distributions, we have

$$(a \sqcap b)\,_{\frac{1}{2}}\!\oplus (a \sqcap c) \xrightarrow{\hat{\tau}} a\,_{\frac{1}{2}}\!\oplus (a \sqcap b\,_{\frac{1}{2}}\!\oplus c).$$

This follows because as a distribution $(a \sqcap b)\,_{\frac{1}{2}}\!\oplus (a \sqcap c)$ may be written as

$$\tfrac{1}{4} \cdot \overline{(a \sqcap b)} + \tfrac{1}{4} \cdot \overline{(a \sqcap b)} + \tfrac{1}{4} \cdot \overline{(a \sqcap c)} + \tfrac{1}{4} \cdot \overline{(a \sqcap c)}$$

and we have the four moves from states to distributions:

$$(a \sqcap b) \xrightarrow{\hat{\tau}} \overline{a} \qquad (a \sqcap b) \xrightarrow{\hat{\tau}} \overline{(a \sqcap b)}$$

$$(a \sqcap c) \xrightarrow{\hat{\tau}} \overline{a} \qquad (a \sqcap c) \xrightarrow{\hat{\tau}} \overline{c}.$$

5.5.1 The Simulation Preorders

Following tradition it would be natural to define simulations as relations between states in a pLTS [13, 14], as we did in Sect. 3.5. However, technically it is more convenient to use relations in $\mathsf{sCSP} \times \mathcal{D}(\mathsf{sCSP})$. One reason may be understood through the example in Fig. 5.5. Although in Example 5.3 we found out that $R_2 \not\sqsubseteq_{\text{pmay}} R_1$, we do have $R_1 \sqsubseteq_{\text{pmay}} R_2$. If we are to relate these processes via a simulation-like relation, then the initial state of R_1 needs to be related to the initial *distribution* of R_2, containing the two states $a.b$ and $a.c$.

Our definition of simulation uses *weak transitions* [15], which have the standard definitions except that they now apply to distributions, and $\xrightarrow{\hat{\tau}}$ is used instead of $\xrightarrow{\tau}$. This reflects the understanding that if a distribution may perform a sequence of internal moves before or after executing a visible action, different parts of the distribution may perform different numbers of internal actions:

- Let $\Delta_1 \overset{\hat{\tau}}{\Longrightarrow} \Delta_2$ whenever $\Delta_1 \xrightarrow{\hat{\tau}}{}^* \Delta_2$.
- Similarly, $\Delta_1 \overset{\hat{a}}{\Longrightarrow} \Delta_2$ denotes $\Delta_1 \xrightarrow{\hat{\tau}}{}^* \xrightarrow{a} \xrightarrow{\hat{\tau}}{}^* \Delta_2$ whenever $a \in \mathsf{Act}$.
- We write $s \overset{A}{\not\longrightarrow}$ with $A \subseteq \mathsf{Act}$ when $\forall \alpha \in A \cup \{\tau\} : s \overset{\alpha}{\not\longrightarrow}$, and also $\Delta \overset{A}{\not\longrightarrow}$ when $\forall s \in \lceil \Delta \rceil : s \overset{A}{\not\longrightarrow}$.
- More generally, write$\Delta \Longrightarrow \overset{A}{\not\longrightarrow}$ if $\Delta \longrightarrow \Delta^{\text{pre}}$ for some Δ^{pre} such that $\Delta^{\text{pre}} \overset{A}{\not\longrightarrow}$.

Definition 5.4 A relation $\mathcal{R} \subseteq S \times \mathcal{D}(S)$ is said to be a *failure simulation* if $s \mathcal{R} \Theta$ implies

- if $s \xrightarrow{\alpha} \Delta$ then there exists some Θ' such that $\Theta \overset{\hat{\alpha}}{\Longrightarrow} \Theta'$ and $\Delta \; \mathcal{R}^\dagger \; \Theta'$
- if $s \overset{A}{\not\longrightarrow}$ then $\Theta \Longrightarrow \overset{A}{\not\longrightarrow}$.

We write $s \triangleleft_{FS} \Theta$ if there is some failure simulation $\mathcal{R}$ such that $s \mathcal{R} \Theta$. Similarly, we define *(forward) simulation* and $s \triangleleft_s \Theta$ by dropping the second clause in Definition 5.4.

Definition 5.5 The *(forward) simulation preorder* $\sqsubseteq_S$ and *failure simulation preorder* $\sqsubseteq_{FS}$ on pCSP are defined as follows:

$$P \sqsubseteq_S Q \text{ iff } [\![Q]\!] \overset{\hat{\tau}}{\Longrightarrow} \Theta \text{ for some } \Theta \text{ with } [\![P]\!] \; (\triangleleft_s)^\dagger \, \Theta$$
$$P \sqsubseteq_{FS} Q \text{ iff } [\![P]\!] \overset{\hat{\tau}}{\Longrightarrow} \Theta \text{ for some } \Theta \text{ with } [\![Q]\!] \; (\triangleleft_s)^\dagger \, \Theta \, .$$

We have chosen the orientation of the preorder symbol to match that of must testing, which goes back to the work of De Nicola and Hennessy [3]. This orientation also matches the one used in CSP [1] and related work, where we have SPECIFICATION $\sqsubseteq$ IMPLEMENTATION. At the same time, we like to stick to the convention popular in the CCS community of writing the simulated process to the left of the preorder symbol and the simulating process (that mimics moves of the simulated one) on the right. This is the reason for the orientation of the symbol $\triangleleft_{FS}$.

$$P_1 = a.(b \,{}_{\frac{1}{2}}\!\oplus (c \,{}_{\frac{1}{2}}\!\oplus c)) \qquad P_2 = a.((b \,{}_{\frac{1}{2}}\!\oplus b) \,{}_{\frac{1}{2}}\!\oplus c)$$

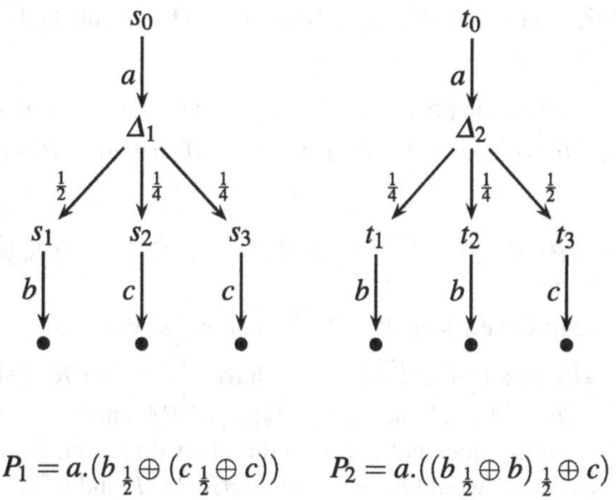

Fig. 5.7 Two simulation equivalent processes

The equivalences generated by $\sqsubseteq_S$ and $\sqsubseteq_{FS}$ are called *(forward) simulation equivalence* and *failure simulation equivalence*, denoted $\simeq_S$ and $\simeq_{FS}$, respectively.

If $P \in \mathsf{sCSP}$, that is if P is a state in the pLTS of pCSP and so $[\![P]\!] = \overline{P}$, then to establish $P \sqsubseteq_S Q$ it is sufficient to exhibit a simulation between the state P and the distribution $[\![Q]\!]$, because trivially $s \lhd_s \Delta$ implies $\overline{s}(\lhd_s)^\dagger \Delta$

Example 5.13   Consider the two processes P_i in Fig. 5.7. To show $P_1 \sqsubseteq_S P_2$ it is sufficient to exhibit a simulation $\mathcal{R}$ such that $s_0 \mathcal{R} \overline{t_0}$. Let $\mathcal{R} \subseteq \mathsf{sCSP} \times \mathcal{D}(\mathsf{sCSP})$ be defined by

$$s_0 \, \mathcal{R} \, \overline{t_0} \qquad s_1 \, \mathcal{R} \, \Delta_t \qquad s_2 \, \mathcal{R} \, \overline{t_3} \qquad s_3 \, \mathcal{R} \, \overline{t_3} \qquad \mathbf{0} \, \mathcal{R} \, \overline{\mathbf{0}}$$

where Δ_t is the two-point distribution mapping both t_1 and t_2 to the probability $\frac{1}{2}$. Then it is straightforward to check that it satisfies the requirements of a simulation: the only nontrivial requirement is that $\Delta_1 \, \mathcal{R}^\dagger \, \Delta_2$. But this follows from the fact that

$$\Delta_1 = \tfrac{1}{2} \cdot \overline{s_1} + \tfrac{1}{4} \cdot \overline{s_2} + \tfrac{1}{4} \cdot \overline{s_3}$$
$$\Delta_2 = \tfrac{1}{2} \cdot \Delta_t + \tfrac{1}{4} \cdot \overline{t_3} + \tfrac{1}{4} \cdot \overline{t_3}.$$

As another example reconsider $R_1 = a.(b \,{}_{\frac{1}{2}}\!\oplus c)$ and $R_2 = a.b \,{}_{\frac{1}{2}}\!\oplus a.c$ from Fig. 5.5, where for convenience we use process terms to denote their semantic interpretations. It is easy to see that $R_1 \sqsubseteq_S R_2$ because of the simulation

$$R_1 \, \mathcal{R} \, [\![R_2]\!] \qquad b \, \mathcal{R} \, \overline{b} \qquad c \, \mathcal{R} \, \overline{c} \qquad \mathbf{0} \, \mathcal{R} \, \overline{\mathbf{0}}$$

The transition $R_1 \xrightarrow{a} (b \,{}_{\frac{1}{2}}\!\oplus c)$ is matched by the transition $R_2 \xrightarrow{a} (b \,{}_{\frac{1}{2}}\!\oplus c)$ with $(b \,{}_{\frac{1}{2}}\!\oplus c) \, \mathcal{R} \, (b \,{}_{\frac{1}{2}}\!\oplus c)$.

Similarly, $(a \,{}_{\frac{1}{2}}\!\oplus c) \sqcap (b \,{}_{\frac{1}{2}}\!\oplus c) \sqsubseteq_S (a \sqcap b) \,{}_{\frac{1}{2}}\!\oplus c$ because it is possible to find a simulation between the state $(a \,{}_{\frac{1}{2}}\!\oplus c) \sqcap (b \,{}_{\frac{1}{2}}\!\oplus c)$ and the distribution $(a \sqcap b) \,{}_{\frac{1}{2}}\!\oplus c$.

In case $P \notin \mathsf{sCSP}$, a statement $P \sqsubseteq_S Q$ cannot always be established by a simulation $\mathcal{R}$ such that $[\![P]\!]\ \mathcal{R}^\dagger\ [\![Q]\!]$.

Example 5.14 Compare the processes $P = a\,{}_{\frac{1}{2}}\!\oplus b$ and $P \sqcap P$. Note that $[\![P]\!]$ is the distribution $\frac{1}{2}\overline{a} + \frac{1}{2}\overline{b}$, whereas $[\![P \sqcap P]\!]$ is the point distribution $\overline{P \sqcap P}$. The relation $\mathcal{R}$ given by

$$(P \sqcap P)\,\mathcal{R}\,(\tfrac{1}{2}\overline{a} + \tfrac{1}{2}\overline{b}) \qquad a\,\mathcal{R}\,\overline{a} \qquad b\,\mathcal{R}\,\overline{b} \qquad \mathbf{0}\,\mathcal{R}\,\overline{\mathbf{0}}$$

is a simulation, because the τ step $P \sqcap P \xrightarrow{\tau} (\frac{1}{2}\overline{a} + \frac{1}{2}\overline{b})$ can be matched by the idle transition $(\frac{1}{2}\overline{a} + \frac{1}{2}\overline{b}) \stackrel{\hat{\tau}}{\Longrightarrow} (\frac{1}{2}\overline{a} + \frac{1}{2}\overline{b})$, and we have $(\frac{1}{2}\overline{a} + \frac{1}{2}\overline{b})\ \mathcal{R}^\dagger\ (\frac{1}{2}\overline{a} + \frac{1}{2}\overline{b})$. Thus $(P \sqcap P) \lhd_s (\frac{1}{2}\overline{a} + \frac{1}{2}\overline{b}) = [\![P]\!]$, hence $[\![P \sqcap P]\!](\lhd_s)^\dagger[\![P]\!]$, and therefore $P \sqcap P \sqsubseteq_S P$.

This type of reasoning does not apply to the other direction. Any simulation $\mathcal{R}$ with $(\frac{1}{2}\overline{a} + \frac{1}{2}\overline{b})\ \mathcal{R}^\dagger\ \overline{P \sqcap P}$ would have to satisfy $a \mathcal{R} \overline{P \sqcap P}$ and $b \mathcal{R} \overline{P \sqcap P}$. However, the move $a \xrightarrow{a} \mathbf{0}$ cannot be matched by the process $\overline{P \sqcap P}$, as the only transition the latter process can do is $\overline{P \sqcap P} \xrightarrow{\tau} (\frac{1}{2}\overline{a} + \frac{1}{2}\overline{b})$, and only half of that distribution can match the a move. As a consequence, no such simulation exists, and we find that $[\![P]\!](\lhd_s)^\dagger[\![P \sqcap P]\!]$ does not hold. Nevertheless, we still have $P \sqsubseteq_S P \sqcap P$. Here, the transition $\stackrel{\hat{\tau}}{\Longrightarrow}$ from Definition 5.5 comes to the rescue. As $[\![P \sqcap P]\!] \stackrel{\hat{\tau}}{\Longrightarrow} [\![P]\!]$ and $[\![P]\!](\lhd_s)^\dagger[\![P]\!]$, we obtain $P \sqsubseteq_S P \sqcap P$.

Example 5.15 Let $P = a\,{}_{\frac{1}{2}}\!\oplus b$ and $Q = P \;\Box\; P$. We can establish that $P \sqsubseteq_S Q$ because $[\![P]\!](\lhd_s)^\dagger[\![Q]\!]$ that comes from the following observations:

1. $[\![P]\!] = \frac{1}{2}\overline{a} + \frac{1}{2}\overline{b}$
2. $[\![Q]\!] = \frac{1}{2}(\frac{1}{2}\overline{a \,\Box\, a} + \frac{1}{2}\overline{a \,\Box\, b}) + \frac{1}{2}(\frac{1}{2}\overline{a \,\Box\, b} + \frac{1}{2}\overline{b \,\Box\, b})$
3. $a \lhd_s (\frac{1}{2}\overline{a \,\Box\, a} + \frac{1}{2}\overline{a \,\Box\, b})$
4. $b \lhd_s (\frac{1}{2}\overline{a \,\Box\, b} + \frac{1}{2}\overline{b \,\Box\, b})$.

This kind of reasoning does not apply to $\lhd_{FS}$. For example, we have

$$a \not\lhd_{FS} (\frac{1}{2}\overline{a \,\Box\, a} + \frac{1}{2}\overline{a \,\Box\, b})$$

because the state on the left hand side can refuse to do action b while the distribution on the right hand side cannot. Indeed, it holds that $Q \not\sqsubseteq_{FS} P$.

Because of the asymmetric use of distributions in the definition of simulations it is not immediately obvious that $\sqsubseteq_S$ and $\sqsubseteq_{FS}$ are actually preorders (reflexive and transitive relations) and hence, $\simeq_S$ and $\simeq_{FS}$ are equivalence relations. In order to show this, we first need to establish some properties of $\lhd_s$ and $\lhd_{FS}$.

Lemma 5.4 *Suppose* $\sum_{i \in I} p_i = 1$ *and* $\Delta_i \stackrel{\hat{\alpha}}{\Longrightarrow} \Phi_i$ *for each* $i \in I$, *with* I *a finite index set. Then*

$$\sum_{i \in I} p_i \cdot \Delta_i \stackrel{\hat{\alpha}}{\Longrightarrow} \sum_{i \in I} p_i \cdot \Phi_i$$

Proof We first prove the case $\alpha = \tau$. For each $i \in I$ there is a number k_i such that $\Delta_i = \Delta_{i0} \xrightarrow{\hat{\tau}} \Delta_{i1} \xrightarrow{\hat{\tau}} \Delta_{i2} \xrightarrow{\hat{\tau}} \cdots \xrightarrow{\hat{\tau}} \Delta_{ik_i} = \Delta'_i$. Let $k = max\{k_i \mid i \in I\}$, using that I is finite. Since we have $\Phi \xrightarrow{\hat{\tau}} \Phi$ for any $\Phi \in \mathcal{D}(S)$, we can add spurious transitions to these sequences, until all k_i equal k. After this preparation the lemma follows by k applications of the linearity of the lifting operation (cf. Definition 5.3), taking $\xrightarrow{\hat{\tau}}$ for $\mathcal{R}$.

The case $\alpha \in \mathsf{Act}$ now follows by one more application of the linearity of lifting operation, this time with $\mathcal{R} = \xrightarrow{a}$, preceded and followed by an application of the case $\alpha = \tau$. □

Lemma 5.5 *Suppose $\Delta(\lhd_s)^\dagger\Phi$ and $\Delta \xrightarrow{\alpha} \Delta'$. Then $\Phi \overset{\hat{\alpha}}{\Longrightarrow} \Phi'$ for some Φ' such that $\Delta'(\lhd_s)^\dagger\Phi'$.*

Proof First $\Delta(\lhd_s)^\dagger\Phi$ means that

$$\Delta = \sum_{i\in I} p_i \cdot \overline{s_i}, \qquad s_i \lhd_s \Phi_i, \qquad \Phi = \sum_{i\in I} p_i \cdot \Phi_i\ ; \tag{5.2}$$

also $\Delta \xrightarrow{\alpha} \Delta'$ means

$$\Delta = \sum_{j\in J} q_j \cdot \overline{t_j}, \qquad t_j \xrightarrow{\alpha} \Theta_j, \qquad \Delta' = \sum_{j\in J} q_j \cdot \Theta_j\ , \tag{5.3}$$

and we can assume without loss of generality that all the coefficients p_i, q_j are nonzero. Now define $I_j = \{ i \in I \mid s_i = t_j\}$ and $J_i = \{ j \in J \mid t_j = s_i\}$, so that trivially

$$\{(i,j) \mid i \in I,\ j \in J_i\} = \{(i,j) \mid j \in J,\ i \in I_j\} \tag{5.4}$$

and note that

$$\Delta(s_i) = \sum_{j\in J_i} q_j \qquad \text{and} \qquad \Delta(t_j) = \sum_{i\in I_j} p_i \tag{5.5}$$

Because of (5.5) we have

$$\begin{aligned} \Phi = \sum_{i\in I} p_i \cdot \Phi_i &= \sum_{i\in I} p_i \cdot \sum_{j\in J_i} \frac{q_j}{\Delta(s_i)} \cdot \Phi_i \\ &= \sum_{i\in I}\sum_{j\in J_i} \frac{p_i \cdot q_j}{\Delta(s_i)} \cdot \Phi_i \end{aligned}$$

Now for each j in J_i we know that in fact $t_j = s_i$, and so from the middle parts of (5.2) and (5.3) we obtain $\Phi_i \overset{\hat{\alpha}}{\Longrightarrow} \Phi_{ij}$ such that $\Theta_j(\lhd_s)^\dagger\Phi_{ij}$. Lemma 5.4 yields

$$\Phi \overset{\hat{\alpha}}{\Longrightarrow} \Phi' = \sum_{i\in I}\sum_{j\in J_i} \frac{p_i \cdot q_j}{\Delta(s_i)} \cdot \Phi_{ij}$$

where within the summations $s_i = t_j$, so that, using (5.4), Φ' can also be written as

$$\sum_{j\in J}\sum_{i\in I_j} \frac{p_i \cdot q_j}{\Delta(t_j)} \cdot \Phi_{ij} \tag{5.6}$$

All that remains is to show that $\Delta'(\lhd_s)^\dagger \Phi'$, which we do by manipulating Δ' so that it takes on a form similar to that in (5.6):

$$\begin{aligned}\Delta' &= \sum_{j\in J} q_j \cdot \Theta_j \\ &= \sum_{j\in J} q_j \cdot \sum_{i\in I_j} \frac{p_i}{\Delta(t_j)} \cdot \Theta_j \qquad \text{using (5.5) again} \\ &= \sum_{j\in J}\sum_{i\in I_j} \frac{p_i \cdot q_j}{\Delta(t_j)} \cdot \Theta_j .\end{aligned}$$

Comparing this with (5.6) above we see that the required result, $\Delta'(\lhd_s)^\dagger \Phi'$, follows from an application of the linearity of lifting operation. □

Lemma 5.6 *Suppose $\Delta(\lhd_s)^\dagger \Phi$ and $\Delta \stackrel{\hat{\alpha}}{\Longrightarrow} \Delta'$. Then $\Phi \stackrel{\hat{\alpha}}{\longrightarrow} \Phi'$ for some Φ' such that $\Delta'(\lhd_s)^\dagger \Phi'$.*

Proof First we consider two claims

(i) If $\Delta(\lhd_s)^\dagger \Phi$ and $\Delta \stackrel{\hat{\tau}}{\longrightarrow} \Delta'$, then $\Phi \stackrel{\hat{\tau}}{\Longrightarrow} \Phi'$ for some Φ' such that $\Delta'(\lhd_s)^\dagger \Phi'$.
(ii) If $\Delta(\lhd_s)^\dagger \Phi$ and $\Delta \stackrel{\hat{\tau}}{\Longrightarrow} \Delta'$, then $\Phi \stackrel{\hat{\tau}}{\Longrightarrow} \Phi'$ for some Φ' such that $\Delta'(\lhd_s)^\dagger \Phi'$.

The proof of claim (i) is similar to the proof of Lemma 5.5. Claim (ii) follows from claim (i) by induction on the length of the derivation of $\stackrel{\hat{\tau}}{\Longrightarrow}$. By combining claim (ii) with Lemma 5.5, we obtain the required result. □

Proposition 5.2 *The relation $(\lhd_s)^\dagger$ is both reflexive and transitive on distributions.*

Proof We leave reflexivity to the reader; it relies on the fact that $s \lhd_s \overline{s}$ for every state s.

For transitivity, let $\mathcal{R} \subseteq \mathsf{sCSP} \times \mathcal{D}(\mathsf{sCSP})$ be given by $s\mathcal{R}\Phi$ iff $s \lhd_s \Delta(\lhd_s)^\dagger \Phi$ for some intermediate distribution Δ. Transitivity follows from the two claims

(i) $\Theta(\lhd_s)^\dagger \Delta(\lhd_s)^\dagger \Phi$ implies $\Theta\ \mathcal{R}^\dagger\ \Phi$
(ii) $\mathcal{R}$ is a simulation, hence $\mathcal{R}^\dagger \subseteq (\lhd_s)^\dagger$.

Claim (ii) is a straightforward application of Lemma 5.6, so let us look at (i). From $\Theta(\lhd_s)^\dagger \Delta$ we have

$$\Theta = \sum_{i\in I} p_i \cdot \overline{s_i}, \qquad s_i \lhd_s \Delta_i, \qquad \Delta = \sum_{i\in I} p_i \cdot \Delta_i .$$

Since $\Delta(\lhd_s)^\dagger \Phi$, in analogy to Proposition 3.3 we can show that $\Phi = \sum_{i\in I} p_i \cdot \Phi_i$ such that $\Delta_i(\lhd_s)^\dagger \Phi_i$. So for each i we have $s_i \mathcal{R} \Phi_i$, from which it follows that $\Theta\ \mathcal{R}^\dagger\ \Phi$. □

Proposition 5.3 *$\sqsubseteq_S$ and $\sqsubseteq_{FS}$ are preorders, i.e. they are reflexive and transitive.*

Proof By combination of Lemma 5.6 and Proposition 5.2, we obtain that $\sqsubseteq_S$ is a preorder. The case for $\sqsubseteq_{FS}$ can be similarly established by proving the counterparts of Lemma 5.6 and Proposition 5.2. □

5.5.2 The Simulation Preorders are Precongruences

In Theorem 5.1 of this section we establish that the pCSP operators are monotone with respect to the simulation preorders, i.e. that both $\sqsubseteq_S$ and $\sqsubseteq_{FS}$ are precongruences for pCSP. This implies that the pCSP operators are compositional for them or, equivalently, that $\simeq_S$ and $\simeq_{FS}$ are congruences for pCSP. The following two lemmas gather some facts we need in the proof of this theorem. Their proofs are straightforward, although somewhat tedious.

Lemma 5.7

(i) If $\Phi \stackrel{\hat{\tau}}{\Longrightarrow} \Phi'$, then $\Phi \,\Box\, \Delta \stackrel{\hat{\tau}}{\Longrightarrow} \Phi' \,\Box\, \Delta$ and $\Delta \,\Box\, \Phi \stackrel{\hat{\tau}}{\Longrightarrow} \Delta \,\Box\, \Phi'$.
(ii) If $\Phi \stackrel{a}{\longrightarrow} \Phi'$, then $\Phi \,\Box\, \Delta \stackrel{a}{\longrightarrow} \Phi'$ and $\Delta \,\Box\, \Phi \stackrel{a}{\longrightarrow} \Phi'$.
(iii) $(\sum_{j\in J} p_j \cdot \Phi_j) \,\Box\, (\sum_{k\in K} q_k \cdot \Delta_k) = \sum_{j\in J} \sum_{k\in K} (p_j \cdot q_k) \cdot (\Phi_j \,\Box\, \Delta_k)$.
(iv) Given two binary relations $\mathcal{R}, \mathcal{R}' \subseteq \mathsf{sCSP} \times \mathcal{D}(\mathsf{sCSP})$ satisfying $s\mathcal{R}'\Delta$ whenever $s = s_1 \,\Box\, s_2$ and $\Delta = \Delta_1 \,\Box\, \Delta_2$ with $s_1\mathcal{R}\Delta_1$ and $s_2\mathcal{R}\Delta_2$. Then $\Phi_i \; \mathcal{R}^\dagger \; \Delta_i$ for $i = 1, 2$ implies $(\Phi_1 \,\Box\, \Phi_2) \; \mathcal{R}'^\dagger \; (\Delta_1 \,\Box\, \Delta_2)$. □

Lemma 5.8

(i) If $\Phi \stackrel{\hat{\tau}}{\Longrightarrow} \Phi'$, then $\Phi \,|_A\, \Delta \stackrel{\hat{\tau}}{\Longrightarrow} \Phi' \,|_A\, \Delta$ and $\Delta \,|_A\, \Phi \stackrel{\hat{\tau}}{\longrightarrow} \Delta \,|_A\, \Phi'$.
(ii) If $\Phi \stackrel{a}{\longrightarrow} \Phi'$ and $a \notin A$, then $\Phi \,|_A\, \Delta \stackrel{a}{\longrightarrow} \Phi' \,|_A\, \Delta$ and $\Delta \,|_A\, \Phi \stackrel{a}{\longrightarrow} \Delta \,|_A\, \Phi'$.
(iii) If $\Phi \stackrel{a}{\longrightarrow} \Phi'$, $\Delta \stackrel{a}{\longrightarrow} \Delta'$ and $a \in A$, then $\Delta \,|_A\, \Phi \stackrel{\tau}{\longrightarrow} \Delta' \,|_A\, \Phi'$.
(iv) $(\sum_{j\in J} p_j \cdot \Phi_j) \,|_A\, (\sum_{k\in K} q_k \cdot \Delta_k) = \sum_{j\in J} \sum_{k\in K} (p_j \cdot q_k) \cdot (\Phi_j \,|_A\, \Delta_k)$.
(v) Let $\mathcal{R}, \mathcal{R}' \subseteq \mathsf{sCSP} \times \mathcal{D}(\mathsf{sCSP})$ be two binary relations satisfying $s\mathcal{R}'\Delta$ whenever $s = s_1 \,|_A\, s_2$ and $\Delta = \Delta_1 \,|_A\, \Delta_2$ with $s_1\mathcal{R}\Delta_1$ and $s_2\mathcal{R}\Delta_2$. Then $\Phi_i \; \mathcal{R}^\dagger \; \Delta_i$ for $i = 1, 2$ implies $(\Phi_1 \,|_A\, \Phi_2) \; \mathcal{R}'^\dagger \; (\Delta_1 \,|_A\, \Delta_2)$. □

Theorem 5.1 *Let $\sqsubseteq \in \{\sqsubseteq_S, \sqsubseteq_{FS}\}$. Suppose $P_i \sqsubseteq Q_i$ for $i = 1, 2$. Then*

1. *$a.P_1 \sqsubseteq a.Q_1$*
2. *$P_1 \sqcap P_2 \sqsubseteq Q_1 \sqcap Q_2$*
3. *$P_1 \,\Box\, P_2 \sqsubseteq Q_1 \,\Box\, Q_2$*
4. *$P_1 \,{}_p\!\oplus\, P_2 \sqsubseteq Q_1 \,{}_p\!\oplus\, Q_2$*
5. *$P_1 \,|_A\, P_2 \sqsubseteq Q_1 \,|_A\, Q_2$.*

Proof We first consider the case for $\sqsubseteq_S$.

1. Since $P_1 \sqsubseteq_S Q_1$, there must be a Δ_1 such that $[\![Q_1]\!] \stackrel{\hat{\tau}}{\Longrightarrow} \Delta_1$ and $[\![P_1]\!](\lhd_s)^\dagger \Delta_1$. It is easy to see that $a.P_1 \lhd_s \overline{a.Q_1}$ because the transition $a.P_1 \stackrel{a}{\longrightarrow} [\![P_1]\!]$ can be matched by $\overline{a.Q_1} \stackrel{a}{\longrightarrow} [\![Q_1]\!] \stackrel{\hat{\tau}}{\Longrightarrow} \Delta_1$. Thus $[\![a.P_1]\!] = \overline{a.P_1}\,(\lhd_s)^\dagger\,\overline{a.Q_1} = [\![a.Q_1]\!]$.
2. Since $P_i \sqsubseteq_S Q_i$, there must be a Δ_i such that $[\![Q_i]\!] \stackrel{\hat{\tau}}{\Longrightarrow} \Delta_i$ and $[\![P_i]\!](\lhd_s)^\dagger \Delta_i$. It is easy to see that $P_1 \sqcap P_2 \lhd_s \overline{Q_1 \sqcap Q_2}$ because the transition $P_1 \sqcap P_2 \stackrel{\tau}{\longrightarrow} [\![P_i]\!]$, for $i = 1$ or $i = 2$, can be matched by $\overline{Q_1 \sqcap Q_2} \stackrel{\tau}{\longrightarrow} [\![Q_i]\!] \stackrel{\hat{\tau}}{\Longrightarrow} \Delta_i$. Thus, we have that $[\![P_1 \sqcap P_2]\!] = \overline{P_1 \sqcap P_2}\,(\lhd_s)^\dagger\,\overline{Q_1 \sqcap Q_2} = [\![Q_1 \sqcap Q_2]\!]$.
3. Let $\mathcal{R} \subseteq \mathsf{sCSP} \times \mathcal{D}(\mathsf{sCSP})$ be defined by $s\mathcal{R}\Delta$, iff either $s \lhd_s \Delta$ or $s = s_1 \,\Box\, s_2$ and $\Delta = \Delta_1 \,\Box\, \Delta_2$ with $s_1 \lhd_s \Delta_1$ and $s_2 \lhd_s \Delta_2$. We show that $\mathcal{R}$ is a simulation. Suppose $s_1 \lhd_s \Delta_1, s_2 \lhd_s \Delta_2$ and $s_1 \,\Box\, s_2 \stackrel{a}{\longrightarrow} \Theta$ with $a \in \mathsf{Act}$. Then $s_i \stackrel{a}{\longrightarrow} \Theta$ for $i = 1$ or $i = 2$. Thus $\Delta_i \stackrel{\hat{a}}{\Longrightarrow} \Delta$ for some Δ with $\Theta(\lhd_s)^\dagger \Delta$, and hence $\Theta\ \mathcal{R}^\dagger\ \Delta$. By Lemma 5.7 we have $\Delta_1 \,\Box\, \Delta_2 \stackrel{\hat{a}}{\Longrightarrow} \Delta$.
   Now suppose that $s_1 \lhd_s \Delta_1$, $s_2 \lhd_s \Delta_2$ and $s_1 \,\Box\, s_2 \stackrel{\tau}{\longrightarrow} \Theta$. Then we have $s_1 \stackrel{\tau}{\longrightarrow} \Phi$ and $\Theta = \Phi \,\Box\, \overline{s_2}$ or $s_2 \stackrel{\tau}{\longrightarrow} \Phi$ and $\Theta = \overline{s_1} \,\Box\, \Phi$. By symmetry we may restrict attention to the first case. Thus $\Delta_1 \stackrel{\hat{\tau}}{\Longrightarrow} \Delta$ for some Δ with $\Phi(\lhd_s)^\dagger \Delta$. By Lemma 5.7 we have $(\Phi \,\Box\, \overline{s_2})\ \mathcal{R}^\dagger\ (\Delta \,\Box\, \Delta_2)$ and $\Delta_1 \,\Box\, \Delta_2 \stackrel{\hat{\tau}}{\longrightarrow} \Delta \,\Box\, \Delta_2$.
   The case that $s \lhd_s \Delta$ is trivial, so we have checked that $\mathcal{R}$ is a simulation indeed.
   Using this, we proceed to show that $P_1 \,\Box\, P_2 \sqsubseteq_S Q_1 \,\Box\, Q_2$.
   Since $P_i \sqsubseteq_S Q_i$, there must be a Δ_i such that $[\![Q_i]\!] \stackrel{\hat{\tau}}{\Longrightarrow} \Delta_i$ and $[\![P_i]\!](\lhd_s)^\dagger \Delta_i$. By Lemma 5.7, we have $[\![P_1 \,\Box\, P_2]\!] = ([\![P_1]\!] \,\Box\, [\![P_2]\!])\ \mathcal{R}^\dagger\ (\Delta_1 \,\Box\, \Delta_2)$. Therefore, it holds that $[\![P_1 \,\Box\, P_2]\!](\lhd_s)^\dagger(\Delta_1 \,\Box\, \Delta_2)$. By Lemma 5.7 we also obtain
   $$[\![Q_1 \,\Box\, Q_2]\!] = [\![Q_1]\!] \,\Box\, [\![Q_2]\!] \stackrel{\hat{\tau}}{\Longrightarrow} \Delta_1 \,\Box\, [\![Q_2]\!] \stackrel{\hat{\tau}}{\Longrightarrow} \Delta_1 \,\Box\, \Delta_2,$$
   so the required result is established.
4. Since $P_i \sqsubseteq_S Q_i$, there must be a Δ_i such that $[\![Q_i]\!] \stackrel{\hat{\tau}}{\Longrightarrow} \Delta_i$ and $[\![P_i]\!](\lhd_s)^\dagger \Delta_i$. Thus $[\![Q_1\,{}_p\!\oplus Q_2]\!] = p \cdot [\![Q_1]\!] + (1-p) \cdot [\![Q_2]\!] \stackrel{\hat{\tau}}{\Longrightarrow} p \cdot \Delta_1 + (1-p) \cdot \Delta_2$ by Lemma 5.4 and $[\![P_1\,{}_p\!\oplus P_2]\!] = p \cdot [\![P_1]\!] + (1-p) \cdot [\![P_2]\!]\,(\lhd_s)^\dagger\, p \cdot \Delta_1 + (1-p) \cdot \Delta_2$ by the linearity of lifting operation. Hence $P_1\,{}_p\!\oplus P_2 \sqsubseteq_S Q_1\,{}_p\!\oplus Q_2$.
5. Let $\mathcal{R} \subseteq \mathsf{sCSP} \times \mathcal{D}(\mathsf{sCSP})$ be defined by $s\mathcal{R}\Delta$, iff $s = s_1 \,|_A\, s_2$ and $\Delta = \Delta_1 \,|_A\, \Delta_2$ with $s_1 \lhd_s \Delta_1$ and $s_2 \lhd_s \Delta_2$. We show that $\mathcal{R}$ is a simulation. There are three cases to consider.
   a) Suppose $s_1 \lhd_s \Delta_1, s_2 \lhd_s \Delta_2$ and $s_1 \,|_A\, s_2 \stackrel{\alpha}{\longrightarrow} \Theta_1 \,|_A\, \overline{s_2}$ because of the transition $s_1 \stackrel{\alpha}{\longrightarrow} \Theta_1$ with $\alpha \notin A$. Then $\Delta_1 \stackrel{\hat{\alpha}}{\Longrightarrow} \Delta_1'$ for some Δ_1' with $\Theta_1(\lhd_s)^\dagger \Delta_1'$. By Lemma 5.8 we have $\Delta_1 \,|_A\, \Delta_2 \stackrel{\hat{\alpha}}{\Longrightarrow} \Delta_1' \,|_A\, \Delta_2$ and also it can be seen that $(\Theta_1 \,|_A\, \overline{s_2})\ \mathcal{R}^\dagger\ (\Delta_1' \,|_A\, \Delta_2)$.
   b) The symmetric case can be similarly analysed.
   c) Suppose $s_1 \lhd_s \Delta_1, s_2 \lhd_s \Delta_2$ and $s_1 \,|_A\, s_2 \stackrel{\tau}{\longrightarrow} \Theta_1 \,|_A\, \Theta_2$ because of the transitions $s_1 \stackrel{a}{\longrightarrow} \Theta_1$ and $s_2 \stackrel{a}{\longrightarrow} \Theta_2$ with $a \in A$. Then for $i = 1$ and $i = 2$ we have

$\Delta_i \stackrel{\hat{\tau}}{\Longrightarrow} \Delta_i' \stackrel{a}{\longrightarrow} \Delta_i'' \stackrel{\hat{\tau}}{\Longrightarrow} \Delta_i'''$ for some $\Delta_i', \Delta_i'', \Delta_i'''$ with $\Theta_i(\lhd_s)^\dagger \Delta_i'''$. By Lemma 5.8 we have $\Delta_1 \mid_A \Delta_2 \stackrel{\hat{\tau}}{\Longrightarrow} \Delta_1' \mid_A \Delta_2' \stackrel{\tau}{\longrightarrow} \Delta_1'' \mid_A \Delta_2'' \stackrel{\hat{\tau}}{\longrightarrow} \Delta_1''' \mid_A \Delta_2'''$ and $(\Theta_1 \mid_A \Theta_2) \; \mathcal{R}^\dagger \; (\Delta_1''' \mid_A \Delta_2''')$.

So we have checked that $\mathcal{R}$ is a simulation.

Since $P_i \sqsubseteq_S Q_i$, there must be a Δ_i such that $[\![Q_i]\!] \stackrel{\hat{\tau}}{\Longrightarrow} \Delta_i$ and $[\![P_i]\!](\lhd_s)^\dagger \Delta_i$. By Lemma 5.8 we have $[\![P_1 \mid_A P_2]\!] = ([\![P_1]\!] \mid_A [\![P_2]\!]) \; \mathcal{R}^\dagger \; (\Delta_1 \mid_A \Delta_2)$. Therefore, we have $[\![P_1 \mid_A P_2]\!](\lhd_s)^\dagger(\Delta_1 \mid_A \Delta_2)$. By Lemma 5.8 we also obtain

$$[\![Q_1 \mid_A Q_2]\!] \;=\; [\![Q_1]\!] \mid_A [\![Q_2]\!] \stackrel{\hat{\tau}}{\Longrightarrow} \Delta_1 \mid_A [\![Q_2]\!] \stackrel{\hat{\tau}}{\Longrightarrow} \Delta_1 \mid_A \Delta_2,$$

which had to be established.

The case for $\sqsubseteq_{FS}$ is analogous. As an example, we show that $\sqsubseteq_{FS}$ is preserved under parallel composition. The key step is to show that the binary relation $\mathcal{R} \subseteq \mathsf{sCSP} \times \mathcal{D}(\mathsf{sCSP})$ defined by

$$\mathcal{R} \;:=\; \{(s_1|_A s_2, \Delta_1|_A\Delta_2) \mid s_1 \lhd_{FS} \Delta_1 \wedge s_2 \lhd_{FS} \Delta_2\}.$$

is a failure simulation.

Suppose $s_i \lhd_{FS} \Delta_i$ for $i = 1,2$ and $s_1 \mid_A s_2 \stackrel{X}{\not\longrightarrow}$ for some $X \subseteq \mathsf{Act}$. For each $a \in X$ there are two possibilities:

- If $a \notin A$, then $s_1 \stackrel{a}{\not\rightarrow}$ and $s_2 \stackrel{a}{\not\rightarrow}$, since otherwise we would have $s_1 \mid_A s_2 \stackrel{a}{\not\rightarrow}$.
- If $a \in A$, then either $s_1 \stackrel{a}{\not\rightarrow}$ or $s_2 \stackrel{a}{\not\rightarrow}$, since otherwise we would have $s_1 \mid_A s_2 \stackrel{\tau}{\not\rightarrow}$.

Hence, we can partition the set X into three subsets: X_0, X_1 and X_2 such that $X_0 = X \backslash A$ and $X_1 \cup X_2 \subseteq A$ with $s_1 \stackrel{X_1}{\not\rightarrow}$ and $s_2 \stackrel{X_2}{\not\rightarrow}$, but allowing $s_1 \stackrel{a}{\not\rightarrow}$ for some $a \in X_2$ and $s_2 \stackrel{a}{\not\rightarrow}$ for some $a \in X_1$. We then have that $s_i \stackrel{X_0 \cup X_i}{\not\longrightarrow}$ for $i = 1,2$. By the assumption that $s_i \lhd_{FS} \Delta_i$ for $i = 1,2$, there is a Δ_i' with $\Delta_i \stackrel{\hat{\tau}}{\Longrightarrow} \Delta_i' \stackrel{X_0 \cup X_i}{\not\longrightarrow}$. Therefore, $\Delta_1'|_A\Delta_2' \stackrel{X}{\not\rightarrow}$ as well. By Lemma 5.8(i) we have $\Delta_1 \mid_A \Delta_2 \stackrel{\hat{\tau}}{\Longrightarrow} \Delta_1' \mid_A \Delta_2'$. Hence $\Delta_1 \mid_A \Delta_2$ can match the failures of $s_1 \mid_A s_2$.

The matching of transitions and the using of $\mathcal{R}$ to prove the preservation property of $\sqsubseteq_{FS}$ under parallel composition are similar to those in the above corresponding proof for simulations, so we omit them. □

5.5.3 *Simulations Are Sound for Testing Preorders*

In this section we show that simulation is sound for may testing and failure simulation is sound for must testing. That is, we aim to prove that (i) $P \sqsubseteq_S Q$ implies $P \sqsubseteq_{\text{pmay}} Q$ and (ii) $P \sqsubseteq_{FS} Q$ implies $P \sqsubseteq_{\text{pmust}} Q$.

Originally, the relation $\sqsubseteq_S$ is defined on pCSP, we now extend it to pCSP^ω, keeping Definition 5.5 unchanged.

Theorem 5.2 *If $P \sqsubseteq_S Q$ then $P \sqsubseteq_{pmay} Q$.*

Proof For any test $T \in \mathsf{pCSP}^\omega$ and process $P \in \mathsf{pCSP}$ the set $\mathbb{V}_f(T \,|_{\mathsf{Act}} P)$ is finite, so

$$P \sqsubseteq_{\text{pmay}} Q, \text{ iff } max(\mathbb{V}_f([\![T \,|_{\mathsf{Act}} P]\!])) \leq max(\mathbb{V}_f([\![T \,|_{\mathsf{Act}} Q]\!])) \text{ for every test } T. \tag{5.7}$$

The following properties for $\Delta_1, \Delta_2 \in \mathsf{pCSP}^\omega$ and $\alpha \in \mathsf{Act}_\tau$ are not hard to establish:

$$\Delta_1 \stackrel{\hat{\alpha}}{\Longrightarrow} \Delta_2 \text{implies } max(\mathbb{V}_f(\Delta_1)) \geq max(\mathbb{V}_f(\Delta_2)). \tag{5.8}$$

$$\Delta_1 (\lhd_s)^\dagger \Delta_2 \text{ implies } max(\mathbb{V}_f(\Delta_1)) \leq max(\mathbb{V}_f(\Delta_2)). \tag{5.9}$$

Now suppose $P \sqsubseteq_S Q$. Since $\sqsubseteq_S$ is preserved by the parallel operator, we have $T \,|_{\mathsf{Act}} P \sqsubseteq_S T \,|_{\mathsf{Act}} Q$ for an arbitrary test T. By definition, this means that there is a distribution Δ such that $[\![T \,|_{\mathsf{Act}} Q]\!] \stackrel{\hat{\tau}}{\Longrightarrow} \Delta$ and $[\![T \,|_{\mathsf{Act}} P]\!] \; (\lhd_s)^\dagger \; \Delta$. By (5.8) and (5.9) we infer that $max(\mathbb{V}_f[\![T \,|_{\mathsf{Act}} P]\!]) \leq max(\mathbb{V}_f[\![T \,|_{\mathsf{Act}} Q]\!])$. The result now follows from (5.7). □

It is tempting to use the same idea to prove that $P \sqsubseteq_{FS} Q$ implies $P \sqsubseteq_{\text{pmust}} Q$, but now using the function *min* in place of *max*. However, the *min*-analogue of Property (5.8) is in general invalid. For example, let R be the process $a \,|_{\mathsf{Act}} (a \,\Box\, \omega)$. We have $min(\mathbb{V}_f(R)) = 1$, yet $R \stackrel{\tau}{\rightarrow} \overline{\mathbf{0} \,|_{\mathsf{Act}} \mathbf{0}}$ and $min(\mathbb{V}_f(\overline{\mathbf{0} \,|_{\mathsf{Act}} \mathbf{0}})) = 0$. Therefore, it is not the case that $\Delta_1 \stackrel{\hat{\tau}}{\Longrightarrow} \Delta_2$ implies $min(\mathbb{V}_f(\Delta_1)) \leq min(\mathbb{V}_f(\Delta_2))$. Examining this example reveals that the culprit is the "scooting" τ transitions, which are τ transitions of a state that can enable an ω action at the same time.

Our strategy is therefore as follows: when comparing two states, "scooting" transitions are purposefully ignored. Write $s \stackrel{\alpha}{\longrightarrow}_\omega \Delta$ if both $s \stackrel{\omega}{\not\longrightarrow}$ and $s \stackrel{\alpha}{\longrightarrow} \Delta$ hold. We define $\stackrel{\hat{\tau}}{\longrightarrow}_\omega$ as $\stackrel{\hat{\tau}}{\longrightarrow}$ using $\stackrel{\tau}{\longrightarrow}_\omega$ in place of $\stackrel{\tau}{\longrightarrow}$. Similarly, we define $\Longrightarrow_\omega$ and $\stackrel{\hat{\alpha}}{\Longrightarrow}_\omega$. Thus, the subscript ω on a transition of any kind indicates that no state is passed through in which ω is enabled. A version of failure simulation adapted to these transition relations is then defined as follows.

Definition 5.6 Let $\lhd_{FS}^O \subseteq \mathsf{sCSP}^\omega \times \mathcal{D}(\mathsf{sCSP}^\omega)$ be the largest binary relation such that $s \lhd_{FS}^O \Theta$ implies

- if $s \stackrel{\alpha}{\longrightarrow}_\omega \Delta$, then there is some Θ' with $\Theta \stackrel{\hat{\alpha}}{\Longrightarrow}_\omega \Theta'$ and $\Delta \; (\lhd_{FS}^O)^\dagger \; \Theta'$
- if $s \stackrel{X}{\not\longrightarrow}$ with $\omega \in X$, then there is some Θ' with $\Theta \stackrel{\hat{\tau}}{\Longrightarrow}_\omega \Theta'$ and $\Theta' \stackrel{X}{\not\longrightarrow}$.

Let $P \sqsubseteq_{FS}^O Q$, iff $[\![P]\!] \stackrel{\hat{\tau}}{\Longrightarrow}_\omega \Theta$ for some Θ with $[\![Q]\!] \; (\lhd_{FS}^O)^\dagger \; \Theta$.
Note that for processes P, Q in pCSP (as opposed to pCSP^ω), we have $P \sqsubseteq_{FS} Q$ iff $P \sqsubseteq_{FS}^O Q$.

Proposition 5.4 *If P, Q are processes in pCSP with $P \sqsubseteq_{FS} Q$ and T is a process in pCSP^ω then $T \,|_{\mathsf{Act}} P \sqsubseteq_{FS}^O T \,|_{\mathsf{Act}} Q$.*

Proof Similar to the proof of Theorem 5.1. □

Proposition 5.5 *The following properties hold, with* $\Delta_1, \Delta_2 \in \mathcal{D}(\mathsf{sCSP}^\omega)$*:*

$$P \sqsubseteq_{pmust} Q \text{ iff } min(\mathbb{V}_f[\![T \mid_{Act} P]\!]) \leq min(\mathbb{V}_f([\![T \mid_{Act} Q]\!])) \text{ for every test } T. \quad (5.10)$$

$$\Delta_1 \stackrel{\hat{\alpha}}{\Longrightarrow}_\omega \Delta_2 \text{ for } \alpha \in \mathsf{Act}_\tau \text{ implies } min(\mathbb{V}_f(\Delta_1)) \leq min(\mathbb{V}_f(\Delta_2)). \quad (5.11)$$

$$\Delta_1 \; (\lhd^O_{FS})^\dagger \; \Delta_2 \text{ implies } min(\mathbb{V}_f(\Delta_1)) \geq min(\mathbb{V}_f(\Delta_2)). \quad (5.12)$$

Proof Property (5.10) is again straightforward, and Property (5.11) can be established just as the proof of (5.8), but with all $\leq$-signs reversed. Property (5.12) follows by structural induction, simultaneously with the property, for $s \in \mathsf{sCSP}^\omega$ and $\Delta \in \mathcal{D}(\mathsf{sCSP}^\omega)$, that

$$s \lhd^O_{FS} \Delta \text{ implies } min(\mathbb{V}_f(s)) \geq min(\mathbb{V}_f(\Delta)) . \quad (5.13)$$

The reduction of Property (5.12) to (5.13) proceeds exactly as that for (5.9). For (5.13) itself we distinguish three cases:

- If $s \stackrel{\omega}{\longrightarrow}$, then $min(\mathbb{V}_f(s)) = 1 \geq min(\mathbb{V}_f(\Delta))$ trivially.
- If $s \stackrel{\omega}{\not\longrightarrow}$ but $s \rightarrow$, then each "non-scooting" transition from s will be matched by a "nonscooting" transition from Θ. Whenever $s \stackrel{\alpha}{\longrightarrow}_\omega \Theta$, for $\alpha \in \mathsf{Act}_\tau$ and $\Theta \in \mathcal{D}(\mathsf{sCSP}^\omega)$, then $s \lhd^O_{FS} \Delta$ implies the existence of some Δ_Θ such that $\Delta \stackrel{\hat{\alpha}}{\Longrightarrow}_\omega \Delta_\Theta$ and $\Theta \; (\lhd^O_{FS})^\dagger \; \Delta_\Theta$. By induction, using (5.12), it follows that

  $$min(\mathbb{V}_f(\Theta)) \geq min(\mathbb{V}_f(\Delta_\Theta)).$$

  Consequently, we have that

  $$\begin{aligned} min(\mathbb{V}_f(s)) &= min(\{min(\mathbb{V}_f(\Theta)) \mid s \stackrel{\alpha}{\longrightarrow} \Theta\}) \\ &\geq min(\{min(\mathbb{V}_f(\Delta_\Theta)) \mid s \stackrel{\alpha}{\longrightarrow} \Theta\}) \\ &\geq min(\{min(\mathbb{V}_f(\Delta)) \mid s \stackrel{\alpha}{\longrightarrow} \Theta\}) \qquad \text{(by (5.11))} \\ &= min(\mathbb{V}_f(\Delta)) . \end{aligned}$$

- If $s \not\rightarrow$, that is $s \stackrel{\mathsf{Act}^\omega}{\not\longrightarrow}$, then there is some Δ' such that $\Delta \stackrel{\hat{\tau}}{\Longrightarrow}_\omega \Delta'$ and $\Delta' \stackrel{\mathsf{Act}^\omega}{\not\longrightarrow}$. By the definition of $\mathbb{V}_f$, $min(\mathbb{V}_f(\Delta')) = 0$. Using (5.11), we have

  $$min(\mathbb{V}_f(\Delta)) \; \leq \; min(\mathbb{V}_f(\Delta')),$$

  so $min(\mathbb{V}_f(\Delta)) = 0$ as well. Thus, also in this case $min(\mathbb{V}_f(s)) \geq min(\mathbb{V}_f(\Delta))$.

□

Theorem 5.3 *If* $P \sqsubseteq_{FS} Q$ *then* $P \sqsubseteq_{pmust} Q$*.*

Proof Similar to the proof of Theorem 5.2, using (5.10)–(5.12). □

The next two sections are devoted to proving the converse of Theorems 5.2 and 5.3. That is, we will establish:

Theorem 5.4 *Let P and Q be* pCSP *processes.*

- $P \sqsubseteq_{pmay} Q$ *implies* $P \sqsubseteq_S Q$ *and*
- $P \sqsubseteq_{pmust} Q$ *implies* $P \sqsubseteq_{FS} Q$.

Because of Theorem 4.7, it will suffice to show that

- $P \sqsubseteq^{\Omega}_{\mathrm{pmay}} Q$ implies $P \sqsubseteq_S Q$ and
- $P \sqsubseteq^{\Omega}_{\mathrm{pmust}} Q$ implies $P \sqsubseteq_{FS} Q$.

This shift from scalar testing to vector-based testing is motivated by the fact that the latter enables us to use more informative tests, allowing us to discover more intensional properties of the processes being tested.

The crucial characteristics of $\mathcal{A}$ needed for the above implications are summarised in Lemmas 5.9 and 5.10. For convenience of presentation, we write $\vec{\omega}$ for the vector in $[0,1]^{\Omega}$ defined by $\vec{\omega}(\omega) = 1$ and $\vec{\omega}(\omega') = 0$ for $\omega' \neq \omega$. We also have the vector $\vec{0} \in [0,1]^{\Omega}$ with $\vec{0}(\omega) = 0$ for all $\omega \in \Omega$. Sometimes we treat a distribution Δ of finite support as the pCSP expression $\bigoplus_{s \in \lceil \Delta \rceil} \Delta(s) \cdot s$, so that $\mathcal{A}(T, \Delta) := \mathrm{Exp}_{\Delta} \mathcal{A}(\mathrm{T}, \_)$.

Lemma 5.9 *Let P be a* pCSP *process, and* T, T_i *be tests.*

1. $o \in \mathcal{A}(\omega, P)$, *iff* $o = \vec{\omega}$.
2. $\vec{0} \in \mathcal{A}(\Box_{a \in X} a.\omega, P)$, *iff* $\exists \Delta : [\![P]\!] \stackrel{\hat{\tau}}{\Longrightarrow} \Delta \stackrel{X}{\not\longrightarrow}$.
3. *Suppose the action* ω *does not occur in the test T. Then* $o \in \mathcal{A}(\tau.\omega \,\Box\, a.T, P)$ *with* $o(\omega) = 0$, *iff there is a* $\Delta \in \mathcal{D}$ (sCSP) *with* $[\![P]\!] \stackrel{\hat{a}}{\Longrightarrow} \Delta$ *and* $o \in \mathcal{A}(T, \Delta)$.
4. $o \in \mathcal{A}(T_1 \,{}_p{\oplus}\, T_2, P)$, *iff* $o = p \cdot o_1 + (1-p) \cdot o_2$ *for some* $o_i \in \mathcal{A}(T_i, P)$.
5. $o \in \mathcal{A}(T_1 \sqcap T_2, P)$ *if there are some probability* $q \in [0,1]$ *and* $\Delta_1, \Delta_2 \in \mathcal{D}$ (sCSP) *such that* $[\![P]\!] \stackrel{\hat{\tau}}{\Longrightarrow} q \cdot \Delta_1 + (1-q) \cdot \Delta_2$ *and* $o = q \cdot o_1 + (1-q) \cdot o_2$ *for some* $o_i \in \mathcal{A}(T_i, \Delta_i)$.

Proof Straightforward, by induction on the structure of P. □

The converse of Lemma 5.9 (5) also holds, as the following lemma says. However, the proof is less straightforward.

Lemma 5.10 *Let P be a* pCSP *process, and* T_i *be tests. If* $o \in \mathcal{A}(\bigsqcap_{i \in I} T_i, P)$ *then for all* $i \in I$ *there are probabilities* $q_i \in [0,1]$ *and* $\Delta_i \in \mathcal{D}$ (sCSP) *with* $\sum_{i \in I} q_i = 1$ *such that* $[\![P]\!] \stackrel{\hat{\tau}}{\Longrightarrow} \sum_{i \in I} q_i \cdot \Delta_i$ *and* $o = \sum_{i \in I} q_i o_i$ *for some* $o_i \in \mathcal{A}(T_i, \Delta_i)$.

Proof Given that the states of our pLTS are sCSP expressions, there exists a well-founded order on the combination of states in sCSP and distributions in $\mathcal{D}$(sCSP), such that $s \stackrel{\alpha}{\longrightarrow} \Delta$ implies that s is larger than Δ, and any distribution is larger than the states in its support. Intuitively, this order corresponds to the usual order on natural numbers if we graphically depict a pCSP process as a finite tree (cf. Sect. 5.2.4) and assign to each node a number to indicate its level in the tree. Let $T = \bigsqcap_{i \in I} T_i$. We prove the following two claims

(a) If s is a state-based process and $o \in \mathcal{A}(T, s)$, then there are some $\{q_i\}_{i \in I}$ with $\sum_{i \in I} q_i = 1$ such that $s \stackrel{\hat{\tau}}{\Longrightarrow} \sum_{i \in I} q_i \cdot \Delta_i$, $o = \sum_{i \in I} q_i o_i$, and $o_i \in \mathcal{A}(T_i, \Delta_i)$.
(b) If $\Delta \in \mathcal{D}(\mathsf{sCSP})$ and $o \in \mathcal{A}(T, \Delta)$, then there are some $\{q_i\}_{i \in I}$ with $\sum_{i \in I} q_i = 1$ such that $\Delta \stackrel{\hat{\tau}}{\Longrightarrow} \sum_{i \in I} q_i \cdot \Delta_i$, $o = \sum_{i \in I} q_i o_i$, and $o_i \in \mathcal{A}(T_i, \Delta_i)$

by simultaneous induction on the order mentioned above, applied to s and Δ.

(a) We have two subcases depending on whether s can make an initial τ move or not.
   - If s cannot make a τ move, that is $s \stackrel{\tau}{\not\longrightarrow}$, then the only possible moves from $T \mid_{\mathsf{Act}} s$ are τ moves originating in T; T has no non-τ moves, and any non-τ moves that might be possible for s on its own are inhibited by the alphabet Act of the composition. Suppose $o \in \mathcal{A}(T, s)$. Then there are some $\{q_i\}_{i \in I}$ with $\sum_{i \in I} q_i = 1$ such that $o = \sum_{i \in I} q_i o_i$ and $o_i \in \mathcal{A}(T_i, s) = \mathcal{A}(T_i, \overline{s})$. Obviously we also have $[\![s]\!] \stackrel{\hat{\tau}}{\Longrightarrow} \sum_{i \in I} q_i \cdot \overline{s}$.
   - If s can make one or more τ moves, then we have $s \stackrel{\tau}{\longrightarrow} \Delta'_j$ for $j \in J$, where without loss of generality J can be assumed to be a nonempty finite set disjoint from I, the index set for T. The possible first moves for $T \mid_{\mathsf{Act}} s$ are τ moves either of T or of s, because T cannot make initial non-τ moves and that prevents a proper synchronisation from occurring on the first step. Suppose that $o \in \mathcal{A}(T, s)$. Then there are some $\{p_k\}_{k \in I \cup J}$ with $\sum_{k \in I \cup J} p_k = 1$ and

$$o = \sum_{k \in I \cup J} p_k o'_k \tag{5.14}$$

$$o'_i \in \mathcal{A}(T_i, s) \qquad \text{for all } i \in I \tag{5.15}$$

$$o'_j \in \mathcal{A}(T, \Delta'_j) \qquad \text{for all } j \in J. \tag{5.16}$$

   For each $j \in J$, we know by the induction hypothesis that

$$\Delta'_j \stackrel{\hat{\tau}}{\Longrightarrow} \sum_{i \in I} p_{ji} \cdot \Delta'_{ji} \tag{5.17}$$

$$o'_j = \sum_{i \in I} p_{ji} o'_{ji} \tag{5.18}$$

$$o'_{ji} \in \mathcal{A}(T_i, \Delta'_{ji}) \tag{5.19}$$

   for some $\{p_{ji}\}_{i \in I}$ with $\sum_{i \in I} p_{ji} = 1$. Let

$$q_i = p_i + \sum_{j \in J} p_j p_{ji}$$

$$\Delta_i = \frac{1}{q_i}(p_i \cdot \overline{s} + \sum_{j \in J} p_j p_{ji} \cdot \Delta'_{ji})$$

$$o_i = \frac{1}{q_i}(p_i o'_i + \sum_{j \in J} p_j p_{ji} o'_{ji})$$

for each $i \in I$, except that Δ_i and o_i are chosen arbitrarily in case $q_i = 0$. It can be checked by arithmetic that q_i, Δ_i, o_i have the required properties, viz. that $\sum_{i \in I} q_i = 1$, that $o = \sum_{i \in I} q_i o_i$ and that

$$\begin{aligned} s &\stackrel{\hat{\tau}}{\Longrightarrow} \sum_{i \in I} p_i \cdot \overline{s} + \sum_{j \in J} p_j \cdot \Delta'_j \\ &\stackrel{\hat{\tau}}{\Longrightarrow} \sum_{i \in I} p_i \cdot \overline{s} + \sum_{j \in J} p_j \cdot \sum_{i \in I} p_{ji} \cdot \Delta'_{ji} \qquad \text{by (5.17) and Lemma 5.4} \\ &= \sum_{i \in I} q_i \cdot \Delta_i \,. \end{aligned}$$

Finally, it follows from (5.15) and (5.19) that $o_i \in \mathcal{A}(T_i, \Delta_i)$ for each $i \in I$.

(b) Let $\lceil \Delta \rceil = \{s_j\}_{j \in J}$ and $r_j = \Delta(s_j)$. Without loss of generality we may assume that J is a nonempty finite set disjoint from I. Using the condition that $\mathcal{A}(T, \Delta) := \text{Exp}_\Delta \mathcal{A}(T, \_)$, if $o \in \mathcal{A}(T, \Delta)$ then

$$o = \sum_{j \in J} r_j o'_j \tag{5.20}$$

$$o'_j \in \mathcal{A}(T, s_j). \tag{5.21}$$

For each $j \in J$, we know by the induction hypothesis that

$$s_j \stackrel{\hat{\tau}}{\Longrightarrow} \sum_{i \in I} q_{ji} \cdot \Delta'_{ji} \tag{5.22}$$

$$o'_j = \sum_{i \in I} q_{ji} o'_{ji} \tag{5.23}$$

$$o'_{ji} \in \mathcal{A}(T_i, \Delta'_{ji}) \tag{5.24}$$

for some $\{q_{ji}\}_{i \in I}$ with $\sum_{i \in I} q_{ji} = 1$. Thus let

$$\begin{aligned} q_i &= \sum_{j \in J} r_j q_{ji} \\ \Delta_i &= \frac{1}{q_i} \sum_{j \in J} r_j q_{ji} \cdot \Delta'_{ji} \\ o_i &= \frac{1}{q_i} \sum_{j \in J} r_j q_{ji} o'_{ji} \end{aligned}$$

again choosing Δ_i and o_i arbitrarily in case $q_i = 0$. As in the first case, it can be shown by arithmetic that the collection r_i, Δ_i, o_i has the required properties.

□

5.6 A Modal Logic

In this section, we present logical characterisations $\sqsubseteq^{\mathcal{L}}$ and $\sqsubseteq^{\mathcal{F}}$ of our testing preorders. Besides their intrinsic interest, these logical preorders also serve as a stepping stone in proving Theorem 5.4. In this section we show that the logical preorders are sound with respect to the simulation and failure simulation preorders, and hence with respect to the testing preorders; in the next section we establish completeness. To start with, we define a set $\mathcal{F}$ of modal formulae, inductively, as follows:

- $\top \in \mathcal{F}$,
- $\mathbf{ref}(X) \in \mathcal{F}$ when $X \subseteq \mathsf{Act}$,
- $\langle a \rangle \varphi \in \mathcal{F}$ when $\varphi \in \mathcal{F}$ and $a \in \mathsf{Act}$,
- $\Delta \models \varphi_1 \wedge \varphi_2$ iff $\Delta \models \varphi_1$ and $\Delta \models \varphi_2$,
- $\varphi_1\, {}_p\!\oplus \varphi_2 \in \mathcal{F}$ when $\varphi_1, \varphi_2 \in \mathcal{F}$ and $p \in [0, 1]$.

We often use the generalised probabilistic choice operator $\bigoplus_{i \in I} p_i \cdot \varphi_i$, where I is a nonempty finite index set, and $\sum_{i \in I} p_i = 1$. This can be expressed in our language by nested use of the binary probabilistic choice.

The *satisfaction relation* $\models\, \subseteq \mathcal{D}(\mathsf{sCSP}) \times \mathcal{F}$ is given by:

- $\Delta \models \top$ for any $\Delta \in \mathcal{D}(S)$,
- $\Delta \models \mathbf{ref}(X)$ iff there is a Δ' with $\Delta \stackrel{\hat{\tau}}{\Longrightarrow} \Delta'$ and $\Delta' \stackrel{X}{\not\longrightarrow}$,
- $\Delta \models \langle a \rangle \varphi$ iff there is a Δ' with $\Delta \stackrel{\hat{a}}{\Longrightarrow} \Delta'$ and $\Delta' \models \varphi$,
- $\Delta \models \varphi_1 \wedge \varphi_2$ iff $\Delta \models \varphi_1$ and $\Delta \models \varphi_2$,
- $\Delta \models \varphi_1\, {}_p\!\oplus \varphi_2$ iff there are $\Delta_1, \Delta_2 \in \mathcal{D}(S)$ with $\Delta_1 \models \varphi_1$ and $\Delta_2 \models \varphi_2$, such that $\Delta \stackrel{\hat{\tau}}{\Longrightarrow} p \cdot \Delta_1 + (1-p) \cdot \Delta_2$.

Let $\mathcal{L}$ be the subclass of $\mathcal{F}$ obtained by skipping the $\mathbf{ref}(X)$ clause. We shall write $P \sqsubseteq^{\mathcal{L}} Q$ just when $[\![P]\!] \models \varphi$ implies $[\![Q]\!] \models \varphi$ for all $\varphi \in \mathcal{L}$, and $P \sqsubseteq^{\mathcal{F}} Q$ just when $[\![P]\!] \models \varphi$ is implied by $[\![Q]\!] \models \varphi$ for all $\varphi \in \mathcal{F}$ (Note the opposing directions).

Remark 5.1 Compared with the two-sorted logic in Sect. 3.6.1, the logic $\mathcal{F}$ and its sublogic $\mathcal{L}$ drop state formulae and only contain distribution formulae. The reason is that we will characterise failure simulation preorder and simulation preorder. Both of them are distribution-based and strictly coarser than the state-based bisimilarity investigated in Chap. 3.

In order to obtain the main result of this section, Theorem 5.5, we introduce the following tool.

Definition 5.7 The $\mathcal{F}$*-characteristic formula* φ_s or φ_Δ of a process $s \in \mathsf{sCSP}$ or $\Delta \in \mathcal{D}(\mathsf{sCSP})$ is defined inductively:

- $\varphi_s := \bigwedge_{s \stackrel{a}{\longrightarrow} \Delta} \langle a \rangle \varphi_\Delta \wedge \mathbf{ref}(\{a \mid s \stackrel{a}{\not\longrightarrow}\})$ if $s \stackrel{\tau}{\not\longrightarrow}$,
- $\varphi_s := \bigwedge_{s \stackrel{a}{\longrightarrow} \Delta} \langle a \rangle \varphi_\Delta \wedge \bigwedge_{s \stackrel{\tau}{\longrightarrow} \Delta} \varphi_\Delta$ otherwise,
- $\varphi_\Delta := \bigoplus_{s \in \lceil \Delta \rceil} \Delta(s) \cdot \varphi_s$.

Here the conjunctions $\bigwedge_{s \xrightarrow{a} \Delta}$ range over suitable pairs a, Δ, and $\bigwedge_{s \xrightarrow{\tau} \Delta}$ ranges over suitable Δ. The $\mathcal{L}$-*characteristic formulae* ψ_s and ψ_Δ are defined likewise, but omitting the conjuncts $\mathbf{ref}(\{a \mid s \stackrel{a}{\not\longrightarrow}\})$.
We write $\varphi \Rrightarrow \psi$ with $\varphi, \psi \in \mathcal{F}$ if for each distribution Δ one has $\Delta \models \varphi$ implies $\Delta \models \psi$. Then it is easy to see that $\varphi_{\overline{s}} \Lleftarrow\!\!\Rrightarrow \varphi_s$ and $\bigwedge_{i \in I} \varphi_i \Rrightarrow \varphi_i$ for any $i \in I$; furthermore, the following property can be established by an easy inductive proof.

Lemma 5.11 *For any $\Delta \in \mathcal{D}(\mathsf{sCSP})$ we have $\Delta \models \varphi_\Delta$, as well as $\Delta \models \psi_\Delta$.* □
This and the following lemma help to establish Theorem 5.5.

Lemma 5.12 *For any processes $P, Q \in \mathsf{pCSP}$ we have that $[\![P]\!] \models \varphi_{[\![Q]\!]}$ implies $P \sqsubseteq_{FS} Q$, and likewise $[\![Q]\!] \models \psi_{[\![P]\!]}$ implies $P \sqsubseteq_S Q$.*

Proof To establish the first statement, we define the relation $\mathcal{R}$ by $s\mathcal{R}\Theta$ iff $\Theta \models \varphi_s$; to show that it is a failure simulation we first prove the following technical result:

$$\Theta \models \varphi_\Delta \text{ implies } \exists \Theta' : \Theta \stackrel{\hat{\tau}}{\Longrightarrow} \Theta' \wedge \Delta\, \mathcal{R}^\dagger\, \Theta'. \tag{5.25}$$

Suppose $\Theta \models \varphi_\Delta$ with $\varphi_\Delta = \bigoplus_{i \in I} p_i \cdot \varphi_{s_i}$, so that we have $\Delta = \sum_{i \in I} p_i \cdot \overline{s_i}$ and for all $i \in I$ there are $\Theta_i \in \mathcal{D}(\mathsf{sCSP})$ with $\Theta_i \models \varphi_{s_i}$ such that $\Theta \stackrel{\hat{\tau}}{\Longrightarrow} \Theta'$ with $\Theta' := \sum_{i \in I} p_i \cdot \Theta_i$. Since $s_i \mathcal{R} \Theta_i$ for all $i \in I$ we have $\Delta\, \mathcal{R}^\dagger\, \Theta'$.
Now we show that $\mathcal{R}$ is a failure simulation. Assume that $s\mathcal{R}\Theta$.

- Suppose $s \stackrel{\tau}{\longrightarrow} \Delta$. Then from Definition 5.7 we have $\varphi_s \Rrightarrow \varphi_\Delta$, so that $\Theta \models \varphi_\Delta$. Applying (5.25) gives us $\Theta \stackrel{\hat{\tau}}{\Longrightarrow} \Theta'$ with $\Delta\, \mathcal{R}^\dagger\, \Theta'$ for some Θ'.
- Suppose $s \stackrel{a}{\longrightarrow} \Delta$ with $a \in \mathsf{Act}$. Then $\varphi_s \Rrightarrow \langle a \rangle \varphi_\Delta$, so $\Theta \models \langle a \rangle \varphi_\Delta$. Hence, there exists some Θ' with $\Theta \stackrel{\hat{a}}{\Longrightarrow} \Theta'$ and $\Theta' \models \varphi_\Delta$. Again apply (5.25).
- Suppose $s \stackrel{X}{\not\longrightarrow}$ with $X \subseteq A$. Then $\varphi_s \Rrightarrow \mathbf{ref}(X)$, so $\Theta \models \mathbf{ref}(X)$. Hence, there exists some Θ' with $\Theta \stackrel{\hat{\tau}}{\Longrightarrow} \Theta'$ and $\Theta' \stackrel{X}{\not\longrightarrow}$.

Thus, $\mathcal{R}$ is indeed a failure simulation. By our assumption $[\![P]\!] \models \varphi_{[\![Q]\!]}$, using (5.25), there exists a Θ' such that $[\![P]\!] \stackrel{\hat{\tau}}{\Longrightarrow} \Theta'$ and $[\![Q]\!]\, \mathcal{R}^\dagger\, \Theta'$, which gives $P \sqsubseteq_{FS} Q$ via Definition 5.5.
To establish the second statement, define the relation $\mathcal{S}$ by $s\mathcal{S}\Theta$ iff $\Theta \models \psi_s$; exactly as above one obtains

$$\Theta \models \psi_\Delta \text{ implies } \exists \Theta' : \Theta \stackrel{\hat{\tau}}{\Longrightarrow} \Theta' \wedge \Delta\, \mathcal{S}^\dagger\, \Theta'. \tag{5.26}$$

Just as above it follows that $\mathcal{S}$ is a simulation. By the assumption $[\![Q]\!] \models \varphi_{[\![P]\!]}$, using (5.26), there exists a Θ' such that $[\![Q]\!] \stackrel{\hat{\tau}}{\Longrightarrow} \Theta'$ and $[\![P]\!]\, \overline{\mathcal{S}}\, \Theta'$. Hence, $P \sqsubseteq_S Q$ via Definition 5.5. □

Theorem 5.5

1. *If $P \sqsubseteq^{\mathcal{L}} Q$ then $P \sqsubseteq_S Q$.*
2. *If $P \sqsubseteq^{\mathcal{F}} Q$ then $P \sqsubseteq_{FS} Q$.*

Proof Suppose that $P \sqsubseteq^{\mathcal{F}} Q$. By Lemma 5.11 we have $[\![Q]\!] \models \varphi_{[\![Q]\!]}$ and therefore $[\![P]\!] \models \varphi_{[\![Q]\!]}$. Lemma 5.12 gives $P \sqsubseteq_{FS} Q$.

Now assume that $P \sqsubseteq^{\mathcal{L}} Q$. We have $[\![P]\!] \models \psi_{[\![P]\!]}$, hence $[\![Q]\!] \models \psi_{[\![P]\!]}$, and thus $P \sqsubseteq_S Q$.

5.7 Characteristic Tests

Our final step towards Theorem 5.4 is taken in this section, where we show that every modal formula φ can be characterised by a vector-based test T_φ with the property that any pCSP process satisfies φ just when it passes the test T_φ.

Lemma 5.13 *For every $\varphi \in \mathcal{F}$ there exists a pair (T_φ, v_φ) with T_φ an Ω test and $v_\varphi \in [0,1]^\Omega$, such that*

$$\Delta \models \varphi \quad \textit{iff} \quad \exists o \in \mathcal{A}(T_\varphi, \Delta) : \ o \leq v_\varphi \tag{5.27}$$

for all $\Delta \in \mathcal{D}(\mathsf{sCSP})$, and in case $\varphi \in \mathcal{L}$ we also have

$$\Delta \models \varphi \quad \textit{iff} \quad \exists o \in \mathcal{A}(T_\varphi, \Delta) : \ o \geq v_\varphi\,. \tag{5.28}$$

T_φ is called a characteristic test of φ and v_φ its target value.

Proof First of all note that if a pair (T_φ, v_φ) satisfies the requirements above, then any pair obtained from (T_φ, v_φ) by bijectively renaming the elements of Ω also satisfies these requirements. Hence a characteristic test can always be chosen in such a way that there is a success action $\omega \in \Omega$ that does not occur in (the finite) T_φ. Moreover, any countable collection of characteristic tests can be assumed to be Ω-disjoint, meaning that no $\omega \in \Omega$ occurs in two different elements of the collection.

The required characteristic tests and target values are obtained as follows.

- Let $\varphi = \top$. Take $T_\varphi := \omega$ for some $\omega \in \Omega$, and $v_\varphi := \vec{\omega}$.
- Let $\varphi = \mathbf{ref}(X)$ with $X \subseteq \mathsf{Act}$. Take $T_\varphi := \Box_{a \in X} a.\omega$ for some $\omega \in \Omega$, and $v_\varphi := \vec{0}$.
- Let $\varphi = \langle a \rangle \psi$. By induction, ψ has a characteristic test T_ψ with target value v_ψ. Take $T_\varphi := \tau.\omega \,\Box\, a.T_\psi$ where $\omega \in \Omega$ does not occur in T_ψ, and $v_\varphi := v_\psi$.
- Let $\varphi = \varphi_1 \wedge \varphi_2$. Choose an Ω-disjoint pair (T_i, v_i) of characteristic tests T_i with target values v_i, for $i = 1, 2$. Furthermore, let $p \in (0, 1]$ be chosen arbitrarily, and take $T_\varphi := T_1 \,{}_p\!\oplus T_2$ and $v_\varphi := p \cdot v_1 + (1-p) \cdot v_2$.
- Let $\varphi = \varphi_1 \,{}_p\!\oplus \varphi_2$. Again choose an Ω-disjoint pair (T_i, v_i) of characteristic tests T_i with target values v_i, $i = 1, 2$, this time ensuring that there are two distinct success actions ω_1, ω_2 that do not occur in any of these tests. Let $T_i' := T_i \,{}_{\frac{1}{2}}\!\oplus \omega_i$ and $v_i' := \frac{1}{2} v_i + \frac{1}{2}\vec{\omega_i}$. Note that for $i = 1, 2$ we have that T_i' is also a characteristic test of φ_i with target value v_i'. Take $T_\varphi := T_1' \sqcap T_2'$ and $v_\varphi := p \cdot v_1' + (1-p) \cdot v_2'$.

Note that $v_\varphi(\omega) = 0$ whenever $\omega \in \Omega$ does not occur in T_φ. By induction on φ we now check (5.27) above.

- Let $\varphi = \top$. For all $\Delta \in \mathcal{D}(\mathsf{sCSP})$ we have $\Delta \models \varphi$ and $\exists o \in \mathcal{A}(T_\varphi, \Delta) : o \leq v_\varphi$, using Lemma 5.9(1).
- Let $\varphi = \mathbf{ref}(X)$ with $X \subseteq \mathsf{Act}$. Suppose $\Delta \models \varphi$. Then there is a Δ' with $\Delta \stackrel{\hat{\tau}}{\Longrightarrow} \Delta'$ and $\Delta' \stackrel{X}{\not\longrightarrow}$. By Lemma 5.9(2), $\vec{0} \in \mathcal{A}(T_\varphi, \Delta)$.
  Now suppose $\exists o \in \mathcal{A}(T_\varphi, \Delta) : o \leq v_\varphi$. This implies $o = \vec{0}$, so by Lemma 5.9(2) there is a Δ' with $\Delta \stackrel{\hat{\tau}}{\Longrightarrow} \Delta'$ and $\Delta' \stackrel{X}{\not\longrightarrow}$. Hence $\Delta \models \varphi$.
- Let $\varphi = \langle a \rangle \psi$ with $a \in \mathsf{Act}$. Suppose $\Delta \models \varphi$. Then there is a Δ' with $\Delta \stackrel{\hat{a}}{\Longrightarrow} \Delta'$ and $\Delta' \models \psi$. By induction, $\exists o \in \mathcal{A}(T_\psi, \Delta') : o \leq v_\psi$. By Lemma 5.9(3), we know that $o \in \mathcal{A}(T_\varphi, \Delta)$.
  Now suppose $\exists o \in \mathcal{A}(T_\varphi, \Delta) : o \leq v_\varphi$. This implies $o(\omega) = 0$, so by using Lemma 5.9(3) we see that there is a Δ' with $\Delta \stackrel{\hat{a}}{\Longrightarrow} \Delta'$ and $o \in \mathcal{A}(T_\psi, \Delta')$. By induction, $\Delta' \models \psi$, so $\Delta \models \varphi$.
- Let $\varphi = \varphi_1 \wedge \varphi_2$ and suppose $\Delta \models \varphi$. Then $\Delta \models \varphi_i$ for all $i = 1, 2$, and hence, by induction, $\exists o_i \in \mathcal{A}(T_i, \Delta) : o_i \leq v_i$. Thus $o := p \cdot o_1 + (1-p) \cdot o_2 \in \mathcal{A}(T_\varphi, \Delta)$ by Lemma 5.9(4), and $o \leq v_\varphi$.
  Now suppose $\exists o \in \mathcal{A}(T_\varphi, \Delta) : o \leq v_\varphi$. Then, using Lemma 5.9(4), we have that $o = p \cdot o_1 + (1-p) \cdot o_2$ for certain $o_i \in \mathcal{A}(T_i, \Delta)$. Note that T_1, T_2 are Ω-disjoint tests. One has $o_i \leq v_i$ for all $i = 1, 2$, for if $o_i(\omega) > v_i(\omega)$ for some $i = 1$ or 2 and $\omega \in \Omega$, then ω must occur in T_i and hence cannot occur in T_{3-i}. This implies $v_{3-i}(\omega) = 0$ and thus $o(\omega) > v_\varphi(\omega)$, in contradiction with the assumption. By induction, $\Delta \models \varphi_i$ for all $i = 1, 2$, and hence $\Delta \models \varphi$.
- Let $\varphi = \varphi_1 \, {}_p\!\oplus \varphi_2$. Suppose $\Delta \models \varphi$. Then for all $i = 1, 2$ there are $\Delta_i \in \mathcal{D}(\mathsf{sCSP})$ with $\Delta_i \models \varphi_i$ such that $\Delta \stackrel{\hat{\tau}}{\Longrightarrow} p \cdot \Delta_1 + (1-p) \cdot \Delta_2$. By induction, for $i = 1, 2$, there are $o_i \in \mathcal{A}(T_i, \Delta_i)$ with $o_i \leq v_i$. Hence, there are $o_i' \in \mathcal{A}(T_i', \Delta_i)$ with $o_i' \leq v_i'$. Thus $o := p \cdot o_1' + (1-p) \cdot o_2' \in \mathcal{A}(T_\varphi, \Delta)$ by Lemma 5.9(5), and $o \leq v_\varphi$.
  Now suppose $\exists o \in \mathcal{A}(T_\varphi, \Delta) : o \leq v_\varphi$. Then, by Lemma 5.10, there are $q \in [0, 1]$ and Δ_1, Δ_2, such that $\Delta \stackrel{\hat{\tau}}{\Longrightarrow} q \cdot \Delta_1 + (1-q) \cdot \Delta_2$ and $o = q \cdot o_1 + (1-q) o_2'$ for some $o_i' \in \mathcal{A}(T_i', \Delta_i)$. Now $\forall i : o_i'(\omega_i) = v_i'(\omega_i) = \frac{1}{2}$, so, using that T_1, T_2 are Ω-disjoint tests, we have $\frac{1}{2} q = q \cdot o_1'(\omega_1) = o(\omega_1) \leq v_\varphi(\omega_1) = p \cdot v_1'(\omega_1) = \frac{1}{2} p$ and $\frac{1}{2}(1-q) = (1-q) \cdot o_2'(\omega_2) = o(\omega_2) \leq v_\varphi(\omega_) = (1-p) \cdot v_2'(\omega_2) = \frac{1}{2}(1-p)$. Together, these inequalities say that $q = p$. Exactly as in the previous case one obtains $o_i' \leq v_i'$ for all $i = 1, 2$. Given that $T_i' = T_i \, {}_{\frac{1}{2}}\!\oplus \omega_i$, using Lemma 5.9(4), it must be that $o' = \frac{1}{2} o_i + \frac{1}{2} \vec{\omega_i}$ for some $o_i \in \mathcal{A}(T_i, \Delta_i)$ with $o_i \leq v_i$. By induction, $\Delta_i \models \varphi_i$ for all $i = 1, 2$, and hence $\Delta \models \varphi$.

In case $\varphi \in \mathcal{L}$, the formula cannot be of the form $\mathbf{ref}(X)$. Then a straightforward induction yields that $\sum_{\omega \in \Omega} v_\varphi(\omega) = 1$ and for all $\Delta \in \mathcal{D}(\mathsf{pCSP})$ and $o \in \mathcal{A}(T_\varphi, \Delta)$ we have $\sum_{\omega \in \Omega} o(\omega) = 1$. Therefore, $o \leq v_\varphi$ iff $o \geq v_\varphi$ iff $o = v_\varphi$, yielding (5.28). □

Theorem 5.6

1. If $P \sqsubseteq^{\Omega}_{pmay} Q$ *then* $P \sqsubseteq^{\mathcal{L}} Q$.

2. *If* $P \sqsubseteq^{\Omega}_{pmust} Q$ *then* $P \sqsubseteq^{\mathcal{F}} Q$.

Proof Suppose $P \sqsubseteq^{\Omega}_{\text{pmust}} Q$ and $[\![Q]\!] \models \varphi$ for some $\varphi \in \mathcal{F}$. Let T_φ be a characteristic test of φ with target value v_φ. Then Lemma 5.13 yields $\exists o \in \mathcal{A}(T_\varphi, [\![Q]\!]) : o \leq v_\varphi$, and hence, given that $P \sqsubseteq^{\Omega}_{\text{pmust}} Q$ and $\mathcal{A}(T_\varphi, [\![R]\!]) = \mathcal{A}(T_\varphi, R)$ for any $R \in$ pCSP, by the Smyth preorder we have $\exists o' \in \mathcal{A}(T_\varphi, [\![P]\!]) : o' \leq v_\varphi$. Thus $[\![P]\!] \models \varphi$.

The may-case goes likewise, via the Hoare preorder. □

Combining Theorems 4.7, 5.6 and 5.5, we obtain Theorem 5.4, the goal we set ourselves in Sect. 5.5.3. Thus, with Theorems 5.2 and 5.3, we have shown that the may preorder coincides with simulation and that the must preorder coincides with failure simulation. These results also imply the converse of both statements in Theorem 5.6, and thus that the logics $\mathcal{L}$ and $\mathcal{F}$ give logical characterisations of the simulation and failure simulation preorders $\sqsubseteq_S$ and $\sqsubseteq_{FS}$. □

5.8 Equational Theories

Having settled the problem of characterising the may preorder in terms of simulation, and the must preorder in terms of failure simulation, we now turn to complete axiomatisations of the preorders.

In order to focus on the essentials we consider just those pCSP processes that do not use the parallel operator $|_A$; we call the resulting sublanguage nCSP.

Let us write $P =_E Q$ for equivalences that can be derived using the equations given in the upper part of Fig. 5.8. Given the way we defined the syntax of pCSP, axiom (**D1**) is merely a case of abbreviation-expansion; thanks to (**D1**) there is no need for (meta-)variables ranging over the subsort of state-based processes anywhere in the axioms. Many of the standard equations for CSP [1] are missing; they are not sound for $\simeq_{FS}$. Typical examples include:

$$a.(P \sqcap Q) = a.P \sqcap a.Q$$
$$P = P \,\Box\, P$$
$$P \,\Box\, (Q \sqcap R) = (P \,\Box\, Q) \sqcap (P \,\Box\, R)$$
$$P \sqcap (Q \,\Box\, R) = (P \sqcap Q) \,\Box\, (P \sqcap R).$$

A detailed discussion of the standard equations for CSP in the presence of probabilistic processes has been given in Sect. 5.3.

Proposition 5.6 *Suppose* $P =_E Q$, *then* $P \simeq_{FS} Q$.

Proof Because of Theorem 5.1, that $\sqsubseteq_{FS}$ is a precongruence, it is sufficient to exhibit appropriate witness failure simulations for the axioms in the upper part of Fig. 5.8. □

As $\simeq_S$ is a less discriminating equivalence than $\simeq_{FS}$ it follows that $P =_E Q$ implies $P \simeq_S Q$.

This equational theory allows us to reduce terms to a form in which the external choice operator is applied to prefix terms only.

Definition 5.8 (Normal Forms) The set of *normal forms* N is given by the following grammar:

$$N ::= N_1\, {}_p\!\oplus N_2 \mid N_1 \sqcap N_2 \mid \Box_{i\in I} a_i.N_i.$$

Proposition 5.7 *For every $P \in$ nCSP, there is a normal form N such that $P =_E N$.*

Proof A fairly straightforward induction, heavily relying on **(D1)**–**(D3)**. □

We can also show that the axioms **(P1)**–**(P3)** and **(D1)** are in some sense all that are required to reason about probabilistic choice. Let $P =_{\text{prob}} Q$ denote that equivalence of P and Q can be derived using those axioms alone. Then we have the following property.

Lemma 5.14 *Let $P, Q \in$ nCSP. Then $[\![P]\!] = [\![Q]\!]$ implies $P =_{prob} Q$.*

Here $[\![P]\!] = [\![Q]\!]$ says that $[\![P]\!]$ and $[\![Q]\!]$ are the very same distributions of state-based processes in sCSP; this is a much stronger prerequisite than P and Q being testing equivalent.

Proof The axioms **(P1)**–**(P3)** and **(D1)** essentially allow any processes to be written in the unique form $\bigoplus_{i\in I} p_i s_i$, where the $s_i \in$ sCSP are all different. □

5.9 Inequational Theories

In order to characterise the simulation preorders, and the associated testing preorders, we introduce *inequations*. We write $P \sqsubseteq_{E_{\text{may}}} Q$ when $P \sqsubseteq Q$ is derivable from the inequational theory obtained by adding the four *may* inequations in the middle part to the upper part of Fig. 5.8. The first three additions, **(May0)**–**(May2)**, are used in the standard testing theory of CSP [1, 3, 4]. For the *must* case, we write $P \sqsubseteq_{E_{\text{must}}} Q$ when $P \sqsubseteq Q$ is derivable from the equations and inequations in the upper and lower parts of Fig. 5.8. In addition to the standard inequation **(Must1)**, we require an inequational schema, **(Must2)**; this uses the notation *inits(P)* to denote the (finite) set of initial actions of P. Formally,

$$\begin{aligned}
inits(0) &= \emptyset \\
inits(a.P) &= \{a\} \\
inits(P\, {}_p\!\oplus Q) &= inits(P) \cup inits(Q) \\
inits(P \Box Q) &= inits(P) \cup inits(Q) \\
inits(P \sqcap Q) &= \{\tau\}.
\end{aligned}$$

Equations :

(P1)	$P\,{}_p\!\oplus P = P$
(P2)	$P\,{}_p\!\oplus Q = Q\,{}_{1-p}\!\oplus P$
(P3)	$(P\,{}_p\!\oplus Q)\,{}_q\!\oplus R = P\,{}_{p\cdot q}\!\oplus (Q\,{}_{\frac{(1-p)\cdot q}{1-p\cdot q}}\!\oplus R)$
(I1)	$P \sqcap P = P$
(I2)	$P \sqcap Q = Q \sqcap P$
(I3)	$(P \sqcap Q) \sqcap R = P \sqcap (Q \sqcap R)$
(E1)	$P \,\Box\, \mathbf{0} = P$
(E2)	$P \,\Box\, Q = Q \,\Box\, P$
(E3)	$(P \,\Box\, Q) \,\Box\, R = P \,\Box\, (Q \,\Box\, R)$
(EI)	$a.P \,\Box\, a.Q = a.P \sqcap a.Q$
(D1)	$P \,\Box\, (Q\,{}_p\!\oplus R) = (P \,\Box\, Q)\,{}_p\!\oplus (P \,\Box\, R)$
(D2)	$a.P \,\Box\, (Q \sqcap R) = (a.P \,\Box\, Q) \sqcap (a.P \,\Box\, R)$
(D3)	$(P_1 \sqcap P_2) \,\Box\, (Q_1 \sqcap Q_2) = (P_1 \,\Box\, (Q_1 \sqcap Q_2)) \sqcap (P_2 \,\Box\, (Q_1 \sqcap Q_2)) \sqcap ((P_1 \sqcap P_2) \,\Box\, Q_1) \sqcap ((P_1 \sqcap P_2) \,\Box\, Q_2)$

May :

(May0)	$a.P \,\Box\, b.Q = a.P \sqcap b.Q$
(May1)	$P \sqsubseteq P \sqcap Q$
(May2)	$\mathbf{0} \sqsubseteq P$
(May3)	$a.(P\,{}_p\!\oplus Q) \sqsubseteq a.P\,{}_p\!\oplus a.Q$

Must :

(Must1)	$P \sqcap Q \sqsubseteq Q$
(Must2)	$R \sqcap \bigsqcap_{i\in I} \bigoplus_{j\in J_i} p_j \cdot (a_i.Q_{ij} \,\Box\, P_{ij}) \sqsubseteq \Box_{i\in I} a_i.\bigoplus_{j\in J_i} p_j \cdot Q_{ij}$, provided $\mathit{inits}(R) \subseteq \{a_i\}_{i\in I}$

Fig. 5.8 Equations and inequations. (©[2007] IEEE. Reprinted, with permission, from [16])

The axiom (**Must2**) can equivalently be formulated as follows:

$$\bigoplus_{k\in K} \Box_{\ell\in L_k} a_{k\ell}.R_{k\ell} \sqcap \bigsqcap_{i\in I} \bigoplus_{j\in J_i} p_j \cdot (a_i.Q_{ij} \Box P_{ij}) \sqsubseteq \Box_{i\in I} a_i.\bigoplus_{j\in J_i} p_j \cdot Q_{ij},$$

$$\text{provided} \qquad \{a_{k\ell} \mid k \in K, \ell \in K_k\} \subseteq \{a_i \mid i \in I\}.$$

This is the case because a term R satisfies $\mathit{inits}(R) \subseteq \{a_i\}_{i\in I}$ iff it can be converted into the form $\bigoplus_{k\in K} \Box_{\ell\in L_k} a_{k\ell}.R_{k\ell}$ by means of axioms (**D1**), (**P1**)–(**P3**) and (**E1**)–(**E3**) of Fig. 5.8. This axiom can also be reformulated in an equivalent but more semantic style:

(**Must2′**)
$$R \sqcap \bigsqcap_{i\in I} P_i \;\sqsubseteq\; \Box_{i\in I} a_i.Q_i,$$
$$\text{provided } [\![P_i]\!] \xrightarrow{a_i} [\![Q_i]\!] \text{ and } [\![R]\!] \stackrel{X}{\not\longrightarrow} \text{ with } X = \mathsf{Act}\backslash\{a_i\}_{i\in I}.$$

This is the case because $[\![P]\!] \xrightarrow{a} [\![Q]\!]$ iff, up to the axioms in Fig. 5.8, P has the form $\bigoplus_{j\in J} p_j \cdot (a.Q_j \,\Box\, P_j)$ and Q has the form $\bigoplus_{j\in J} p_j \cdot Q_j$ for certain P_j, Q_j and p_j, for $j \in J$.

Note that (**Must2**) can be used, together with (**I1**), to derive the dual of (**May3**) via the following inference:

$$\begin{aligned} a.P\,{}_p\!\oplus a.Q &=_E (a.P\,{}_p\!\oplus a.Q) \sqcap (a.P\,{}_p\!\oplus a.Q) \\ &\sqsubseteq_{E_{\text{must}}} a.(P\,{}_p\!\oplus Q). \end{aligned}$$

An important inequation that follows from (**May1**) and (**P1**) is

$$\textbf{(May4)} \quad P\,{}_p\!\oplus Q \sqsubseteq_{E_{\text{may}}} P \sqcap Q$$

saying that any probabilistic choice can be simulated by an internal choice. It is derived as follows:

$$\begin{aligned} P\,{}_p\!\oplus Q &\sqsubseteq_{E_{\text{may}}} (P \sqcap Q)\,{}_p\!\oplus (P \sqcap Q) \\ &=_E (P \sqcap Q). \end{aligned}$$

Likewise, we have

$$P \sqcap Q \sqsubseteq_{E_{\text{must}}} P\,{}_p\!\oplus Q\ .$$

Theorem 5.7 *For* $P,\ Q$ *in nCSP, it holds that*

(i) $P \sqsubseteq_S Q$ *iff* $P \sqsubseteq_{E_{may}} Q$
(ii) $P \sqsubseteq_{FS} Q$ *iff* $P \sqsubseteq_{E_{must}} Q$.

Proof For one direction it is sufficient to check that the inequations, and the inequational schema in Fig. 5.8 are sound. The converse, completeness, is established in the next section. □

5.10 Completeness

The completeness proof of Theorem 5.7 depends on the following variation on the *Derivative lemma* of [15]:

Lemma 5.15 *(Derivative Lemma) Let* $P, Q \in$ *nCSP.*

(i) If $[\![P]\!] \stackrel{\hat{\tau}}{\Longrightarrow} [\![Q]\!]$ *then* $P \sqsubseteq_{E_{must}} Q$ *and* $Q \sqsubseteq_{E_{may}} P$.
(ii) If $[\![P]\!] \stackrel{a}{\Longrightarrow} [\![Q]\!]$ *then* $a.Q \sqsubseteq_{E_{may}} P$.

Proof The proof of (i) proceeds in four stages. We only deal with $\sqsubseteq_{E_{\text{may}}}$, as the proof for $\sqsubseteq_{E_{\text{must}}}$ is entirely analogous.

First we show by structural induction on $s \in$ sCSP $\cap$ nCSP that $s \stackrel{\tau}{\longrightarrow} [\![Q]\!]$ implies $Q \sqsubseteq_{E_{\text{may}}} s$. So suppose $s \stackrel{\tau}{\longrightarrow} [\![Q]\!]$. In case s has the form $P_1 \sqcap P_2$ it follows by the operational semantics of pCSP that $Q = P_1$ or $Q = P_2$. Hence, $Q \sqsubseteq_{E_{\text{may}}} s$ by

(**May1**). The only other possibility is that s has the form $s_1 \Box s_2$. In that case there must be a distribution Δ such that either $s_1 \xrightarrow{\tau} \Delta$ and $[\![Q]\!] = \Delta \Box s_2$, or $s_2 \xrightarrow{\tau} \Delta$ and $[\![Q]\!] = s_1 \Box \Delta$. Using symmetry, we may restrict attention to the first case. Let R be a term such that $[\![R]\!] = \Delta$. Then $[\![R \Box s_2]\!] = \Delta \Box s_2 = [\![Q]\!]$, so Lemma 5.14 yields $Q =_{\text{prob}} R \Box s_2$. By induction we have $R \sqsubseteq_{E_{\text{may}}} s_1$, hence $R \Box s_2 \sqsubseteq_{E_{\text{may}}} s_1 \Box s_2$, and thus $Q \sqsubseteq_{E_{\text{may}}} s$.

Now we show that $s \xrightarrow{\hat{\tau}} [\![Q]\!]$ implies $Q \sqsubseteq_{E_{\text{may}}} s$. This follows because $s \xrightarrow{\hat{\tau}} [\![Q]\!]$ means that either $s \xrightarrow{\tau} [\![Q]\!]$ or $[\![Q]\!] = \overline{s}$, and in the latter case Lemma 5.14 yields $Q =_{\text{prob}} s$.

Next we show that $[\![P]\!] \xrightarrow{\hat{\tau}} [\![Q]\!]$ implies $Q \sqsubseteq_{E_{\text{may}}} P$. So suppose $[\![P]\!] \xrightarrow{\hat{\tau}} [\![Q]\!]$, that is

$$[\![P]\!] = \sum_{i \in I} p_i \cdot \overline{s_i} \qquad s_i \xrightarrow{\hat{\tau}} [\![Q_i]\!] \qquad [\![Q]\!] = \sum_{i \in I} p_i \cdot [\![Q_i]\!]$$

for some I, $p_i \in (0, 1]$, $s_i \in \mathsf{sCSP} \cap \mathsf{nCSP}$ and $Q_i \in \mathsf{nCSP}$. Now

1. $[\![P]\!] = [\![\bigoplus_{i \in I} p_i \cdot s_i]\!]$. By Lemma 5.14 we have $P =_{\text{prob}} \bigoplus_{i \in I} p_i \cdot s_i$.
2. $[\![Q]\!] = [\![\bigoplus_{i \in I} p_i \cdot Q_i]\!]$. Again Lemma 5.14 yields $Q =_{\text{prob}} \bigoplus_{i \in I} p_i \cdot Q_i$.
3. $s_i \xrightarrow{\hat{\tau}} [\![Q_i]\!]$ implies $Q_i \sqsubseteq_{E_{\text{may}}} s_i$. Therefore, $\bigoplus_{i \in I} p_i \cdot Q_i \sqsubseteq_{E_{\text{may}}} \bigoplus_{i \in I} p_i \cdot s_i$.

Combining (1), (2) and (3) we obtain $Q \sqsubseteq_{E_{\text{may}}} P$.

Finally, the general case, when $[\![P]\!] \stackrel{\hat{\tau}}{\Longrightarrow} \Delta$, is now a simple inductive argument on the length of the derivation.

The proof of (ii) is similar: first we treat the case when $s \xrightarrow{a} [\![Q]\!]$ by structural induction, using (**May2**); then the case $[\![P]\!] \xrightarrow{a} [\![Q]\!]$, exactly as above; and finally use part (i) to derive the general case. □

The completeness result now follows from the following two propositions.

Proposition 5.8 *Let P and Q be in* nCSP. *Then $P \sqsubseteq_S Q$ implies $P \sqsubseteq_{E_{may}} Q$.*

Proof The proof is by structural induction on P and Q, and we may assume that both P and Q are in normal form because of Proposition 5.7. So take $P, Q \in \mathsf{pCSP}$ and suppose the claim has been established for all subterms P' of P and Q' of Q, of which at least one of the two is a strict subterm. We start by proving that if $P \in \mathsf{sCSP}$ then we have

$$P \lhd_s [\![Q]\!] \quad \text{implies} \quad P \sqsubseteq_{E_{\text{may}}} Q. \tag{5.29}$$

There are two cases to consider.

1. P has the form $P_1 \sqcap P_2$. Since $P_i \sqsubseteq_{E_{\text{may}}} P$ we know $P_i \sqsubseteq_S P \sqsubseteq_S Q$. We use induction to obtain $P_i \sqsubseteq_{E_{\text{may}}} Q$, from which the result follows using (**I1**).
2. P has the form $\Box_{i \in I} a_i . P_i$. If I contains two or more elements then P may also be written as $\sqcap_{i \in I} a_i . P_i$, using (**May0**) and (**D2**), and we may proceed as in case (1) above. If I is empty, that is P is $\mathbf{0}$, then we can use (**May2**). So we are left

with the possibility that P is $a.P'$. Thus suppose that $a.P' \lhd_s [\![Q]\!]$. We proceed by a case analysis on the structure of Q.

- Q is $a.Q'$. We know from $a.P' \lhd_s [\![a.Q']\!]$ that $[\![P']\!](\lhd_s)^\dagger \Theta$ for some Θ with $[\![Q']\!] \stackrel{\hat{\tau}}{\Longrightarrow} \Theta$, thus $P' \sqsubseteq_S Q'$. Therefore, we have $P' \sqsubseteq_{E_{\text{may}}} Q'$ by induction. It follows that $a.P' \sqsubseteq_{E_{\text{may}}} a.Q'$.
- Q is $\Box_{j\in J} a_j.Q_j$ with at least two elements in J. We use (**May0**) and then proceed as in the next case.
- Q is $Q_1 \sqcap Q_2$. We know from $a.P' \lhd_s [\![Q_1 \sqcap Q_2]\!]$ that $[\![P']\!](\lhd_s)^\dagger \Theta$ for some Θ such that one of the following two conditions holds
  a) $[\![Q_i]\!] \stackrel{a}{\Longrightarrow} \Theta$ for $i = 1$ or 2. In this case, $a.P' \lhd_s [\![Q_i]\!]$, hence $a.P' \sqsubseteq_S Q_i$. By induction we have $a.P' \sqsubseteq_{E_{\text{may}}} Q_i$; then we apply (**May1**).
  b) $[\![Q_1]\!] \stackrel{a}{\Longrightarrow} \Theta_1$ and $[\![Q_2]\!] \stackrel{a}{\Longrightarrow} \Theta_2$ such that $\Theta = p \cdot \Theta_1 + (1-p) \cdot \Theta_2$ for some $p \in (0, 1)$. Let $\Theta_i = [\![Q'_i]\!]$ for $i = 1, 2$. By the Derivative Lemma, we have $a.Q'_1 \sqsubseteq_{E_{\text{may}}} Q_1$ and $a.Q'_2 \sqsubseteq_{E_{\text{may}}} Q_2$. Clearly, $[\![Q'_1 \, _p\!\oplus Q'_2]\!] = \Theta$, thus we have $P' \sqsubseteq_S Q'_1 \, _p\!\oplus Q'_2$. By induction, we infer that $P' \sqsubseteq_{E_{\text{may}}} Q'_1 \, _p\!\oplus Q'_2$. So

$$
\begin{array}{rlr}
a.P' & \sqsubseteq_{E_{\text{may}}} a.(Q'_1 \, _p\!\oplus Q'_2) & \\
& \sqsubseteq_{E_{\text{may}}} a.Q'_1 \, _p\!\oplus a.Q'_2 & \textbf{(May3)} \\
& \sqsubseteq_{E_{\text{may}}} Q_1 \, _p\!\oplus Q_2 & \\
& \sqsubseteq_{E_{\text{may}}} Q_1 \sqcap Q_2 & \textbf{(May4)}
\end{array}
$$

- Q is $Q_1 \, _p\!\oplus Q_2$. We know from $a.P' \lhd_s [\![Q_1 \, _p\!\oplus Q_2]\!]$ that $[\![P']\!](\lhd_s)^\dagger \Theta$ for some Θ such that $[\![Q_1 \, _p\!\oplus Q_2]\!] \stackrel{a}{\Longrightarrow} \Theta$. From Lemma 5.4 we know that Θ must take the form $p \cdot [\![Q'_1]\!] + (1-p) \cdot [\![Q'_2]\!]$, where $[\![Q_i]\!] \stackrel{a}{\Longrightarrow} [\![Q'_i]\!]$ for $i = 1, 2$. Hence $P' \sqsubseteq_S Q'_1 \, _p\!\oplus Q'_2$, and by induction we get $P' \sqsubseteq_{E_{\text{may}}} Q'_1 \, _p\!\oplus Q'_2$. Then we can derive $a.P' \sqsubseteq_{E_{\text{may}}} Q_1 \, _p\!\oplus Q_2$ as in the previous case.

Now we use (5.29) to show that $P \sqsubseteq_S Q$ implies $P \sqsubseteq_{E_{\text{may}}} Q$. Suppose $P \sqsubseteq_S Q$. Applying Definition 5.5 with the understanding that any distribution $\Theta \in \mathcal{D}(\text{sCSP})$ can be written as $[\![Q']\!]$ for some $Q' \in \text{pCSP}$, this basically means that $[\![P]\!](\lhd_s)^\dagger [\![Q']\!]$ for some $[\![Q]\!] \stackrel{\hat{\tau}}{\Longrightarrow} [\![Q']\!]$. The Derivative Lemma yields $Q' \sqsubseteq_{E_{\text{may}}} Q$. So it suffices to show $P \sqsubseteq_{E_{\text{may}}} Q'$. We know that $[\![P]\!](\lhd_s)^\dagger [\![Q']\!]$ means that

$$
[\![P]\!] = \sum_{k\in K} r_k \cdot \overline{t_k} \qquad t_k \lhd_s [\![Q'_k]\!] \qquad [\![Q']\!] = \sum_{k\in K} r_k \cdot [\![Q'_k]\!]
$$

for some K, $r_k \in (0, 1]$, $t_k \in \text{sCSP}$ and $Q'_k \in \text{pCSP}$. Now

1. $[\![P]\!] = [\![\bigoplus_{k\in K} r_k \cdot t_k]\!]$. By Lemma 5.14 we have $P =_{\text{prob}} \bigoplus_{k\in K} r_k \cdot t_k$.
2. $[\![Q']\!] = [\![\bigoplus_{k\in K} r_k \cdot Q'_k]\!]$. Again Lemma 5.14 yields $Q' =_{\text{prob}} \bigoplus_{k\in K} r_k \cdot Q'_k$.
3. $t_k \lhd_s [\![Q'_k]\!]$ implies $t_k \sqsubseteq_{E_{\text{may}}} Q'_k$ by (5.29). Therefore, $\bigoplus_{k\in K} r_k \cdot t_k \sqsubseteq_{E_{\text{may}}} \bigoplus_{k\in K} r_k \cdot Q'_k$.

Combining (1), (2) and (3) above we obtain $P \sqsubseteq_{E_{\text{may}}} Q'$, hence $P \sqsubseteq_{E_{\text{may}}} Q$. □

Proposition 5.9 *Let P and Q be in* nCSP*. Then* $P \sqsubseteq_{FS} Q$ *implies* $P \sqsubseteq_{E_{must}} Q$.

Proof Similar to the proof of Proposition 5.8, but using a reversed orientation of the preorders. The only real difference is the case (2), which we consider now. So assume $Q \lhd_{FS} [\![P]\!]$, where Q has the form $\Box_{i\in I} a_i.Q_i$. Let X be any set of actions such that $X \cap \{a_i\}_{i\in I} = \emptyset$; then $\Box_{i\in I} a_i.Q_i \stackrel{X}{\not\longrightarrow}$. Therefore, there exists a P' such that $[\![P]\!] \stackrel{\hat{\tau}}{\Longrightarrow} [\![P']\!] \stackrel{X}{\not\longrightarrow}$. By the Derivative lemma,

$$P \sqsubseteq_{E_{\mathrm{must}}} P'. \tag{5.30}$$

Since $\Box_{i\in I} a_i.Q_i \stackrel{a_i}{\longrightarrow} [\![Q_i]\!]$, there exist P_i, P_i', P_i'' such that

$$[\![P]\!] \stackrel{\hat{\tau}}{\Longrightarrow} [\![P_i]\!] \stackrel{a_i}{\longrightarrow} [\![P_i']\!] \stackrel{\hat{\tau}}{\Longrightarrow} [\![P_i'']\!] \text{ and } [\![Q_i]\!] (\lhd_s)^\dagger [\![P_i'']\!].$$

Now

$$P \sqsubseteq_{E_{\mathrm{must}}} P_i \tag{5.31}$$

using the Derivative lemma, and $P_i' \sqsubseteq_{FS} Q_i$, by Definition 5.5. By induction, we have $P_i' \sqsubseteq_{E_{\mathrm{must}}} Q_i$, hence

$$\Box_{i\in I} a_i.P_i' \sqsubseteq_{E_{\mathrm{must}}} \Box_{i\in I} a_i.Q_i \tag{5.32}$$

The desired result is now obtained as follows:

$$\begin{aligned} P &\sqsubseteq_{E_{\mathrm{must}}} P' \sqcap \textstyle\bigsqcap_{i\in I} P_i && \text{by } (\mathbf{I1}), (5.30) \text{ and } (5.31)\\ &\sqsubseteq_{E_{\mathrm{must}}} \Box_{i\in I} a_i.P_i' && \text{by } (\mathbf{Must2'})\\ &\sqsubseteq_{E_{\mathrm{must}}} \Box_{i\in I} a_i.Q_i && \text{by } (5.32) \end{aligned}$$

□

Propositions 5.8 and 5.9 give us the completeness result stated in Theorem 5.7.

5.11 Bibliographic Notes

In this chapter we have studied three different aspects of may and must testing preorders for finite processes: (i) we have shown that the may preorder can be characterised as a coinductive simulation relation, and the must preorder as a failure simulation relation; (ii) we have given a characterisation of both preorders in a finitary modal logic; and (iii) we have also provided complete axiomatisations for both preorders over a probabilistic version of recursion-free CSP. Although we omitted our parallel operator $|_A$ from the axiomatisations, it and similar CSP and CCS-like parallel operators can be handled using standard techniques, in the must case at the expense of introducing auxiliary operators. A generalisation of results (i) and (ii) above to a probabilistic π-calculus, with similar proof techniques of characteristic formulae and characteristic tests, has appeared in [17].

5.11.1 Probabilistic Equivalences

Whereas the testing semantics explored in the present chapter is based on the idea that processes should be *distinguished* only when there is a compelling reason to do so, (strong) bisimulation semantics [15] is based on the idea that processes should be *identified* only when there is a compelling reason to do so. It has been extended to reactive probabilistic processes in [18], to generative ones in [19], and to processes combining nondeterminism and probability in [12]. The latter paper also features a complete axiomatisation of a probabilistic extension of recursion-free CCS.

Weak and branching bisimulation [15, 20] are versions of strong bisimulation that respect the hidden nature of the internal action τ. Generalisations of these notions to nondeterministic probabilistic processes appear, amongst others, in [7, 14, 21–25], with complete axiomatisations reported in [23, 24, 26, 27]. The authors of these paper tend to distinguish whether they work in an *alternating* [21, 22, 25, 27] or a *nonalternating* model of probabilistic processes [7, 14, 24, 26], the two approaches being compared in [23]. The nonalternating model stems from [14] and is similar to our model of Sect. 5.2.2. The alternating model is attributed to [12], and resembles our graphical representation of processes in Sect. 5.2.4. It is easy to see that mathematically the alternating and nonalternating model can be translated into each other without loss of information [23]. The difference between the two is one of interpretation. In the alternating interpretation, the nodes of form $\circ$ in our graphical representations are interpreted as actual states a process can be in, whereas in the nonalternating representation they are not. Take for example the process $R_1 = a.(b\,_{\frac{1}{2}}\!\oplus c)$ depicted in Fig. 5.5. In the alternating representation this process passes through a state in which a has already happened, but the probabilistic choice between b and c has not yet been made. In the nonalternating interpretation on the other hand the execution of a is what constitutes this probabilistic choice; after doing a there is a fifty–fifty chance of ending up in either state. Although in strong bisimulation semantics the alternating and nonalternating interpretation lead to the same semantic equivalence, in weak and branching bisimulation semantics the resulting equivalences are different, as illustrated in [21, 23, 25]. Our testing and simulation preorders as presented here can be classified as nonalternating; however, we believe that an alternating approach would lead to the very same preorders.

Early additions of probability to CSP include work by Lowe [28], Seidel [29] and Morgan et al. [30]; but all of them are forced to make compromises of some kind in order to address the potentially complicated interactions between the three forms of choice. The last [30] for example applies the Jones/Plotkin probabilistic powerdomain [31] directly to the failures model of CSP [10], the resulting compromise being that probability distributed outwards through all other operators; one controversial result of that is that internal choice is no longer idempotent, and that it is "clairvoyant" in the sense that it can adapt to probabilistic-choice outcomes that have not yet occurred. Mislove addresses this problem in [32] by presenting a denotational model in which internal choice distributes outwards through probabilistic choice. However, the distributivities of both [30] and [32] constitute identifications that cannot be justified by our testing approach; see [6].

In Jou and Smolka [33], as in [28, 29], probabilistic equivalences based on traces, failures and readies are defined. These equivalences are coarser than $\simeq_{\text{pmay}}$. For example, let

$$P := a.((b.d \Box c.e)\,{}_{\frac{1}{2}}\!\oplus (b.f \Box c.g))$$

$$Q := a.((b.d \Box c.g)\,{}_{\frac{1}{2}}\!\oplus (b.f \Box c.e)).$$

The two processes cannot be distinguished by the equivalences of [28, 29, 33]. However, we can tell them apart by the test:

$$T := a.((b.d.\omega\,{}_{\frac{1}{2}}\!\oplus c.e.\omega) \sqcap (b.f.\omega\,{}_{\frac{1}{2}}\!\oplus c.g.\omega))$$

since $\mathcal{A}(T, P) = \{0, \frac{1}{2}, 1\}$ and $\mathcal{A}(T, Q) = \{\frac{1}{2}\}$, that is, $P \not\sqsubseteq_{\text{pmay}} Q$.

5.11.2 Probabilistic Simulations

Four different notions of simulation for probabilistic processes occur in the literature, each a generalisation of the well-known concept of simulation for nondeterministic processes [34]. The most straightforward generalisation [35] is defined as in Definition 3.5. This simulation induces a preorder strictly finer than $\sqsubseteq_S$ and $\sqsubseteq_{\text{pmay}}$. For example, it does not satisfy the law

$$a.(P\,{}_p\!\oplus Q) \sqsubseteq a.P \Box a.Q$$

that holds in probabilistic *may* testing semantics. The reason is that the process $a.P \Box a.Q$ can answer the initial a move of $a.(P\,{}_p\!\oplus Q)$ by taking either the a move to P, or the a move to Q, but not by a probabilistic combination of the two. Such probabilistic combinations are allowed in the probabilistic simulation of [14], which induces a coarser preorder on processes, satisfying the above law. In our terminology it can be defined by changing the requirement above into

$$\text{if } s \mathcal{R}\, t \text{ and } s \xrightarrow{\alpha} \Theta \text{ then there is a } \Delta' \text{ with } \overline{t} \xrightarrow{\alpha} \Delta' \text{ and } \Theta\ \mathcal{R}^\dagger\ \Delta'.$$

A *weak* version of this probabilistic simulation, abstracting from the internal action τ, weakens this requirement into

$$\text{if } s \mathcal{R}\, t \text{ and } s \xrightarrow{\alpha} \Theta \text{ then there is a } \Delta' \text{ with } \overline{t} \stackrel{\hat{\alpha}}{\Longrightarrow} \Delta' \text{ and } \Theta\ \mathcal{R}^\dagger\ \Delta'.$$

Nevertheless, also this probabilistic simulation does not satisfy all the laws we have shown to hold for probabilistic may testing. In particular, it does not satisfy the law (**May3**). Consider for instance the processes $R_1 = a.b.c.(d\,{}_{\frac{1}{2}}\!\oplus e)$ and $R_2 = a.(b.c.d\,{}_{\frac{1}{2}}\!\oplus b.c.e)$. The law (**May3**), which holds for probabilistic *may* testing, would yield $R_1 \sqsubseteq R_2$. If we are to relate these processes via a probabilistic simulation à la

[14], the state $c.(d_{\frac{1}{2}}\oplus e)$ of R_1, reachable after an a and a b step, needs to be related to the *distribution* $(c.d_{\frac{1}{2}}\oplus c.e)$ of R_2, containing the two states $a.b$ and $a.c$. This relation cannot be obtained through lifting, as this would entail relating the single state $c.(d_{\frac{1}{2}}\oplus e)$ to each of the states $c.d$ and $c.e$. Such a relation would not be sound, because $c.(d_{\frac{1}{2}}\oplus e)$ is able to perform the sequence of actions ce half of the time, whereas the process $c.d$ cannot mimic this.

In [36], another notion of simulation is proposed, whose definition is too complicated to explain in few sentences. They show for a class of probabilistic processes that do not contain τ actions, that probabilistic *may* testing is captured exactly by their notion of simulation. Nevertheless, their notion of simulation makes strictly more identifications than ours. As an example, let us consider the processes $R_1 = a_{\frac{1}{2}}\oplus(b\ \square\ c)$ and $R_3 = (a\ \square\ b)_{\frac{1}{2}}\oplus(a\ \square\ c)$ of Example 5.11, which also appear in Sect. 5 of [36]. There it is shown that $R_1 \sqsubseteq R_3$ holds in their semantics. However, in our framework we have $R_1 \not\sqsubseteq_{\text{pmay}} R_3$, as demonstrated in Example 5.11. The difference can only be explained by the circumstance that in [36] processes, and hence also tests, may not have internal actions. So this example shows that tests with internal moves can distinguish more processes than tests without internal moves, even when applied to processes that have no internal moves themselves.

Our notion of forward simulation first appears in [7], although the preorder $\sqsubseteq_S$ of Definition 5.5 is new. Segala has no expressions that denote distributions and consequently is only interested in the restriction of the simulation preorder to states (*automata* in his framework). It turns out that for states s and t (or expressions in the set sCSP in our framework) we have $s \sqsubseteq_S t$ iff $s \vartriangleleft_s \bar{t}$, so on their common domain of definition, the simulation preorder of [7] agrees with ours. This notion of simulation is strictly more discriminating than the simulation of [36], and strictly less discriminating than the ones from [14] and [35].

References

1. Hoare, C.A.R.: Communicating Sequential Processes. Prentice Hall, Englewood Cliffs (1985)
2. Segala, R.: Testing probabilistic automata. Proceedings of the 7th International Conference on Concurrency Theory. Lecture Notes in Computer Science, vol. 1119, pp. 299–314. Springer, Heidelberg (1996)
3. De Nicola, R., Hennessy, M.: Testing equivalences for processes. Theor. Comput. Sci. **34**, 83–133 (1984)
4. Hennessy, M.: An Algebraic Theory of Processes. The MIT Press, Cambridge (1988)
5. Yi, W., Larsen, K.G.: Testing probabilistic and nondeterministic processes. Proceedings of the IFIP TC6/WG6.1 12th International Symposium on Protocol Specification, Testing and Verification, *IFIP Transactions*, vol. C-8, pp. 47–61. North-Holland (1992)
6. Deng, Y., van Glabbeek, R., Hennessy, M., Morgan, C.C., Zhang, C.: Remarks on testing probabilistic processes. Electron. Notes Theor. Comput. Sci. **172**, 359–397 (2007)
7. Segala, R.: Modeling and verification of randomized distributed real-time systems. Tech. Rep. MIT/LCS/TR-676. PhD thesis, MIT, Dept. of EECS (1995)
8. Lynch, N., Segala, R., Vaandrager, F.W.: Observing branching structure through probabilistic contexts. SIAM J. Comput. **37**(4), 977–1013 (2007)

9. van Glabbeek, R.: The linear time—branching time spectrum II; the semantics of sequential systems with silent moves. Proceedings of the 4th International Conference on Concurrency Theory. Lecture Notes in Computer Science, vol. 715, pp. 66–81. Springer, Heidelberg (1993)
10. Brookes, S., Hoare, C., Roscoe, A.: A theory of communicating sequential processes. J. ACM **31**(3), 560–599 (1984)
11. Olderog, E.R., Hoare, C.: Specification-oriented semantics for communicating processes. Acta Inf. **23**, 9–66 (1986)
12. Hansson, H., Jonsson, B.: A calculus for communicating systems with time and probabilities. Proceedings of IEEE Real-Time Systems Symposium, pp. 278–287. IEEE Computer Society Press (1990)
13. Jonsson, B., Yi, W., Larsen, K.G.: Probabilistic extensions of process algebras. Handbook of Process Algebra, Chapter. 11, pp. 685–710. Elsevier (2001)
14. Segala, R., Lynch, N.: Probabilistic simulations for probabilistic processes. Proceedings of the 5th International Conference on Concurrency Theory. Lecture Notes in Computer Science, vol. 836, pp. 481–496. Springer, Heidelberg (1994)
15. Milner, R.: Communication and Concurrency. Prentice Hall, New York (1989)
16. Deng, Y., van Glabbeek, R., Hennessy, M., Morgan, C.C., Zhang, C.: Characterising testing preorders for finite probabilistic processes. Proceedings of the 22nd Annual IEEE Symposium on Logic in Computer Science, pp. 313–325. IEEE Computer Society (2007)
17. Deng, Y., Tiu, A.: Characterisations of testing preorders for a finite probabilistic π-calculus. Form. Asp. Comput. **24**(4–6), 701–726 (2012)
18. Larsen, K.G., Skou, A.: Bisimulation through probabilistic testing. Inf. Comput. **94**(1), 1–28 (1991)
19. van Glabbeek, R., Smolka, S.A., Steffen, B., Tofts, C.: Reactive, generative, and stratified models of probabilistic processes. Proceedings of the 5th Annual IEEE Symposium on Logic in Computer Science, pp. 130–141. Computer Society Press (1990)
20. van Glabbeek, R., Weijland, W.: Branching time and abstraction in bisimulation semantics. J. ACM **43**(3), 555–600 (1996)
21. Philippou, A., Lee, I., Sokolsky, O.: Weak bisimulation for probabilistic systems. Proceedings of the 11th International Conference on Concurrency Theory. Lecture Notes in Computer Science, vol. 1877, pp. 334–349. Springer, Heidelberg (2000)
22. Andova, S., Baeten, J.C.: Abstraction in probabilistic process algebra. Proceedings of the 7th International Conference on Tools and Algorithms for the Construction and Analysis of Systems. Lecture Notes in Computer Science, vol. 2031, pp. 204–219. Springer, Heidelberg (2001)
23. Bandini, E., Segala, R.: Axiomatizations for probabilistic bisimulation. In: Proceedings of the 28th International Colloquium on Automata, Languages and Programming. Lecture Notes in Computer Science, vol. 2076, pp. 370–381. Springer (2001)
24. Deng, Y., Palamidessi, C.: Axiomatizations for probabilistic finite-state behaviors. Proceedings of the 8th International Conference on Foundations of Software Science and Computation Structures. Lecture Notes in Computer Science, vol. 3441, pp. 110–124. Springer, Heidelberg (2005)
25. Andova, S., Willemse, T.A.: Branching bisimulation for probabilistic systems: Characteristics and decidability. Theor. Comput. Sci. **356**(3), 325–355 (2006)
26. Deng, Y., Palamidessi, C., Pang, J.: Compositional reasoning for probabilistic finite-state behaviors. Processes, Terms and Cycles: Steps on the Road to Infinity, Essays Dedicated to Jan Willem Klop, on the Occasion of His 60th Birthday. Lecture Notes in Computer Science, vol. 3838, pp. 309–337. Springer, Heidelberg (2005)
27. Andova, S., Baeten, J.C., Willemse, T.A.: A complete axiomatisation of branching bisimulation for probabilistic systems with an application in protocol verification. Proceedings of the 17th International Conference on Concurrency Theory. Lecture Notes in Computer Science, vol. 4137, pp. 327–342. Springer, Heidelberg (2006)
28. Lowe, G.: Representing nondeterminism and probabilistic behaviour in reactive processes. Technical Report TR-11-93, Computing laboratory, Oxford University (1993)

29. Seidel, K.: Probabilistic communicating processes. Theor. Comput. Sci. **152**(2), 219–249 (1995)
30. Morgan, C.C., McIver, A.K., Seidel, K., Sanders, J.: Refinement-oriented probability for CSP. Form. Asp. Comput. **8**(6), 617–647 (1996)
31. Jones, C., Plotkin, G.: A probabilistic powerdomain of evaluations. Proceedings of the 4th Annual IEEE Symposium on Logic in Computer Science, pp. 186–195. Computer Society Press (1989)
32. Mislove, M.W.: Nondeterminism and probabilistic choice: Obeying the laws. Proceedings of the 11th International Conference on Concurrency Theory. Lecture Notes in Computer Science, vol. 1877, pp. 350–364. Springer, Heidelberg (2000)
33. Jou, C.C., Smolka, S.A.: Equivalences, congruences, and complete axiomatizations for probabilistic processes. Proceedings of the 1st International Conference on Concurrency Theory. Lecture Notes in Computer Science, vol. 458, pp. 367–383. Springer, Heidelberg (1990)
34. Park, D.: Concurrency and automata on infinite sequences. Proceedings of the 5th GI Conference. Lecture Notes in Computer Science, vol. 104, pp. 167–183. Springer, Heidelberg (1981)
35. Jonsson, B., Larsen, K.G.: Specification and refinement of probabilistic processes. Proceedings of the 6th Annual IEEE Symposium on Logic in Computer Science, pp. 266–277. Computer Society Press (1991)
36. Jonsson, B., Yi, W.: Testing preorders for probabilistic processes can be characterized by simulations. Theor. Comput. Sci. **282**(1), 33–51 (2002)

Chapter 6
Testing Finitary Probabilistic Processes

Abstract In this chapter we extend the results of Chap. 5 from finite to finitary processes, that is, finite-state and finitely branching processes. Testing preorders can still be characterised as simulation preorders and admit modal characterisations. However, to prove these results demands more advanced techniques. A new notion of weak derivation is introduced; some of its fundamental properties are established. Of particular importance is the finite generability property of the set of derivatives from any distribution, which enables us to approximate coinductive simulation relations by stratified inductive relations. This opens the way to characterise the behaviour of finitary process by a finite modal logic. Therefore, if two processes are related it suffices to construct a simulation relation, otherwise a finite test can be constructed to tell them apart.

We also introduce a notion of real-reward testing that allows for negative rewards. Interestingly, for finitary convergent processes, real-reward testing preorder coincides with nonnegative-reward testing preorder.

Keywords Finitary processes · Testing preorder · Simulation preorder · Modal logic · Weak derivation · Finite generability · Real-reward testing

6.1 Introduction

In order to describe infinite behaviour of processes, we extend the language pCSP to a version of rpCSP with recursive process descriptions; we add a construct $\mathsf{rec}\, x.P$ for recursion and extend the intensional semantics of Fig. 5.1 in a straightforward manner. We restrict ourselves to *finitary* rpCSP processes, having finitely many states and displaying finite branching.

The simulation relations $\sqsubseteq_S$ and $\sqsubseteq_{FS}$ given in Sect. 5.5.1 were defined in terms of weak transitions $\stackrel{\hat{\tau}}{\Longrightarrow}$ between distributions, obtained as the transitive closure of a relation $\stackrel{\hat{\tau}}{\longrightarrow}$ between distributions that allows one part of a distribution to make a τ-move with the other part remaining in place. This definition is, however, inadequate for processes that can do an unbounded number of τ-steps. The problem is highlighted by the process $Q_1 = \mathsf{rec}\, x.(\tau.x \,{}_{\frac{1}{2}}\!\oplus a.\mathbf{0})$ illustrated in Fig. 6.1a. Process Q_1 is indistinguishable, using tests, from the simple process $a.\mathbf{0}$: we have $Q_1 \simeq_{\mathrm{pmay}} a.\mathbf{0}$ and $Q_1 \simeq_{\mathrm{pmust}} a.\mathbf{0}$. This is because the process Q_1 will eventually perform the action

DOI 10.1007/978-3-662-45198-4_6

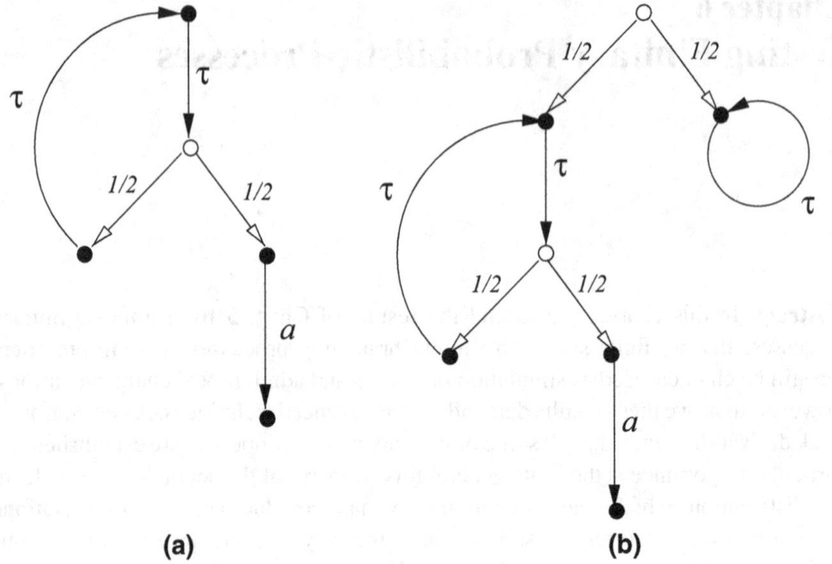

Fig. 6.1 The pLTSs of processes Q_1 and Q_2. Reprinted from [1], with kind permission from Springer Science+Business Media

a with probability 1. However, the action $[\![a.\mathbf{0}]\!] \xrightarrow{a} [\![\mathbf{0}]\!]$ cannot be simulated by a corresponding move $[\![Q_1]\!] \xRightarrow{\hat{\tau}} \xrightarrow{a}$. No matter which distribution Δ we obtain from executing a finite sequence of internal moves $[\![Q_1]\!] \xRightarrow{\hat{\tau}} \Delta$, still part of it is unable to subsequently perform the action a.

To address this problem, we propose a new relation $\Delta \Longrightarrow \Theta$, to indicate that Θ can be derived from Δ by performing an unbounded sequence of internal moves; we call Θ a *weak derivative* of Δ. For example $[\![a.\mathbf{0}]\!]$ will turn out to be a weak derivative of $[\![Q_1]\!]$, that is, $[\![Q_1]\!] \Longrightarrow [\![a.\mathbf{0}]\!]$, via the infinite sequence of internal moves

$$[\![Q_1]\!] \xrightarrow{\hat{\tau}} [\![Q_1 {}_{\frac{1}{2}}\!\oplus a.\mathbf{0}]\!] \xrightarrow{\hat{\tau}} [\![Q_1 {}_{\frac{1}{2^2}}\!\oplus a.\mathbf{0}]\!] \xrightarrow{\hat{\tau}} \ldots \xrightarrow{\hat{\tau}} [\![Q_1 {}_{\frac{1}{2^n}}\!\oplus a.\mathbf{0}]\!] \xrightarrow{\hat{\tau}} \ldots .$$

Here, we make a significant use of "subdistributions" that sum to *no more than* one [2, 3]. For example, the empty subdistribution ε elegantly represents the chaotic behaviour of processes that in communicating sequential processes (CSP) and in must-testing semantics is tantamount to divergence, because we have $\varepsilon \xrightarrow{\alpha} \varepsilon$ for any action α, and a process like $\mathsf{rec}\,x.x$ that diverges via an infinite τ path gives rise to the weak transition $\mathsf{rec}\,x.x \Longrightarrow \varepsilon$. So the process $Q_2 = Q_1 {}_{\frac{1}{2}}\!\oplus \mathsf{rec}\,x.x$ illustrated in Fig. 6.1b will enable the weak transition $[\![Q_2]\!] \Longrightarrow \frac{1}{2}[\![a.\mathbf{0}]\!]$, where intuitively the latter is a proper subdistribution mapping the state $a.\mathbf{0}$ to the probability $\frac{1}{2}$. Our weak transition relation $\Longrightarrow$ can be regarded as an extension of the *weak hyper-transition* from [4] to partial distributions; the latter, although defined in a very different way, can be represented in terms of ours by requiring weak derivatives to be total distributions.

We end this introduction with a brief glimpse at our proof strategy. In Chap. 5, the characterisations for finite rpCSP processes were obtained using a probabilistic extension of the Hennessy–Milner logic. Moving to recursive processes, we know that process behaviour can be captured by a finite modal logic only if the underlying labelled transition system (LTS) is finitely branching, or at least image-finite [5]. Thus to take advantage of a finite probabilistic Hennessy–Milner logic, we need a property of pLTSs corresponding to finite branching in LTSs; this is topological compactness, whose relevance we now sketch.

Subdistributions over (derivatives of) finitary rpCSP processes inherit the standard (complete) Euclidean metric. One of our key results is that

Theorem 6.1 *For every finitary rpCSP process P, the set $\{ \Delta \mid [\![P]\!] \Longrightarrow \Delta\}$ is convex and compact.*

Indeed, using techniques from the theory of Markov decision processes [6], we can show that the potentially uncountable set $\{ \Delta \mid [\![P]\!] \Longrightarrow \Delta\}$ is nevertheless the convex closure of a *finite* set of subdistributions, from which Theorem 6.1 follows.

This key result allows an *inductive* characterisation of the simulation preorders $\sqsubseteq_S$ and $\sqsubseteq_{FS}$, here defined using our novel weak derivation relation $\Longrightarrow$.[1] We first construct a sequence of approximations $\sqsubseteq_S^k$ for $k \geq 0$ and, using Theorem 6.1, we prove

Theorem 6.2 *For every finitary rpCSP process P, and for every $k \in \mathbb{N}$, the set*

$$\{ \Delta \mid [\![P]\!] \sqsubseteq_S^k \Delta\}$$

is convex and compact.

This in turn enables us to use the *Finite Intersection Property* of compact sets to prove

Theorem 6.2 *For finitary* rpCSP *processes we have $P \sqsubseteq_S Q$ iff $P \sqsubseteq_S^k Q$ for all $k \geq 0$.*

Our main characterisation results can then be obtained by extending the probabilistic modal logic used in Sect. 5.6, so that, for example,

- it characterises $\sqsubseteq_S^k$ for every $k \geq 0$, and therefore it also characterises $\sqsubseteq_S$;
- every probabilistic modal formula can be mimicked by a may-test.

Similar results accrue for must testing: details are given in Sect. 6.7.

[1] When restricted to finite processes, the new definitions of simulation and failure simulation preorders degenerate into the preorders in Sect. 5.5.1. So the extension is conservative and justifies our use of the same symbols $\sqsubseteq_S$ and $\sqsubseteq_{FS}$ in this chapter.

$$a.P \xrightarrow{a} [\![P]\!] \qquad \mathsf{rec}x.P \xrightarrow{\tau} [\![P[x \mapsto \mathsf{rec}x.P]]\!]$$

$$P \sqcap Q \xrightarrow{\tau} [\![P]\!] \qquad P \sqcap Q \xrightarrow{\tau} [\![Q]\!]$$

$$\frac{s_1 \xrightarrow{a} \Delta}{s_1 \,\Box\, s_2 \xrightarrow{a} \Delta} \qquad \frac{s_2 \xrightarrow{a} \Delta}{s_1 \,\Box\, s_2 \xrightarrow{a} \Delta}$$

$$\frac{s_1 \xrightarrow{\tau} \Delta}{s_1 \,\Box\, s_2 \xrightarrow{\tau} \Delta \,\Box\, s_2} \qquad \frac{s_2 \xrightarrow{\tau} \Delta}{s_1 \,\Box\, s_2 \xrightarrow{\tau} s_1 \,\Box\, \Delta}$$

$$\frac{s_1 \xrightarrow{\alpha} \Delta \quad \alpha \notin A}{s_1 \mid_A s_2 \xrightarrow{\alpha} \Delta \mid_A s_2} \qquad \frac{s_2 \xrightarrow{\alpha} \Delta \quad \alpha \notin A}{s_1 \mid_A s_2 \xrightarrow{\alpha} s_1 \mid_A \Delta}$$

$$\frac{s_1 \xrightarrow{a} \Delta_1, s_2 \xrightarrow{a} \Delta_2 \ a \in A}{s_1 \mid_A s_2 \xrightarrow{\tau} \Delta_1 \mid_A \Delta_2}$$

Fig. 6.2 Operational semantics of rpCSP

6.2 The Language rpCSP

Let Act be a set of visible actions that a process can perform, and let Var be an infinite set of variables. The language rpCSP of probabilistic CSP processes is given by the following two-sorted syntax, in which $p \in [0, 1]$, $a \in \mathsf{Act}$ and $A \subseteq \mathsf{Act}$:

$$P ::= S \mid P\,{}_p{\oplus} P$$
$$S ::= \mathbf{0} \mid x \in \mathsf{Var} \mid a.P \mid P \sqcap P \mid S \,\Box\, S \mid S \mid_A S \mid \mathsf{rec}\, x.P$$

This is essentially the finite language pCSP given in Sect. 5.2 plus the recursive construct $\mathsf{rec}\,x.P$ in which x is a variable and P a term. The notions of free and bound variables are standard; by $Q[x \mapsto P]$ we indicate substitution of term P for variable x in Q, with renaming if necessary. We write rpCSP for the set of closed P-terms defined by this grammar, and still use sCSP for its *state-based* subset of closed S-terms.

The operational semantics of rpCSP is defined in terms of a particular pLTS $\langle \mathsf{sCSP}, \mathsf{Act}_\tau, \rightarrow \rangle$, in which sCSP is the set of states and Act_τ is the set of transition labels. We interpret rpCSP processes P as distributions $[\![P]\!] \in \mathcal{D}(\mathsf{sCSP})$, as we did for pCSP processes in Sect. 5.2.2.

The transition relation $\rightarrow$ is defined in Fig. 6.2, where A ranges over subsets of Act, and actions a, α are elements of Act, Act_τ respectively. This is a slight extension of the rules in Fig. 5.1 for finite processes; one new rule is required to interpret recursive processes. The process $\mathsf{rec}\; x P$ performs an internal action when unfolding. As our testing semantics will abstract from internal actions, these τ-steps are harmless and merely simplify the semantics.

We graphically depict the operational semantics of a rpCSP expression P by drawing the part of the pLTS reachable from $[\![P]\!]$ as a directed graph with states represented by filled nodes • and distributions by open nodes ∘, as described in Sect. 3.2.

Note that for each $P \in \mathsf{rpCSP}$ the distribution $[\![P]\!]$ has finite support. Moreover, our pLTS is *finitely branching* in the sense that for each state $s \in \mathsf{sCSP}$, there are only finitely many pairs $(\alpha, \Delta) \in \mathsf{Act}_\tau \times \mathcal{D}(\mathsf{sCSP})$ with $s \xrightarrow{\alpha} \Delta$. In spite of $[\![P]\!]$'s finite support and the finite branching of our pLTS, it is possible for there to be infinitely many states reachable from $[\![P]\!]$; when there are only finitely many, then P is said to be finitary.

Definition 6.1 A subdistribution $\Delta \in \mathcal{D}_{sub}(S)$ in a pLTS $\langle S, L, \rightarrow \rangle$ is *finitary* if only finitely many states are reachable from Δ; a rpCSP expression P is *finitary* if $[\![P]\!]$ is.

6.3 A General Definition of Weak Derivations

In this section, we develop a new definition of what it means for a recursive process to evolve by silent activity into another process; it allows the simulation and failure-simulation preorders given in Definition 5.5 to be adapted to characterise the testing preorders for at least finitary probabilistic processes.

Recall, for example, the process Q_1 depicted in Fig. 6.1a. It turns out that in our testing framework this process is indistinguishable from $a.\mathbf{0}$; both processes can do nothing else than an a-action, possibly after some internal moves, and in both cases the probability that the process will never do the a-action is 0. In Sect. 5.5, where we did not deal with recursive processes like Q_1, we defined a weak transition relation $\stackrel{\hat{a}}{\Longrightarrow}$ in such a way that $P \stackrel{\hat{a}}{\Longrightarrow}$ if there is a finite number of τ-moves after which the entire distribution $[\![P]\!]$ will have done an a-action. Lifting this definition verbatim to a setting with recursion would create a difference between $a.\mathbf{0}$ and Q_1, for only the former admits such a weak transition $\stackrel{\hat{a}}{\Longrightarrow}$. The purpose of this section is to propose a new definition of weak transitions, with which we can capture the intuition that the process Q_1 can perform the action a with probability 1, provided it is allowed to run for an unbounded amount of time.

We construct our generalised definition of weak move by revising what it means for a probabilistic process to execute an indefinite sequence of (internal) τ moves. The key technical innovation is to change the focus from distributions to *subdistributions* that enable us to express divergence very conveniently.

First some relatively standard terminology. For any subset X of $\mathcal{D}_{sub}(S)$, with S a set, recall that $\updownarrow X$ stands for the convex closure of X. As the smallest convex set containing X, it satisfies that $\Delta \in \updownarrow X$ if and only if $\Delta = \sum_{i \in I} p_i \cdot \Delta_i$, where $\Delta_i \in X$ and $p_i \in [0,1]$, for some index set I such that $\sum_{i \in I} p_i = 1$.

In case S is a finite set, it makes no difference whether we restrict I to being finite or not; in fact, index sets of size 2 will suffice. However, in general they do not.

Example 6.1 Let $S = \{s_i \mid i \in \mathbb{N}\}$. Then $\updownarrow\{\overline{s_i} \mid i \in \mathbb{N}\}$ consists of all total distributions whose support is included in S. However, with a definition of convex closure that requires only binary interpolations of distributions to be included, $\updownarrow\{\overline{s_i} \mid i \in \mathbb{N}\}$ would merely consist of all such distributions with finite support.

Convex closure is a closure operator in the standard sense, in that it satisfies

- $X \subseteq \updownarrow X$
- $X \subseteq Y$ implies $\updownarrow X \subseteq \updownarrow Y$
- $\updownarrow\updownarrow X = \updownarrow X$.

We say a binary relation $\mathcal{R} \subseteq Y \times \mathcal{D}_{sub}(S)$ is convex whenever the set $\{\Delta \mid y\mathcal{R}\Delta\}$ is convex for every y in Y, and let $\updownarrow\mathcal{R}$ denote the smallest convex relation containing $\mathcal{R}$.

6.3.1 Lifting Relations

In a pLTS, actions are only performed by states, in that, actions are given by relations from states to distributions. But rpCSP processes in general correspond to distributions over states, so in order to define what it means for a process to perform an action, we need to lift these relations so that they also apply to distributions. In fact, we will find it convenient to lift them to subdistributions by a straightforward generalisation of Definition 5.3.

Definition 6.2 Let $(S, L, \rightarrow)$ be a pLTS and $\mathcal{R} \subseteq S \times \mathcal{D}_{sub}(S)$ be a relation from states to subdistributions. Then $\mathcal{R}^\dagger \subseteq \mathcal{D}_{sub}(S) \times \mathcal{D}_{sub}(S)$ is the smallest relation that satisfies:

(1) $s\mathcal{R}\Theta$ implies $\overline{s}\ \mathcal{R}^\dagger\ \Theta$ and
(2) (Linearity) $\Delta_i\ \mathcal{R}^\dagger\ \Theta_i$ for all $i \in I$ implies $(\sum_{i\in I} p_i \cdot \Delta_i)\ \mathcal{R}^\dagger\ (\sum_{i\in I} p_i \cdot \Theta_i)$, where I is a finite index set and $\sum_{i\in I} p_i \leq 1$.

Remark 6.1 By construction $\mathcal{R}^\dagger$ is convex. Moreover, because $s(\updownarrow\mathcal{R})\Theta$ implies $\overline{s}\ \mathcal{R}^\dagger\ \Theta$ we have $\mathcal{R}^\dagger = (\updownarrow\mathcal{R}^\dagger)$, which means that when considering a lifted relation we can without loss of generality assume the original relation to have been convex. In fact when $\mathcal{R}$ is indeed convex, we have that $\overline{s}\ \mathcal{R}^\dagger\ \Theta$ and $s\mathcal{R}\Theta$ are equivalent.

An application of this notion is when the relation is $\xrightarrow{\alpha}$ for $\alpha \in \mathsf{Act}_\tau$; in that case we also write $\xrightarrow{\alpha}$ for $(\xrightarrow{\alpha})^\dagger$. Thus, as source of a relation $\xrightarrow{\alpha}$ we now also allow distributions, and even subdistributions. A subtlety of this approach is that for any action α, we have

$$\varepsilon \xrightarrow{\alpha} \varepsilon \tag{6.1}$$

simply by taking $I = \emptyset$ or $\sum_{i\in I} p_i = 0$ in Definition 6.2. That will turn out to make ε especially useful for modelling the "chaotic" aspects of divergence, in particular that in the must-case a divergent process can simulate any other.

We have the following variant of Lemma 5.3.

Lemma 6.1 *$\Delta\ \mathcal{R}^\dagger\ \Theta$ if and only if there is a collection of states $\{s_i\}_{i\in I}$, a collection of subdistributions $\{\Theta_i\}_{i\in I}$, and a collection of probabilities $\{p_i\}_{i\in I}$, for some finite index set I, such that $\sum_{i\in I} p_i \leq 1$ and Δ, Θ can be decomposed as follows:*

1. *$\Delta = \sum_{i\in I} p_i \cdot \overline{s_i}$;*
2. *$\Theta = \sum_{i\in I} p_i \cdot \Theta_i$;*
3. *For each $i \in I$ we have $s_i \mathcal{R} \Theta_i$.* □

A simple but important property of this lifting operation is the following:

Lemma 6.2 *Suppose $\Delta\ \mathcal{R}^\dagger\ \Theta$, where $\mathcal{R}$ is any relation in $S \times \mathcal{D}_{sub}(S)$. Then*

(i) $|\Delta| \geq |\Theta|$.
(ii) If $\mathcal{R}$ is a relation in $S \times \mathcal{D}(S)$ then $|\Delta| = |\Theta|$.

Proof This follows immediately from the characterisation in Lemma 6.1. □

So, for example, if $\varepsilon\ \mathcal{R}^\dagger\ \Theta$ then $0 = |\varepsilon| \geq |\Theta|$, whence Θ is also ε.

Remark 6.2 From Lemma 6.1 it also follows that lifting enjoys the following two properties:

(i) (Scaling) If $\Delta\ \mathcal{R}^\dagger\ \Theta$, $p \in \mathbb{R}_{\geq 0}$ and $|p \cdot \Delta| \leq 1$ then $p \cdot \Delta\ \mathcal{R}^\dagger\ p \cdot \Theta$.
(ii) (Additivity) If $\Delta_i\ \mathcal{R}^\dagger\ \Theta_i$ for $i \in I$ and $|\sum_{i\in I} \Delta_i| \leq 1$ then $(\sum_{i\in I} \Delta_i)\ \mathcal{R}^\dagger\ (\sum_{i\in I} \Theta_i)$, where I is a finite index set.

In fact, we could have presented Definition 6.2 using scaling and additivity instead of linearity.

The lifting operation has yet another characterisation, this time in terms of *choice functions*.

Definition 6.3 Let $\mathcal{R} \subseteq S \times \mathcal{D}_{sub}(S)$ be a binary relation from states to subdistributions. Then $f : S \to \mathcal{D}_{sub}(S)$ is a *choice function for* $\mathcal{R}$ if $s\mathcal{R}f(s)$ for every $s \in dom(\mathcal{R})$. We write $\mathbf{Ch}(\mathcal{R})$ for the set of all choice functions of $\mathcal{R}$.

Note that if f is a choice function of $\mathcal{R}$ then f behaves properly at each state s in the domain of R, but for each state s outside the domain of $\mathcal{R}$, the value $f(s)$ can be arbitrarily chosen.

Proposition 6.1 *Suppose $\mathcal{R} \subseteq S \times \mathcal{D}_{sub}(S)$ is a convex relation. Then for any subdistribution $\Delta \in \mathcal{D}_{sub}(S)$, $\Delta\ \mathcal{R}^\dagger\ \Theta$ if and only if there is some choice function $f \in \mathbf{Ch}(\mathcal{R})$ such that $\Theta = Exp_\Delta(f)$.*

Proof First suppose $\Theta = \text{Exp}_\Delta(f)$ for some choice function $f \in \mathbf{Ch}(\mathcal{R})$, that is $\Theta = \sum_{s\in\lceil\Delta\rceil} \Delta(s) \cdot f(s)$. It now follows from Lemma 6.1 that $\Delta\ \mathcal{R}^\dagger\ \Theta$ since $s\mathcal{R}f(s)$ for each $s \in dom(\mathcal{R})$.

Conversely, suppose $\Delta\ \mathcal{R}^\dagger\ \Theta$. We have to find a choice function $f \in \mathbf{Ch}(\mathcal{R})$ such that $\Theta = \text{Exp}_\Delta(f)$. Applying Lemma 6.1 we know that

(i) $\Delta = \sum_{i\in I} p_i \cdot \overline{s_i}$, for some index set I, with $\sum_{i\in I} p_i \leq 1$
(ii) $\Theta = \sum_{i\in I} p_i \cdot \Theta_i$ for some Θ_i satisfying $s_i \mathcal{R} \Theta_i$.

Now let us define the function $f : S \to \mathcal{D}_{sub}(S)$ as follows:

- if $s \in \lceil\Delta\rceil$ then $f(s) = \sum_{\{i \in I \mid s_i = s\}} \frac{p_i}{\Delta(s)} \cdot \Theta_i$;
- if $s \in dom(\mathcal{R}) \backslash \lceil\Delta\rceil$ then $f(s) = \Theta'$ for any Θ' with $s\mathcal{R}\Theta'$;
- otherwise, $f(s) = \varepsilon$, where ε is the empty subdistribution.

Note that if $s \in \lceil\Delta\rceil$, then $\Delta(s) = \sum_{\{i \in I \mid s_i = s\}} p_i$ and therefore by convexity $s\mathcal{R}f(s)$; so f is a choice function for $\mathcal{R}$ as $s\mathcal{R}f(s)$ for each $s \in dom(\mathcal{R})$. Moreover, a simple calculation shows that $\text{Exp}_\Delta(f) = \sum_{i \in I} p_i \cdot \Theta_i$, which by (ii) above is Θ. □

An important further property is the following:

Proposition 6.2 (Left-Decomposable) *If* $(\sum_{i \in I} p_i \cdot \Delta_i)\mathcal{R}^\dagger\Theta$ *then* $\Theta = \sum_{i \in I} p_i \cdot \Theta_i$ *for some subdistributions* Θ_i *such that* $\Delta_i\mathcal{R}^\dagger\Theta_i$ *for each* $i \in I$.

Proof It is possible to adapt the proof of Proposition 3.3. But here we provide another proof that takes advantage of choice functions.

Let $\Delta\ \mathcal{R}^\dagger\ \Theta$, where $\Delta = \sum_{i \in I} p_i \cdot \Delta_i$. By Proposition 6.1, using that $\mathcal{R}^\dagger = (\updownarrow R)^\dagger$, there is a choice function $f \in \mathbf{Ch}(\updownarrow\mathcal{R})$ such that $\Theta = \text{Exp}_\Delta(f)$. Take $\Theta_i := \text{Exp}_{\Delta_i}(f)$ for $i \in I$. Using that $\lceil\Delta_i\rceil \subseteq \lceil\Delta\rceil$, Proposition 6.1 yields $\Delta_i\ \mathcal{R}^\dagger\ \Theta_i$ for $i \in I$. Finally,

$$\begin{aligned}
\textstyle\sum_{i \in I} p_i \cdot \Theta_i &= \textstyle\sum_{i \in I} p_i \cdot \sum_{s \in \lceil\Delta_i\rceil} \Delta_i(s) \cdot f(s) \\
&= \textstyle\sum_{s \in \lceil\Delta\rceil} \sum_{i \in I} p_i \cdot \Delta_i(s) \cdot f(s) \\
&= \textstyle\sum_{s \in \lceil\Delta\rceil} \Delta(s) \cdot f(s) \\
&= \text{Exp}_\Delta(f) \\
&= \Theta.
\end{aligned}$$
□

The converse to the above is not true in general; from $\Delta\ \mathcal{R}^\dagger\ (\sum_{i \in I} p_i \cdot \Theta_i)$ it does not follow that Δ can correspondingly be decomposed. For example, we have $\overline{a.(b\,{}_{\frac{1}{2}}\!\oplus c)} \xrightarrow{a} \frac{1}{2} \cdot \overline{b} + \frac{1}{2} \cdot \overline{c}$, yet $\overline{a.(b\,{}_{\frac{1}{2}}\!\oplus c)}$ cannot be written as $\frac{1}{2} \cdot \Delta_1 + \frac{1}{2} \cdot \Delta_2$ such that $\Delta_1 \xrightarrow{a} b$ and $\Delta_2 \xrightarrow{a} c$.

In fact, a simplified form of Proposition 6.2 holds for unlifted relations, provided they are convex:

Corollary 6.1 *If* $(\sum_{i \in I} p_i \cdot \overline{s_i})\ \mathcal{R}^\dagger\ \Theta$ *and* $\mathcal{R}$ *is convex, then* $\Theta = \sum_{i \in I} p_i \cdot \Theta_i$ *for subdistributions* Θ_i *with* $s_i\mathcal{R}\Theta_i$ *for* $i \in I$.

Proof Take Δ_i to be $\overline{s_i}$ in Proposition 6.2, whence $\Theta = \sum_{i \in I} p_i \cdot \Theta_i$ for some subdistributions Θ_i such that $\overline{s_i}\ \mathcal{R}^\dagger\ \Theta_i$ for $i \in I$. Because $\mathcal{R}$ is convex, we then have $s_i\mathcal{R}\Theta_i$ from Remark 6.1. □

As we have seen in Proposition 3.4, the lifting operation is monotone, that is $\mathcal{R}_1 \subseteq \mathcal{R}_2$ implies $\mathcal{R}_1{}^\dagger \subseteq \mathcal{R}_2{}^\dagger$, and satisfies the following monadic property with respect to composition.

Lemma 6.3 *Let* $\mathcal{R}_1, \mathcal{R}_2 \subseteq S \times \mathcal{D}_{sub}(S)$. *Then the forward relational composition* $\mathcal{R}_1{}^\dagger \cdot \mathcal{R}_2{}^\dagger$ *is equal to the lifted composition* $(\mathcal{R}_1 \cdot \mathcal{R}_2)^\dagger$.

Proof Suppose $\Delta\ \mathcal{R}_1{}^{\dagger} \cdot \mathcal{R}_2{}^{\dagger}\ \Phi$. Then, there is some Θ such that $\Delta\ \mathcal{R}_1{}^{\dagger}\ \Theta\ \mathcal{R}_2{}^{\dagger}\ \Phi$. By Lemma 6.1 we have the decomposition $\Delta = \sum_{i\in I} p_i \cdot \overline{s_i}$ and $\Theta = \sum_{i\in I} p_i \cdot \Theta_i$ with $s_i \mathcal{R}_1 \Theta_i$ for each $i \in I$. By Proposition 6.2, we obtain $\Phi = \sum_{i\in I} p_i \cdot \Phi_i$ and for each $i \in I$ we have $\Theta_i\ \mathcal{R}_2{}^{\dagger}\ \Phi_i$. It follows that $s_i \mathcal{R}_1 \cdot \mathcal{R}_2{}^{\dagger}\ \Phi_i$, and thus $\Delta\ (\mathcal{R}_1 \cdot \mathcal{R}_2{}^{\dagger})^{\dagger}\ \Phi$. So we have shown that $\mathcal{R}_1{}^{\dagger} \cdot \mathcal{R}_2{}^{\dagger} \subseteq (\mathcal{R}_1 \cdot \mathcal{R}_2{}^{\dagger})^{\dagger}$. The other direction can be proved similarly. □

Notice that in the second part of Definition 6.2, we have required the index set I to be finite. In fact, this constraint is unnecessary if the state space S is finite. Then we say that the lifting operation has a property of *infinite linearity*. To show the property we need two technical lemmas.

Let us write $\updownarrow_\omega X$ for the set of subdistributions of the form $\sum_{i\geq 0} p_i \cdot \Delta_i$, where $\Delta_i \in X$ and $\sum_{i\geq 0} p_i = 1$.

Lemma 6.4 *If the set S is finite then $\updownarrow X = \updownarrow_\omega X$ for any subset X of $\mathcal{D}_{sub}(S)$.*

Proof It is clear that $\updownarrow X \subseteq \updownarrow_\omega X$, so we prove the inverse inclusion, $\updownarrow_\omega X \subseteq \updownarrow X$. The basic idea is to view a subdistribution over S as a point in Euclidean space of dimension $|S|$ and give a geometric proof, by induction on the size of S. More specifically we prove, by induction on k, that if X is a subset in a space of dimension k, then $\updownarrow X = \updownarrow_\omega X$. The base case, when $|S| = 1$ is trivial. Let us consider the inductive case, where the dimension is $(k+1)$.

Suppose there is a point $x \in \updownarrow_\omega X$ but $x \notin \updownarrow X$. Then, by the Separation theorem (cf. Theorem 2.7) there exists a hyperplane H that separates x from $\updownarrow X$. If h is the normal of H, we can assume without loss of generality that there is a constant c satisfying

$$h \cdot x \geq c \text{ and } h \cdot x' \leq c \text{ for all } x' \in X,$$

where with a slight abuse of notation we write $\cdot$ for dot product of two vectors of dimension $(k+1)$. Note that the separation here may not be strict because $\updownarrow X$ is convex but not necessarily Cauchy closed.

Since $x \in \updownarrow_\omega X$, there is a sequence of probabilities p_i with $\sum_{i\geq 0} p_i = 1$ and a sequence of points $x_i \in X$ such that $x = \sum_{i\geq 0} p_i \cdot x_i$. We then have

(i) $c \leq h \cdot x = \sum_{i\geq 0} p_i \cdot (h \cdot x_i)$ and
(ii) $h \cdot x_i \leq c$ for all $i \geq 0$.

It follows from (i) and (ii) that actually $h \cdot x_i = c$, for all $i \geq 0$. In other words, it must be the case that $h \cdot x_i = c$ for all i, which means that all the points x_i lie in H; in other words the separation of x from $\updownarrow X$ cannot be strict. Therefore, we have that $x \in \updownarrow_\omega (X \cap H)$ since $\updownarrow_\omega \{x_i \mid i \geq 0\} \subseteq \updownarrow_\omega (X \cap H)$.

On the other hand, since $x \notin \updownarrow X$ we have $x \notin \updownarrow (X \cap H)$. However, $X \cap H$ can be described as a subset in a space of one dimension lower than X, that is of dimension k. We have now contradicted the induction hypothesis. □

In order to use the above lemma, we need to rephrase the lifting operation in terms of the closure operator $\updownarrow(\_)$. To this end let us use $\mathcal{R}(s)$ to denote the set $\{\Delta \in \mathcal{D}_{sub}(S) \mid s\mathcal{R}\Delta\}$, for any $\mathcal{R} \subseteq S \times \mathcal{D}_{sub}(S)$.

Lemma 6.5 *For subdistributions over a finite set S, $\Delta\ \mathcal{R}^\dagger\ \Theta$ if and only if Θ can be written in the form $\sum_{s\in\lceil\Delta\rceil}\Delta(s)\cdot\Theta_s$ where each $\Theta_s \in \updownarrow\mathcal{R}(s)$.*

Proof Suppose $\Theta = \sum_{s\in\lceil\Delta\rceil}\Delta(s)\cdot\Theta_s$ with $\Theta_s \in \updownarrow\mathcal{R}(s)$. To show that $\Delta\ \mathcal{R}^\dagger\ \Theta$, it suffices to prove that $\overline{s}\ \mathcal{R}^\dagger\ \Theta_s$ for each $s \in \lceil\Delta\rceil$, as $\mathcal{R}^\dagger$ is linear. Since $\Theta_s \in \updownarrow\mathcal{R}(s)$, we can rewrite Θ_s as $\Theta_s = \sum_{i\in I} p_i \cdot \Theta_{i_s}$ where $\Theta_{i_s} \in \mathcal{R}(s)$ for some finite index set I. The fact that $\overline{s} = \sum_{i\in I} p_i \cdot \overline{s}$ and $s\mathcal{R}\Theta_{i_s}$ yields that $\overline{s}\ \mathcal{R}^\dagger\ \Theta_s$.

Conversely, suppose $\Delta\ \mathcal{R}^\dagger\ \Theta$. By Lemma 6.1 we have that

$$\Delta = \sum_{i\in I} p_i \cdot \overline{s_i} \qquad s_i\mathcal{R}\Theta_i \qquad \Theta = \sum_{i\in I} p_i \cdot \Theta_i. \tag{6.2}$$

For each $s \in \lceil\Delta\rceil$, let $I_s = \{i \in I \mid s_i = s\}$. Note that $\Delta(s) = \sum_{i\in I_s} p_i$. Hence, we can rewrite Θ as follows:

$$\begin{aligned}\Theta &= \textstyle\sum_{s\in\lceil\Delta\rceil}\sum_{i\in I_s} p_i \cdot \Theta_i\\ &= \textstyle\sum_{s\in\lceil\Delta\rceil}\Delta(s)\cdot(\sum_{i\in I_s}\frac{p_i}{\Delta(s)}\cdot\Theta_i)\end{aligned}$$

Since the subdistribution $\sum_{i\in I_s}\frac{p_i}{\Delta(s)}\cdot\Theta_i$ is a convex combination of $\{\Theta_i \mid i \in I_s\}$, it must be in $\updownarrow\mathcal{R}(s)$ due to (6.2), and the result follows. □

Theorem 6.4 (Infinite Linearity) *Suppose $\mathcal{R}$ is a relation over $S\times\mathcal{D}_{sub}(S)$, where S is finite and $\sum_{i\ge 0} p_i = 1$. Then $\Delta_i\ \mathcal{R}^\dagger\ \Theta_i$ implies $(\sum_{i\ge 0} p_i\cdot\Delta_i)\ \mathcal{R}^\dagger\ (\sum_{i\ge 0} p_i\cdot\Theta_i)$.*

Proof Suppose $\sum_{i\ge 0} p_i = 1$ and $\Delta_i\ \mathcal{R}^\dagger\ \Theta_i$ for each $i \ge 0$. Let Δ, Θ denote $\sum_{i\ge 0} p_i\cdot\Delta_i$ and $\sum_{i\ge 0} p_i\cdot\Theta_i$, respectively. We have to show $\Delta\ \mathcal{R}^\dagger\ \Theta$. By Lemma 6.5 it is sufficient to show

$$\Theta = \sum_{s\in\lceil\Delta\rceil}\Delta(s)\cdot\Gamma_s, \tag{6.3}$$

where $\Gamma_s \in \updownarrow\mathcal{R}(s)$ for each $s \in \lceil\Delta\rceil$.

By the same lemma we know that for each $i \ge 0$, since $\Delta_i\ \mathcal{R}^\dagger\ \Theta_i$,

$$\Theta_i = \sum_{s\in\lceil\Delta_i\rceil}\Delta_i(s)\cdot\Theta_{i_s} \text{ with } \Theta_{i_s} \in \updownarrow\mathcal{R}(s). \tag{6.4}$$

Therefore,

$$\begin{aligned}\Theta &= \textstyle\sum_{i\ge 0} p_i\cdot(\sum_{s\in\lceil\Delta_i\rceil}\Delta_i(s)\cdot\Theta_{i_s})\\ &= \textstyle\sum_{s\in\lceil\Delta\rceil}\sum_{i\ge 0}(p_i\cdot\Delta_i(s))\cdot\Theta_{i_s}\end{aligned}$$

Let w_i^s denote $p_i\cdot\Delta_i(s)$ and note that $\Delta(s)$ is the infinite sum $\sum_{i\ge 0} w_i^s$. Therefore, we can continue:

$$\begin{aligned}\Theta &= \textstyle\sum_{s\in\lceil\Delta\rceil}\sum_{i\ge 0} w_i^s\cdot\Theta_{i_s}\\ &= \textstyle\sum_{s\in\lceil\Delta\rceil}\Delta(s)\cdot(\sum_{i\ge 0}\frac{w_i^s}{\Delta(s)}\cdot\Theta_{i_s}).\end{aligned}$$

The required (6.3) above will follow if we can show $(\sum_{i\geq 0}\frac{w_i^s}{\Delta(s)}\cdot\Theta_{i_s})\in\updownarrow\mathcal{R}(s)$ for each $s\in\lceil\Delta\rceil$.

From (6.4) we know $\Theta_{i_s}\in\updownarrow\mathcal{R}(s)$, and therefore, by construction we have that $(\sum_{i\geq 0}\frac{w_i^s}{\Delta(s)}\cdot\Theta_{i_s})\in\updownarrow_\omega\updownarrow\mathcal{R}(s)$. But now an application of Lemma 6.4 yields $\updownarrow_\omega\updownarrow\mathcal{R}(s)=\updownarrow\updownarrow\mathcal{R}(s)$, and this coincides with $\updownarrow\mathcal{R}(s)$ because $\updownarrow(\_)$ is a closure operator. □

Consequently, for finite-state pLTSs we can freely use the infinite linearity property of the lifting operation.

6.3.2 *Weak Transitions*

We now formally define a notion of weak derivatives.

Definition 6.4 (Weak τ Moves to Derivatives) Suppose we have subdistributions $\Delta, \Delta_k^{\rightarrow}, \Delta_k^{\times}$, for $k\geq 0$, with the following properties:

$$\begin{aligned}\Delta &= \Delta_0^{\rightarrow}+\Delta_0^{\times}\\ \Delta_0^{\rightarrow} &\xrightarrow{\tau}\Delta_1^{\rightarrow}+\Delta_1^{\times}\\ &\vdots\\ \Delta_k^{\rightarrow} &\xrightarrow{\tau}\Delta_{k+1}^{\rightarrow}+\Delta_{k+1}^{\times}\,.\\ &\vdots\end{aligned}$$

The $\xrightarrow{\tau}$ moves above with subdistribution sources are lifted in the sense of the previous section. Then we call $\Delta':=\sum_{k=0}^{\infty}\Delta_k^{\times}$ a *weak derivative* of Δ, and write $\Delta\Longrightarrow\Delta'$ to mean that Δ can make a *weak τ move* to its derivative Δ'.

There is always at least one derivative of any distribution (the distribution itself) and there can be many. Using Lemma 6.2 it is easily checked that Definition 6.4 is well-defined in that derivatives do not sum to more than one.

Example 6.2 Let $\Delta\xrightarrow{\tau}{}^{\star}$ denote the reflexive transitive closure of the relation $\xrightarrow{\tau}$ over subdistributions. By the judicious use of the empty distribution ε in the definition of $\Longrightarrow$ and property (6.1) above, it is easy to see that

$$\Delta\xrightarrow{\tau}{}^{\star}\Theta\quad\text{implies}\quad\Delta\Longrightarrow\Theta$$

because $\Delta\xrightarrow{\tau}{}^{\star}\Theta$ means the existence of a finite sequence of subdistributions such that $\Delta=\Delta_0,\ \Delta_1,\ldots,\Delta_k=\Theta,\ k\geq 0$ for which we can write

$$
\begin{array}{lll}
\Delta & = & \Delta_0 + \varepsilon \\
\Delta_0 & \xrightarrow{\tau} & \Delta_1 + \varepsilon \\
\vdots & & \vdots \\
\Delta_{k-1} & \xrightarrow{\tau} & \varepsilon + \Delta_k \\
\varepsilon & \xrightarrow{\tau} & \varepsilon + \varepsilon \\
 & & \vdots \\
\hline
\text{In total:} & & \Theta
\end{array}
$$

This implies that $\Longrightarrow$ is indeed a generalisation of the standard notion for non-probabilistic transition systems of performing an indefinite sequence of internal τ moves.

In Sect. 5.5, we wrote $s \xrightarrow{\hat{\tau}} \Delta$, if either $s \xrightarrow{\tau} \Delta$ or $\Delta = \overline{s}$. Hence, the lifted relation $\xrightarrow{\hat{\tau}}$ satisfies $\Delta \xrightarrow{\hat{\tau}} \Delta'$ if and only if there are $\Delta^{\rightarrow}$, $\Delta^{\times}$ and Δ_1 with $\Delta = \Delta^{\rightarrow} + \Delta^{\times}$, $\Delta^{\rightarrow} \xrightarrow{\tau} \Delta_1$ and $\Delta' = \Delta_1 + \Delta^{\times}$. Clearly, $\Delta \xrightarrow{\hat{\tau}} \Delta'$ implies $\Delta \Longrightarrow \Delta'$. With a little effort, one can also show that $\Delta \xrightarrow{\hat{\tau}}{}^{\star} \Delta'$ implies $\Delta \Longrightarrow \Delta'$. In fact, this follows directly from the reflexivity and transitivity of $\Longrightarrow$; the latter will be established in Theorem 6.6.

Conversely, in Sect. 5.5.1 we dealt with recursion-free rpCSP processes P, and these have the property that in a sequence as in Definition 6.4 with $\Delta = [\![P]\!]$ we necessarily have that $\Delta_k = \varepsilon$ for some $k \geq 0$. On such processes we have that the relations $\xrightarrow{\hat{\tau}}{}^{\star}$ and $\Longrightarrow$ coincide.

In Definition 6.4, we can see that $\Delta' = \varepsilon$ iff $\Delta_k^{\times} = \varepsilon$ for all k. Thus $\Delta \Longrightarrow \varepsilon$ iff there is an infinite sequence of subdistributions Δ_k such that $\Delta = \Delta_0$ and $\Delta_k \xrightarrow{\tau} \Delta_{k+1}$, that is Δ can give rise to a divergent computation.

Example 6.3 Consider the process $\mathsf{rec}\, x.x$, which recall is a state, and for which we have $\mathsf{rec}\, x.x \xrightarrow{\tau} [\![\mathsf{rec}\, x.x]\!]$ and thus $[\![\mathsf{rec}\, x.x]\!] \xrightarrow{\tau} [\![\mathsf{rec}\, x.x]\!]$. Then $[\![\mathsf{rec}\, x.x]\!] \Longrightarrow \varepsilon$.

Example 6.4 Recall the process $Q_1 = \mathsf{rec}\, x.(\tau.x\,{}_{\frac{1}{2}}\!\oplus a.\mathbf{0})$ from Sect. 6.1. We have $[\![Q_1]\!] \Longrightarrow [\![a.\mathbf{0}]\!]$ because

$$
\begin{aligned}
[\![Q_1]\!] &= [\![Q_1]\!] + \varepsilon \\
[\![Q_1]\!] &\xrightarrow{\tau} \frac{1}{2} \cdot [\![\tau.Q_1]\!] + \frac{1}{2} \cdot [\![a.\mathbf{0}]\!] \\
\frac{1}{2} \cdot [\![\tau.Q_1]\!] &\xrightarrow{\tau} \frac{1}{2} \cdot [\![Q_1]\!] + \varepsilon \\
\frac{1}{2} \cdot [\![Q_1]\!] &\xrightarrow{\tau} \frac{1}{2^2} \cdot [\![\tau.Q_1]\!] + \frac{1}{2^2} \cdot [\![a.\mathbf{0}]\!] \\
&\ldots
\end{aligned}
$$

$$\frac{1}{2^k} \cdot [\![Q_1]\!] \xrightarrow{\tau} \frac{1}{2^{k+1}} \cdot [\![\tau.Q_1]\!] + \frac{1}{2^{k+1}} \cdot [\![a.\mathbf{0}]\!]$$

$$\cdots$$

which means that by definition we have

$$[\![Q_1]\!] \Longrightarrow \varepsilon + \sum_{k\geq 1} \frac{1}{2^k} \cdot [\![a.\mathbf{0}]\!]$$

thus generating the weak derivative $[\![a.\mathbf{0}]\!]$ as claimed.

Example 6.5 Consider the (infinite) collection of states s_k and probabilities p_k for $k \geq 2$ such that

$$s_k \xrightarrow{\tau} [\![a.\mathbf{0}]\!]_{p_k} \oplus \overline{s_{k+1}} \,,$$

where we choose p_k so that starting from any s_k the probability of eventually taking a left-hand branch, and so reaching $[\![a.\mathbf{0}]\!]$ ultimately, is just $\frac{1}{k}$ in total. Thus p_k must satisfy $\frac{1}{k} = p_k + (1-p_k)\frac{1}{k+1}$, whence by arithmetic we have that $p_k := \frac{1}{k^2}$ will do. Therefore, in particular $s_2 \Longrightarrow \frac{1}{2}[\![a.\mathbf{0}]\!]$, with the remaining $\frac{1}{2}$ lost in divergence.

Our final example demonstrates that derivatives of (interpretations of) rpCSP processes may have infinite support, and hence that we can have $[\![P]\!] \Longrightarrow \Delta'$ such that there is no $P' \in$ rpCSP with $[\![P']\!] = \Delta'$.

Example 6.6 Let P denote the process $\mathsf{rec}\, x.(b.\mathbf{0}\,{}_{\frac{1}{2}}\!\oplus (x\,|_{\emptyset}\mathbf{0}))$. Then we have the derivation:

$$[\![P]\!] = [\![P]\!] + \varepsilon$$

$$[\![P]\!] \xrightarrow{\tau} \frac{1}{2} \cdot [\![P|_{\emptyset}\mathbf{0}^1]\!] + \frac{1}{2} \cdot [\![b.\mathbf{0}]\!]$$

$$\frac{1}{2} \cdot [\![P|_{\emptyset}\mathbf{0}^1]\!] \xrightarrow{\tau} \frac{1}{2^2} \cdot [\![P|_{\emptyset}\mathbf{0}^2]\!] + \frac{1}{2^2} \cdot [\![b.\mathbf{0}|_{\emptyset}\mathbf{0}^1]\!]$$

$$\cdots$$

$$\frac{1}{2^k} \cdot [\![P|_{\emptyset}\mathbf{0}^k]\!] \xrightarrow{\tau} \frac{1}{2^{k+1}} \cdot [\![P|_{\emptyset}\mathbf{0}^{k+1}]\!] + \frac{1}{2^{k+1}} \cdot [\![b.\mathbf{0}|_{\emptyset}\mathbf{0}^k]\!]$$

$$\cdots$$

where $\mathbf{0}^k$ represents k instances of $\mathbf{0}$ running in parallel. This implies that

$$[\![P]\!] \Longrightarrow \Theta,$$

where

$$\Theta = \sum_{k\geq 0} \frac{1}{2^{k+1}} \cdot [\![b|_{\emptyset}\mathbf{0}^k]\!],$$

a distribution with infinite support.

6.3.3 Properties of Weak Transitions

Here, we develop some properties of the weak move relation $\Longrightarrow$ that will be important later on. We wish to use weak derivation as much as possible in the same way as the lifted action relations $\xrightarrow{\alpha}$, and therefore, we start with showing that $\Longrightarrow$ enjoys two of the most crucial properties of $\xrightarrow{\alpha}$: linearity of Definition 6.2 and the decomposition property of Proposition 6.2. To this end, we first establish that weak derivations do not increase the mass of distributions and are preserved under scaling.

Lemma 6.6 *For any subdistributions Δ, Θ, Γ, Λ, Π we have*

(i) If $\Delta \Longrightarrow \Theta$ then $|\Delta| \geq |\Theta|$.
(ii) If $\Delta \Longrightarrow \Theta$ and $p \in \mathbb{R}_{\geq 0}$ such that $|p \cdot \Delta| \leq 1$, then $p \cdot \Delta \Longrightarrow p \cdot \Theta$.
(iii) If $\Gamma + \Lambda \Longrightarrow \Pi$ then $\Pi = \Pi^\Gamma + \Pi^\Lambda$ with $\Gamma \Longrightarrow \Pi^\Gamma$ and $\Lambda \Longrightarrow \Pi^\Lambda$.

Proof By definition $\Delta \Longrightarrow \Theta$ means that some $\Delta_k, \Delta_k^\times, \Delta_k^\rightarrow$ exist for all $k \geq 0$ such that

$$\Delta = \Delta_0, \qquad \Delta_k = \Delta_k^\times + \Delta_k^\rightarrow, \qquad \Delta_k^\rightarrow \xrightarrow{\tau} \Delta_{k+1}, \qquad \Theta = \sum_{k=0}^{\infty} \Delta_k^\times.$$

A simple inductive proof shows that

$$|\Delta| = |\Delta_i^\rightarrow| + \sum_{k\leq i} |\Delta_k^\times| \text{ for any } i \geq 0. \tag{6.5}$$

The sequence $\{\sum_{k\leq i} |\Delta_k|\}_{i=0}^{\infty}$ is nondecreasing and by (6.5) each element of the sequence is not greater than $|\Delta|$. Therefore, the limit of this sequence is bounded by $|\Delta|$. That is,

$$|\Delta| \geq \lim_{i\to\infty} \sum_{k\leq i} |\Delta_k^\times| = |\Theta|.$$

Now suppose $p \in \mathbb{R}_{\geq 0}$ such that $|p \cdot \Delta| \leq 1$. From Remark 6.2(i) it follows that

$$p\cdot\Delta = p\cdot\Delta_0, \quad p\cdot\Delta_k = p\cdot\Delta_k^\rightarrow + p\cdot\Delta_k^\times, \quad p\cdot\Delta_k^\rightarrow \xrightarrow{\tau} p\cdot\Delta_{k+1}, \quad p\cdot\Theta = \sum_k p\cdot\Delta_k^\times.$$

Hence, Definition 6.4 yields $p \cdot \Delta \Longrightarrow p \cdot \Theta$.

Next suppose $\Gamma + \Lambda \Longrightarrow \Pi$. By Definition 6.4 there are subdistributions Π_k, $\Pi_k^\rightarrow$, $\Pi_k^\times$ for $k \in \mathbb{N}$ such that

$$\Gamma + \Lambda = \Pi_0, \qquad \Pi_k = \Pi_k^\rightarrow + \Pi_k^\times, \qquad \Pi_k^\rightarrow \xrightarrow{\tau} \Pi_{k+1}, \qquad \Pi = \sum_k \Pi_k^\times.$$

For any $s \in S$, define

$$\begin{aligned} \Gamma_0^\rightarrow(s) &:= min(\Gamma(s), \Pi_0^\rightarrow(s)) \\ \Gamma_0^\times(s) &:= \Gamma(s) - \Gamma_0^\rightarrow(s) \\ \Lambda_0^\times(s) &:= min(\Lambda(s), \Pi_0^\times(s)) \\ \Lambda_0^\rightarrow(s) &:= \Lambda(s) - \Lambda_0^\times(s), \end{aligned} \tag{6.6}$$

and check that $\Gamma_0^{\rightarrow} + \Gamma_0^{\times} = \Gamma$ and $\Lambda_0^{\rightarrow} + \Lambda_0^{\times} = \Lambda$. To show that $\Lambda_0^{\rightarrow} + \Gamma_0^{\rightarrow} = \Pi_0^{\rightarrow}$ and $\Lambda_0^{\times} + \Gamma_0^{\times} = \Pi_0^{\times}$ we fix a state s and distinguish two cases: either (a) $\Pi_0^{\rightarrow}(s) \geq \Gamma(s)$ or (b) $\Pi_0^{\rightarrow}(s) < \Gamma(s)$. In Case (a) we have $\Pi_0^{\times}(s) \leq \Lambda(s)$ and the definitions (6.6) simplify to $\Gamma_0^{\rightarrow}(s) = \Gamma(s)$, $\Gamma_0^{\times}(s) = 0$, $\Lambda_0^{\times}(s) = \Pi_0^{\times}(s)$ and $\Lambda_0^{\rightarrow}(s) = \Lambda(s) - \Pi_0^{\times}(s)$, whence immediately $\Gamma_0^{\rightarrow}(s) + \Lambda_0^{\rightarrow}(s) = \Pi_0^{\rightarrow}(s)$ and $\Gamma_0^{\times}(s) + \Lambda_0^{\times}(s) = \Pi_0^{\times}(s)$. Case (b) is similar.

Since $\Lambda_0^{\rightarrow} + \Gamma_0^{\rightarrow} \xrightarrow{\tau} \Pi_1$, by Proposition 6.2 we find Γ_1, Λ_1 with $\Gamma_0^{\rightarrow} \xrightarrow{\tau} \Gamma_1$ and $\Lambda_0^{\rightarrow} \xrightarrow{\tau} \Lambda_1$ and $\Pi_1 = \Gamma_1 + \Lambda_1$. Being now in the same position with Π_1 as we were with Π_0, we can continue this procedure to find Λ_k, Γ_k, $\Lambda_k^{\rightarrow}$, $\Gamma_k^{\rightarrow}$, $\Lambda_k^{\times}$ and $\Gamma_k^{\times}$ with

$$\Gamma = \Gamma_0, \qquad \Gamma_k = \Gamma_k^{\rightarrow} + \Gamma_k^{\times}, \qquad \Gamma_k^{\rightarrow} \xrightarrow{\tau} \Gamma_{k+1},$$

$$\Lambda = \Lambda_0, \qquad \Lambda_k = \Lambda_k^{\rightarrow} + \Lambda_k^{\times}, \qquad \Lambda_k^{\rightarrow} \xrightarrow{\tau} \Lambda_{k+1},$$

$$\Gamma_k + \Lambda_k = \Pi_k, \quad \Gamma_k^{\rightarrow} + \Lambda_k^{\rightarrow} = \Pi_k^{\rightarrow}, \quad \Gamma_k^{\times} + \Lambda_k^{\times} = \Pi_k^{\times}.$$

Let $\Pi^{\Gamma} := \sum_k \Gamma_k^{\times}$ and $\Pi^{\Lambda} := \sum_k \Lambda_k^{\times}$. Then $\Pi = \Pi^{\Gamma} + \Pi^{\Lambda}$ and Definition 6.4 yields $\Gamma \Longrightarrow \Pi^{\Gamma}$ and $\Lambda \Longrightarrow \Pi^{\Lambda}$. □

Together, Lemma 6.6(ii) and (iii) imply the binary counterpart of the decomposition property of Proposition 6.2. We now generalise this result to infinite (but still countable) decomposition and also establish linearity.

Theorem 6.5 *Let $p_i \in [0, 1]$ for $i \in I$ with $\sum_{i\in I} p_i \leq 1$. Then*

(i) If $\Delta_i \Longrightarrow \Theta_i$ for all $i \in I$ then $\sum_{i\in I} p_i \cdot \Delta_i \Longrightarrow \sum_{i\in I} p_i \cdot \Theta_i$.
(ii) If $\sum_{i\in I} p_i \cdot \Delta_i \Longrightarrow \Theta$ then $\Theta = \sum_{i\in I} p_i \cdot \Theta_i$ for some subdistributions Θ_i such that $\Delta_i \Longrightarrow \Theta_i$ for all $i \in I$.

Proof (i) Suppose $\Delta_i \Longrightarrow \Theta_i$ for all $i \in I$. By Definition 6.4 there are subdistributions $\Delta_{ik}, \Delta_{ik}^{\rightarrow}, \Delta_{ik}^{\times}$ such that

$$\Delta_i = \Delta_{i0}, \qquad \Delta_{ik} = \Delta_{ik}^{\rightarrow} + \Delta_{ik}^{\times}, \qquad \Delta_{ik}^{\rightarrow} \xrightarrow{\tau} \Delta_{i(k+1)}, \qquad \Theta_i = \sum_k \Delta_{ik}^{\times}.$$

We compose relevant subdistributions and obtain that $\sum_{i\in I} p_i \cdot \Delta_i = \sum_{i\in I} p_i \cdot \Delta_{i0}$, $\sum_{i\in I} p_i \cdot \Delta_{ik} = \sum_{i\in I} p_i \cdot \Delta_{ik}^{\rightarrow} + \sum_{i\in I} p_i \cdot \Delta_{ik}^{\times}$, $\sum_{i\in I} p_i \cdot \Delta_{ik}^{\rightarrow} \xrightarrow{\tau} \sum_{i\in I} p_i \cdot \Delta_{i(k+1)}$ by Theorem 6.4, and moreover $\sum_{i\in I} p_i \cdot \Theta_i = \sum_{i\in I} p_i \cdot \sum_k \Delta_{ik}^{\times} = \sum_k (\sum_{i\in I} p_i \cdot \Delta_{ik}^{\times})$. By Definition 6.4 we obtain $\sum_{i\in I} p_i \cdot \Delta_i \Longrightarrow \sum_{i\in I} p_i \cdot \Theta_i$.

(ii) In the light of Lemma 6.6(ii) it suffices to show that if $\sum_{i=0}^{\infty} \Delta_i \Longrightarrow \Theta$ then $\Theta = \sum_{i=0}^{\infty} \Theta_i$ for subdistributions Θ_i such that $\Delta_i \Longrightarrow \Theta_i$ for all $i \geq 0$.
Since $\sum_{i=0}^{\infty} \Delta_i = \Delta_0 + \sum_{i\geq 1} \Delta_i$ and $\sum_{i=0}^{\infty} \Delta_i \Longrightarrow \Theta$, by Lemma 6.6(iii) there are $\Theta_0, \Theta_1^{\geq}$ such that

$$\Delta_0 \Longrightarrow \Theta_0, \qquad \sum_{i\geq 1} \Delta_i \Longrightarrow \Theta_1^{\geq}, \qquad \Theta = \Theta_0 + \Theta_1^{\geq}.$$

Using Lemma 6.6(iii) once more, we have $\Theta_1, \Theta_2^{\geq}$ such that

$$\Delta_1 \Longrightarrow \Theta_1, \quad \sum_{i\geq 2} \Delta_i \Longrightarrow \Theta_2^{\geq}, \quad \Theta_1^{\geq} = \Theta_1 + \Theta_2^{\geq},$$

thus in combination $\Theta = \Theta_0 + \Theta_1 + \Theta_2^{\geq}$. Continuing this process we have that

$$\Delta_i \Longrightarrow \Theta_i, \quad \sum_{j\geq i+1} \Delta_j^{\times} \Longrightarrow \Theta_{i+1}^{\geq}, \quad \Theta = \sum_{j=0}^{i} \Theta_j + \Theta_{i+1}^{\geq}$$

for all $i \geq 0$. Lemma 6.6(i) ensures that $|\sum_{j\geq i+1} \Delta_j| \geq |\Theta_{i+1}^{\geq}|$ for all $i \geq 0$. But since $\sum_{i=0}^{\infty} \Delta_i$ is a subdistribution, we know that the tail sum $\sum_{j\geq i+1} \Delta_j$ converges to ε when i approaches ∞, and therefore that $\lim_{i\to\infty} \Theta_i^{\geq} = \varepsilon$. Thus by taking that limit we conclude that $\Theta = \sum_{i=0}^{\infty} \Theta_i$.

□

With Theorem 6.5, the relation $\Longrightarrow \subseteq \mathcal{D}_{sub}(S) \times \mathcal{D}_{sub}(S)$ can be obtained as the lifting of a relation $\Longrightarrow_S$ from S to $\mathcal{D}_{sub}(S)$, which is defined by writing $s \Longrightarrow_S \Theta$ just when $\overline{s} \Longrightarrow \Theta$.

Proposition 6.3 $(\Longrightarrow_S)^{\dagger} = (\Longrightarrow)$.

Proof That $\Delta\ (\Longrightarrow_S)^{\dagger}\ \Theta$ implies $\Delta \Longrightarrow \Theta$ is a simple application of Part (i) of Theorem 6.5. For the other direction, suppose $\Delta \Longrightarrow \Theta$. Given that $\Delta = \sum_{s\in\lceil\Delta\rceil} \Delta(s)\cdot\overline{s}$, Part (ii) of the same theorem enables us to decompose Θ into $\sum_{s\in\lceil\Delta\rceil} \Delta(s)\cdot\Theta_s$, where $\overline{s} \Longrightarrow \Theta_s$ for each s in $\lceil\Delta\rceil$. But the latter actually means that $s \Longrightarrow_S \Theta_s$, and so by definition this implies $\Delta\ (\Longrightarrow_S)^{\dagger}\ \Theta$. □

It is immediate that the relation $\Longrightarrow$ is convex because of its being a lifting.

We proceed with the important properties of reflexivity and transitivity of weak derivations. First note that reflexivity is straightforward; in Definition 6.4 it suffices to take $\Delta_0^{\rightarrow}$ to be the empty subdistribution ε.

Theorem 6.6 (Transitivity of $\Longrightarrow$) *If $\Delta \Longrightarrow \Theta$ and $\Theta \Longrightarrow \Lambda$ then $\Delta \Longrightarrow \Lambda$.*

Proof By definition $\Delta \Longrightarrow \Theta$ means that some $\Delta_k, \Delta_k^{\times}, \Delta_k^{\rightarrow}$ exist for all $k \geq 0$ such that

$$\Delta = \Delta_0, \quad \Delta_k = \Delta_k^{\times} + \Delta_k^{\rightarrow}, \quad \Delta_k^{\rightarrow} \xrightarrow{\tau} \Delta_{k+1}, \quad \Theta = \sum_{k=0}^{\infty} \Delta_k^{\times}. \tag{6.7}$$

Since $\Theta = \sum_{k=0}^{\infty} \Delta_k^{\times}$ and $\Theta \Longrightarrow \Lambda$, by Theorem 6.5(ii) there are Λ_k for $k \geq 0$ such that $\Lambda = \sum_{k=0}^{\infty} \Lambda_k$ and $\Delta_k^{\times} \Longrightarrow \Lambda_k$ for all $k \geq 0$.

Now for each $k \geq 0$, we know that $\Delta_k^{\times} \Longrightarrow \Lambda_k$ gives us some $\Delta_{kl}, \Delta_{kl}^{\times}, \Delta_{kl}^{\rightarrow}$ for $l \geq 0$ such that

$$\Delta_k^{\times} = \Delta_{k0}, \quad \Delta_{kl} = \Delta_{kl}^{\times} + \Delta_{kl}^{\rightarrow}, \quad \Delta_{kl}^{\rightarrow} \xrightarrow{\tau} \Delta_{k,l+1} \quad \Lambda_k = \sum_{l\geq 0} \Delta_{kl}^{\times}. \tag{6.8}$$

Therefore, we can put all this together with

$$\Lambda = \sum_{k=0}^{\infty} \Lambda_k = \sum_{k,l\geq 0} \Delta_{kl}^{\times} = \sum_{i\geq 0} \left(\sum_{k,l|k+l=i} \Delta_{kl}^{\times} \right), \tag{6.9}$$

where the last step is a straightforward diagonalisation.

Now from the decompositions above we recompose an alternative trajectory of Δ_i''s to take Δ via $\Longrightarrow$ to Λ directly. Define

$$\Delta_i' = \Delta_i'^{\times} + \Delta_i'^{\rightarrow}, \qquad \Delta_i'^{\times} = \sum_{k,l|k+l=i} \Delta_{kl}^{\times}, \qquad \Delta_i'^{\rightarrow} = \left(\sum_{k,l|k+l=i} \Delta_{kl}^{\rightarrow} \right) + \Delta_i^{\rightarrow}, \tag{6.10}$$

so that from (6.9) we have immediately that

$$\Lambda \;=\; \sum_{i\geq 0} \Delta_i'^{\times}. \tag{6.11}$$

We now show that

(i) $\Delta = \Delta_0'$,

(ii) $\Delta_i'^{\rightarrow} \xrightarrow{\tau} \Delta_{i+1}'$,

from which, with (6.11), we will have $\Delta \Longrightarrow \Lambda$ as required. For (i) we observe that

$$\begin{array}{lll}
 & \Delta & \\
= & \Delta_0 & (6.7)\\
= & \Delta_0^{\times} + \Delta_0^{\rightarrow} & (6.7)\\
= & \Delta_{00} + \Delta_0^{\rightarrow} & (6.8)\\
= & \Delta_{00}^{\times} + \Delta_{00}^{\rightarrow} + \Delta_0^{\rightarrow} & (6.8)\\
= & (\sum_{k,l|k+l=0} \Delta_{kl}^{\times}) + (\sum_{k,l|k+l=0} \Delta_{kl}^{\rightarrow}) + \Delta_0^{\rightarrow} & \text{index arithmetic}\\
= & \Delta_0'^{\times} + \Delta_0'^{\rightarrow} & (6.10)\\
= & \Delta_0'\,. & (6.10)
\end{array}$$

For (ii) we observe that

$$\begin{array}{lll}
 & \Delta_i'^{\rightarrow} & \\
= & (\sum_{k,l|k+l=i} \Delta_{kl}^{\rightarrow}) + \Delta_i^{\rightarrow} & (6.10)\\
\xrightarrow{\tau} & (\sum_{k,l|k+l=i} \Delta_{k,l+1}) + \Delta_{i+1} & (6.7), (6.8), \text{additivity}\\
= & (\sum_{k,l|k+l=i} (\Delta_{k,l+1}^{\times} + \Delta_{k,l+1}^{\rightarrow})) + \Delta_{i+1}^{\times} + \Delta_{i+1}^{\rightarrow} & (6.7), (6.8)\\
= & (\sum_{k,l|k+l=i} \Delta_{k,l+1}^{\times}) + \Delta_{i+1}^{\times} + (\sum_{k,l|k+l=i} \Delta_{k,l+1}^{\rightarrow}) + \Delta_{i+1}^{\rightarrow} & \text{rearrange}\\
= & (\sum_{k,l|k+l=i} \Delta_{k,l+1}^{\times}) + \Delta_{i+1,0} + (\sum_{k,l|k+l=i} \Delta_{k,l+1}^{\rightarrow}) + \Delta_{i+1}^{\rightarrow} & (6.8)\\
= & (\sum_{k,l|k+l=i} \Delta_{k,l+1}^{\times}) + \Delta_{i+1,0}^{\times} + \Delta_{i+1,0}^{\rightarrow} + (\sum_{k,l|k+l=i} \Delta_{k,l+1}^{\rightarrow}) + \Delta_{i+1}^{\rightarrow} & (6.8)\\
= & (\sum_{k,l|k+l=i+1} \Delta_{kl}^{\times}) + (\sum_{k,l|k+l=i+1} \Delta_{kl}^{\rightarrow}) + \Delta_{i+1}^{\rightarrow} & \text{index arithmetic}\\
= & \Delta_{i+1}'^{\times} + \Delta_{i+1}'^{\rightarrow} & (6.10)\\
= & \Delta_{i+1}'\,, & (6.10)
\end{array}$$

which concludes the proof. □

Finally, we need a property that is the converse of transitivity. If one executes a given weak derivation partly, by stopping more often and moving on less often, one makes another weak transition that can be regarded as an initial segment of the given one. We need the property that after executing such an initial segment, it is still possible to complete the given derivation.

Definition 6.5 A weak derivation $\Phi \Longrightarrow \Gamma$ is called an *initial segment* of a weak derivation $\Phi \Longrightarrow \Psi$ if for $k \geq 0$ there are $\Gamma_k, \Gamma_k^{\rightarrow}, \Gamma_k^{\times}, \Psi_k, \Psi_k^{\rightarrow}, \Psi_k^{\times} \in \mathcal{D}_{sub}(S)$ such that $\Gamma_0 = \Psi_0 = \Phi$ and

$$\begin{array}{lll} \Gamma_k = \Gamma_k^{\rightarrow} + \Gamma_k^{\times} & \Psi_k = \Psi_k^{\rightarrow} + \Psi_k^{\times} & \Gamma_k^{\rightarrow} \leq \Psi_k^{\rightarrow} \\ \Gamma_k^{\rightarrow} \xrightarrow{\tau} \Gamma_{k+1} & \Psi_k^{\rightarrow} \xrightarrow{\tau} \Psi_{k+1} & \Gamma_k \leq \Psi_k \\ \Gamma = \sum_{i=0}^{\infty} \Gamma_k^{\times} & \Psi = \sum_{i=0}^{\infty} \Psi_k^{\times} & (\Psi_k^{\rightarrow} - \Gamma_k^{\rightarrow}) \xrightarrow{\tau} (\Psi_{k+1} - \Gamma_{k+1}). \end{array}$$

Intuitively, in the derivation $\Phi \Longrightarrow \Psi$, for each $k \geq 0$, we only allow a portion of $\Psi_k^{\rightarrow}$ to make τ moves, and the rest remains unmoved even if it can enable τ moves, so as to obtain an initial segment $\Phi \Longrightarrow \Gamma$. Accordingly, each $\Gamma_k^{\times}$ includes the corresponding unmoved part of $\Psi_k^{\rightarrow}$, which is eventually collected in Γ. Now from Γ if we let those previously unmoved parts perform exactly the same τ moves as in $\Phi \Longrightarrow \Psi$, we will end up with a derivation leading to Ψ. This is formulated in the following proposition.

Proposition 6.4 *If $\Phi \Longrightarrow \Gamma$ is an initial segment of $\Phi \Longrightarrow \Psi$, then $\Gamma \Longrightarrow \Psi$.*

Proof For any subdistributions $\Delta, \Theta \in \mathcal{D}_{sub}(S)$, we define two new subdistributions $\Delta \cap \Theta \in \mathcal{D}_{sub}(S)$ by letting $(\Delta \cap \Theta)(s) := min(\Delta(s), \Theta(s))$ and $\Delta - \Theta \in \mathcal{D}_{sub}(S)$ by $(\Delta - \Theta)(s) := max(\Delta(s) - \Theta(s), 0)$. So we have $\Delta - \Theta = \Delta - (\Delta \cap \Theta)$. Observe that in case $\Theta \leq \Delta$, and only then, we have that $(\Delta - \Theta) + \Theta = \Delta$.

Let $\Gamma_k, \Gamma_k^{\rightarrow}, \Gamma_k^{\times}, \Psi_k, \Psi_k^{\rightarrow}, \Psi_k^{\times} \in \mathcal{D}_{sub}(S)$ be as in Definition 6.5. By induction on $k \leq 0$ we define Δ_{ki}, $\Delta_{ki}^{\rightarrow}$ and $\Delta_{ki}^{\times}$, for $0 \leq i \leq k$, such that

(i) $\Delta_{k0} = \Gamma_k^{\times}$
(ii) $\Psi_k = \sum_{i=0}^{k} \Delta_{ki} + \Gamma_k^{\rightarrow}$
(iii) $\Psi_k^{\times} = \sum_{i=0}^{k} \Delta_{ki}^{\times}$
(iv) $\Delta_{ki} = \Delta_{ki}^{\rightarrow} + \Delta_{ki}^{\times}$
(v) $\Delta_{ki}^{\rightarrow} \xrightarrow{\tau} \Delta_{(k+1)(i+1)}$

Induction Base: Let $\Delta_{00} := \Gamma_0^{\times} = \Gamma_0 - \Gamma_0^{\rightarrow} = \Psi_0 - \Gamma_0^{\rightarrow}$. This way the first two equations are satisfied for $k = 0$. All other statements will be dealt with fully by the induction step.

Induction Step: Suppose Δ_{ki} for $0 \leq i \leq k$ are already known, and moreover we have $\Psi_k = \sum_{i=0}^{k} \Delta_{ki} + \Gamma_k^{\rightarrow}$. With induction on i we define $\Delta_{ki}^{\times} := \Delta_{ki} \cap (\Psi_k^{\times} - \sum_{j=0}^{i-1} \Delta_{kj}^{\times})$ and establish $\sum_{j=0}^{i} \Delta_{kj}^{\times} \leq \Psi_k^{\times}$. Namely, writing Θ_{ki} for $\sum_{j=0}^{i-1} \Delta_{kj}^{\times}$, surely $\Theta_{k0} = \varepsilon \leq \Psi_k^{\times}$, and when assuming that $\Theta_{ki} \leq \Psi_k^{\times}$ and defining

$\Delta_{ki}^{\times} := \Delta_{ki} \cap (\Psi_k^{\times} - \Theta_{ki})$ we obtain $\Theta_{k(i+1)} = \Delta_{ki}^{\times} + \Theta_{ki} \leq (\Psi_k^{\times} - \Theta_{ki}) + \Theta_{ki} = \Psi_k^{\times}$. So in particular $\sum_{i=0}^{k} \Delta_{ki}^{\times} \leq \Psi_k^{\times}$.

Using that $\Gamma_k^{\rightarrow} \leq \Psi_k^{\rightarrow}$, we find

$$\Delta_{kk} = (\Psi_k - \Gamma_k^{\rightarrow}) - \sum_{i=0}^{k-1} \Delta_{ki} = (\Psi_k^{\times} + (\Psi_k^{\rightarrow} - \Gamma_k^{\rightarrow})) - \sum_{i=0}^{k-1} \Delta_{ki} \geq \Psi_k^{\times} - \sum_{i=0}^{k-1} \Delta_{ki},$$

hence, $\Delta_{kk}^{\times} = \Delta_{kk} \cap (\Psi_k^{\times} - \sum_{i=0}^{k-1} \Delta_{ki}^{\times}) = \Psi_k^{\times} - \sum_{i=0}^{k-1} \Delta_{ki}$ and thus $\Psi_k^{\times} = \sum_{i=0}^{k} \Delta_{ki}$.

Now define $\Delta_{ki}^{\rightarrow} := \Delta_{ki} - \Delta_{ki}^{\times}$. This yields $\Delta_{ki} = \Delta_{ki}^{\rightarrow} + \Delta_{ki}^{\times}$ and thereby

$$\Psi_k^{\rightarrow} = \Psi_k - \Psi_k^{\times} = \left(\sum_{i=0}^{k} \Delta_{ki} + \Gamma_k^{\rightarrow} \right) - \sum_{i=0}^{k} \Delta_{ki}^{\times} = \sum_{i=0}^{k} \Delta_{ki}^{\rightarrow} + \Gamma_k^{\rightarrow}.$$

Since $\sum_{i=0}^{k} \Delta_{ki}^{\rightarrow} = (\Psi_k^{\rightarrow} - \Gamma_k^{\rightarrow}) \xrightarrow{\tau} (\Psi_{k+1} - \Gamma_{k+1})$, by Proposition 6.2 we have $\Psi_{k+1} - \Gamma_{k+1} = \sum_{i=0}^{k} \Delta_{(k+1)(i+1)}$ for some subdistributions $\Delta_{(k+1)(i+1)}$ such that they form the transitions $\Delta_{ki}^{\rightarrow} \xrightarrow{\tau} \Delta_{(k+1)(i+1)}$ for $i = 0, \ldots, k$. Furthermore, let us define $\Delta_{(k+1)0} := \Gamma_{k+1}^{\times} = \Gamma_{k+1} - \Gamma_{k+1}^{\rightarrow}$. It follows that

$$\Psi_{k+1} = \sum_{i=0}^{k} \Delta_{(k+1)(i+1)} + \Gamma_{k+1} = \sum_{i=1}^{k+1} \Delta_{(k+1)i} + (\Delta_{(k+1)0} + \Gamma_{k+1}^{\rightarrow}) = \sum_{i=0}^{k+1} \Delta_{(k+1)i} + \Gamma_{k+1}^{\rightarrow}.$$

This ends the inductive definition and proof. Now let us define $\Theta_i := \sum_{k=i}^{\infty} \Delta_{ki}$, $\Theta_i^{\rightarrow} := \sum_{k=i}^{\infty} \Delta_{ki}^{\rightarrow}$ and $\Theta_i^{\times} := \sum_{k=i}^{\infty} \Delta_{ki}^{\times}$. We have that $\Theta_0 = \sum_{k=0}^{\infty} \Delta_{k0} = \sum_{k=0}^{\infty} \Gamma_k^{\times} = \Gamma$, $\Theta_i = \Theta_i^{\rightarrow} + \Theta_i^{+}$, and, using Remark 6.2(ii), $\Theta_i^{\rightarrow} \xrightarrow{\tau} \Theta_{i+1}$. Moreover,

$$\sum_{i=0}^{\infty} \Theta_i^{\times} = \sum_{i=0}^{\infty} \sum_{k=i}^{\infty} \Delta_{ki}^{\times} = \sum_{k=0}^{\infty} \sum_{i=0}^{k} \Delta_{ki}^{\times} = \sum_{k=0}^{\infty} \Psi_k^{\times} = \Psi.$$

Definition 6.4 yields $\Gamma \Longrightarrow \Psi$. □

6.4 Testing rpCSP Processes

Applying a test to a process results in a nondeterministic, but possibly probabilistic, computation structure. The main conceptual issue is how to associate outcomes with these nondeterministic structures. In Sect. 4.2, we have seen an approach of testing, in which we explicitly associate with a nondeterministic structure a set of deterministic computations called resolutions, each of which determines a possible outcome. In this section, we describe an alternative approach, in which intuitively the nondeterministic choices are resolved implicitly in a dynamic manner. We show that although these approaches are formally quite different, they lead to exactly the same testing outcomes.

6.4.1 Testing with Extremal Derivatives

A *test* is simply a process in the language rpCSP, except that it may in addition use special *success* actions for reporting outcomes; these are drawn from a set Ω of fresh actions not already in Act_τ. We refer to the augmented language as rpCSP^Ω. Formally a test T is some process from that language, and to apply test T to process P we form the process $T|_{\mathsf{Act}}P$, in which *all* visible actions of P must synchronise with T. The resulting composition is a process whose only possible actions are τ and the elements of Ω. We will define the result $\mathcal{A}^{\mathrm{d}}(T,P)$ of applying the test T to the process P to be a set of testing outcomes, exactly one of which results from each resolution of the choices in $T|_{\mathsf{Act}}P$. Each *testing outcome* is an Ω-tuple of real numbers in the interval [0,1], that is, a function $o : \Omega \to [0,1]$, and its ω-component $o(\omega)$, for $\omega \in \Omega$, gives the probability that the resolution in question will reach an *ω-success state*, one in which the success action ω is possible.

We will now give a definition of $\mathcal{A}^{\mathrm{d}}(T,P)$, which is intended to be an alternative of $\mathcal{A}(T,P)$. Our definition has three ingredients. First of all, to simplify the presentation we normalise our pLTS by removing all τ-transitions that leave a success state. This way an ω-success state will only have outgoing transitions labelled ω.

Definition 6.6 (ω-Respecting) Let $\langle S, L, \rightarrow\rangle$ be a pLTS such that the set of labels L includes Ω. It is said to be *ω-respecting* whenever $s \xrightarrow{\omega}$, for any $\omega \in \Omega$, $s \stackrel{\tau}{\not\longrightarrow}$.

It is straightforward to modify an arbitrary pLTS so that it is ω-respecting. Here we outline how this is done for our pLTS for rpCSP.

Definition 6.7 (Pruning) Let $[\cdot]$ be the unary operator on Ω-test states given by the operational rules

$$\frac{s \xrightarrow{\omega} \Delta}{[s] \xrightarrow{\omega} [\Delta]}\ (\omega \in \Omega) \qquad\qquad \frac{s \stackrel{\omega}{\not\longrightarrow}\ (\text{for all } \omega \in \Omega),\ \ s \xrightarrow{\alpha} \Delta}{[s] \xrightarrow{\alpha} [\Delta]}\ (\alpha \in \mathsf{Act}_\tau)\,.$$

Just as $\Box$ and $|_A$, this operator extends as syntactic sugar to Ω-tests by distributing $[\cdot]$ over ${}_p\oplus$; likewise, it extends to distributions by $[\Delta]([s]) = \Delta(s)$. Clearly, this operator does nothing else than removing all outgoing transitions of a success state other than the ones labelled with $\omega \in \Omega$.

Next, using Definition 6.4, we get a collection of subdistributions Θ reachable from $[[\![T|_{\mathsf{Act}}P]\!]]$. Then, we isolate a class of special weak derivatives called *extreme derivatives*.

Definition 6.8 (Extreme Derivatives) A state s in a pLTS is called *stable* if $s \stackrel{\tau}{\not\longrightarrow}$, and a subdistribution Θ is called *stable* if every state in its support is stable. We write $\Delta \Longrightarrow\!\!\succ \Theta$ whenever $\Delta \Longrightarrow \Theta$ and Θ is stable, and call Θ an *extreme* derivative of Δ.

Referring to Definition 6.4, we see this means that in the extreme derivation of Θ from Δ at every stage a state must move on if it can, so that every stopping component can contain only states that *must* stop: for $s \in \lceil \Delta_k^{\rightarrow} + \Delta_k^{\times} \rceil$ we have that $s \in \lceil \Delta_k^{\times} \rceil$ if *and now also* only if $s \stackrel{\tau}{\not\longrightarrow}$. Moreover, if the pLTS is ω-respecting then whenever

$s \in \lceil \Delta_k^{\rightarrow} \rceil$, that is whenever it marches on, it is not successful, $s \stackrel{\omega}{\not\longrightarrow}$ for every $\omega \in \Omega$.

Lemma 6.7 **(Existence of Extreme Derivatives)**

(i) For every subdistribution Δ there exists some (stable) Δ' such that $\Delta \Longrightarrow\!\!\succ \Delta'$.
(ii) In a deterministic pLTS, if $\Delta \Longrightarrow\!\!\succ \Delta'$ and $\Delta \Longrightarrow\!\!\succ \Delta''$ then $\Delta' = \Delta''$.

Proof We construct a derivation as in Definition 6.4 of a stable Δ' by defining the components Δ_k, $\Delta_k^{\times}$ and $\Delta_k^{\rightarrow}$ using induction on k. Let us assume that the subdistribution Δ_k has been defined; in the base case $k = 0$ this is simply Δ. The decomposition of this Δ_k into the components $\Delta_k^{\times}$ and $\Delta_k^{\rightarrow}$ is carried out by defining the former to be precisely those states that must stop, that is, those s for which $s \stackrel{\tau}{\not\longrightarrow}$. Formally $\Delta_k^{\times}$ is determined by:

$$\Delta_k^{\times}(s) = \begin{cases} \Delta_k(s) & \text{if } s \stackrel{\tau}{\not\longrightarrow} \\ 0 & \text{otherwise} \end{cases}$$

Then $\Delta_k^{\rightarrow}$ is given by the *remainder* of Δ_k, namely those states that can perform a τ action:

$$\Delta_k^{\rightarrow}(s) = \begin{cases} \Delta_k(s) & \text{if } s \stackrel{\tau}{\longrightarrow} \\ 0 & \text{otherwise} \end{cases}$$

Note that these definitions divide the support of Δ_k into two disjoints sets, namely the support of $\Delta_k^{\times}$ and the support of $\Delta_k^{\rightarrow}$. Moreover, by construction we know that $\Delta_k^{\rightarrow} \stackrel{\tau}{\longrightarrow} \Theta$, for some Θ; we let Δ_{k+1} be an arbitrary such Θ.

This completes our definition of an extreme derivative as in Definition 6.4 and so we have established (i).

For (ii), we observe that in a deterministic pLTS the above choice of Δ_{k+1} is unique, so that the whole derivative construction becomes unique. □

It is worth pointing out that the use of subdistributions, rather than distributions, is essential here. If Δ diverges, that is, if there is an infinite sequence of derivations $\Delta \stackrel{\tau}{\longrightarrow} \Delta_1 \stackrel{\tau}{\longrightarrow} \ldots \Delta_k \stackrel{\tau}{\longrightarrow} \ldots$, then one extreme derivative of Δ is the empty subdistribution ε. For example, the only transition of $\mathsf{rec}\, x.x$ is $\mathsf{rec}\, x.x \stackrel{\tau}{\longrightarrow} \overline{\mathsf{rec}\, x.x}$, and therefore $\overline{\mathsf{rec}\, x.x}$ diverges; consequently its unique extreme derivative is ε.

The final ingredient in the definition of the set of outcomes $\mathcal{A}^{\mathrm{d}}(T, P)$ is to use this notion of extreme derivative to formalise the subdistributions that can be reached from $[\![T|_{\mathsf{Act}}P]\!]$. Note that all states $s \in \lceil \Theta \rceil$ in the support of an extreme derivative either satisfy $s \stackrel{\omega}{\longrightarrow}$ for a unique $\omega \in \Omega$, or have $s \not\longrightarrow$.

Definition 6.9 (Outcomes) The outcome $\$\Theta \in [0,1]^{\Omega}$ of a stable subdistribution Θ is given by $\$\Theta(\omega) := \sum_{s \in \lceil \Theta \rceil,\, s \stackrel{\omega}{\longrightarrow}} \Theta(s)$.

Putting all three ingredients together, we arrive at a definition of $\mathcal{A}^{\mathrm{d}}(T, P)$:

Definition 6.10 Let P be a rpCSP process and T an Ω-test. Then

$$\mathcal{A}^{\mathrm{d}}(T,P) := \{\$\Theta \mid [[[T|_{\mathsf{Act}}P]]] \Longrightarrow\!\!\succ \Theta\}.$$

The role of pruning in the above definition can be seen via the following example.

Example 6.7 Let $P = a.b.\mathbf{0}$ and $T = a.(b.\mathbf{0} \,\Box\, \omega.\mathbf{0})$. The pLTS generated by applying T to P can be described by the process $\tau.(\tau.\mathbf{0} \,\Box\, \omega.\mathbf{0})$. Then $[\![T|_{\mathsf{Act}}P]\!]$ has a unique extreme derivation $[\![T|_{\mathsf{Act}}P]\!] \Longrightarrow\!\!\succ [\![\mathbf{0}]\!]$, and $[[\![T|_{\mathsf{Act}}P]\!]]$ also has a unique extreme derivation $[[\![T|_{\mathsf{Act}}P]\!]] \Longrightarrow\!\!\succ [\![\omega.\mathbf{0}]\!]$. The outcome in $\mathcal{A}^{\mathrm{d}}(T,P)$ shows that process P passes test T with probability 1, which is what we expect for state-based testing, which we use in this book. Without pruning we would get an outcome saying that P passes T with probability 0, which would be what is expected for action-based testing.

Using the two standard methods for comparing two sets of outcomes, the Hoare- and Smyth preorders, we define the may- and must-testing preorders; they are decorated with $\cdot^{\Omega}$ for the repertoire Ω of testing actions they employ.

Definition 6.11

1. $P \sqsubseteq^{\Omega}_{\mathrm{pmay}} Q$ if for every Ω-test T, $\mathcal{A}^{\mathrm{d}}(T,P) \leq_{\mathrm{Ho}} \mathcal{A}^{\mathrm{d}}(T,Q)$.
2. $P \sqsubseteq^{\Omega}_{\mathrm{pmust}} Q$ if for every Ω-test T, $\mathcal{A}^{\mathrm{d}}(T,P) \leq_{\mathrm{Sm}} \mathcal{A}^{\mathrm{d}}(T,Q)$.

These preorders are abbreviated to $P \sqsubseteq_{\mathrm{pmay}} Q$ and $P \sqsubseteq_{\mathrm{pmust}} Q$, when $|\Omega| = 1$.

Example 6.8 Consider the process $Q_1 = \mathsf{rec}\, x.(\tau.x\, {}_{\frac{1}{2}}\!\oplus a.\mathbf{0})$, which was already discussed in Sect. 6.1. When we apply the test $T = a.\omega.\mathbf{0}$ to it we get the pLTS in Fig. 4.3b, which is deterministic and unaffected by pruning; from part (ii) of Lemma 6.7 it follows that $\overline{s_0}$ has a unique extreme derivative Θ. Moreover, Θ can be calculated to be

$$\sum_{k\geq 1} \frac{1}{2^k} \cdot \overline{s_3},$$

which simplifies to the distribution $\overline{s_3}$. Thus it gives the same set of results $\{\vec{\omega}\}$ gained by applying T to $a.\mathbf{0}$ on its own; and in fact it is possible to show that this holds for all tests, giving

$$Q_1 \simeq_{\mathrm{pmay}} a.\mathbf{0} \qquad\qquad Q_1 \simeq_{\mathrm{pmust}} a.\mathbf{0}.$$

Example 6.9 Consider the process $Q_2 = \mathsf{rec}\, x.((x\, {}_{\frac{1}{2}}\!\oplus a.\mathbf{0}) \sqcap (\mathbf{0}\, {}_{\frac{1}{2}}\!\oplus a.\mathbf{0}))$ and the application of the same test $T = a.\omega.\mathbf{0}$ to it, as outlined in Fig. 4.4a and b.

Consider any extreme derivative Δ' from $[[\![T|_{\mathsf{Act}}Q_2]\!]]$, which we have abbreviated to $\overline{s_0}$; note that here again pruning actually has no effect. Using the notation of Definition 6.4, it is clear that $\Delta_0^{\times}$ and $\Delta_0^{\rightarrow}$ must be ε and $\overline{s_0}$ respectively. Similarly, $\Delta_1^{\times}$ and $\Delta_1^{\rightarrow}$ must be ε and $\overline{s_1}$ respectively. But s_1 is a nondeterministic state, having two possible transitions:

(i) $s_1 \xrightarrow{\tau} \Theta_0$, where Θ_0 has support $\{s_0, s_2\}$ and assigns each of them the weight $\frac{1}{2}$
(ii) $s_1 \xrightarrow{\tau} \Theta_1$, where Θ_1 has the support $\{s_3, s_4\}$, again diving the mass equally among them.

So there are many possibilities for Δ_2; Lemma 6.1 shows that in fact Δ_2 can be of the form

$$p \cdot \Theta_0 + (1 - p) \cdot \Theta_1 \tag{6.12}$$

for any choice of $p \in [0, 1]$.

Let us consider one possibility, an extreme one where p is chosen to be 0; only the transition (ii) above is used. Here $\Delta_2^{\rightarrow}$ is the subdistribution $\frac{1}{2}\overline{s_4}$, and $\Delta_k^{\rightarrow} = \varepsilon$ whenever $k > 2$. A simple calculation shows that in this case the extreme derivative generated is $\Theta_1^e = \frac{1}{2}\overline{s_3} + \frac{1}{2}\overline{\omega.\mathbf{0}}$ which implies that $\frac{1}{2}\vec{\omega} \in \mathcal{A}^{\mathrm{d}}(T, Q_2)$.

Another possibility for Δ_2 is Θ_0, corresponding to the choice of $p = 1$ in (6.12) above. Continuing with this derivation leads to Δ_3 being $\frac{1}{2} \cdot \overline{s_1} + \frac{1}{2} \cdot \overline{\omega.\mathbf{0}}$; in other words, $\Delta_3^{\times} = \frac{1}{2} \cdot \overline{\omega.\mathbf{0}}$ and $\Delta_3^{\rightarrow} = \frac{1}{2} \cdot \overline{s_1}$. Now in the generation of Δ_4 from $\Delta_3^{\rightarrow}$ once more we have to resolve a transition from the nondeterministic state s_1, by choosing some arbitrary $p \in [0, 1]$ in (6.12). Suppose that each time this arises, we systematically choose $p = 1$, that is, we ignore completely the transition (ii) above. Then it is easy to see that the extreme derivative generated is

$$\Theta_0^e = \sum_{k \geq 1} \frac{1}{2^k} \cdot \overline{\omega.\mathbf{0}},$$

which simplifies to the distribution $\overline{\omega.\mathbf{0}}$. This in turn means that $\vec{\omega} \in \mathcal{A}^{\mathrm{d}}(T, Q_2)$.

We have seen two possible derivations of extreme derivatives from $\overline{s_0}$. But there are many others. In general, whenever $\Delta_k^{\rightarrow}$ is of the form $q \cdot \overline{s_1}$, we have to resolve the nondeterminism by choosing a $p \in [0, 1]$ in (6.12) above; moreover, each such choice is independent. However, it will follow from later results, specifically, Corollary 6.4, that every extreme derivative Δ' of $\overline{s_0}$ is of the form

$$q \cdot \Theta_0^e + (1 - q)\Theta_1^e$$

for some choice of $q \in [0, 1]$; this is explained in Example 6.11. Consequently, it follows that $\mathcal{A}^{\mathrm{d}}(T, Q_2) = \{q\vec{\omega} \mid q \in [\frac{1}{2}, 1]\}$.

Since $\mathcal{A}^{\mathrm{d}}(T, a.\mathbf{0}) = \{\vec{\omega}\}$ it follows that

$$\mathcal{A}^{\mathrm{d}}(T, a.\mathbf{0}) \leq_{\mathrm{Ho}} \mathcal{A}^{\mathrm{d}}(T, Q_2) \qquad \mathcal{A}^{\mathrm{d}}(T, Q_2) \leq_{\mathrm{Sm}} \mathcal{A}^{\mathrm{d}}(T, a.\mathbf{0})\,.$$

Again it is possible to show that these inequalities result from any test T and that therefore we have

$$a.\mathbf{0} \sqsubseteq_{\mathrm{pmay}} Q_2 \qquad Q_2 \sqsubseteq_{\mathrm{pmust}} a.\mathbf{0}$$

6.4.2 Comparison with Resolution-Based Testing

The derivation of extreme derivatives, via the schema in Definition 6.4, involves the systematic dynamic resolution of nondeterministic states, in each transition from $\Delta_k^{\rightarrow}$ to Δ_{k+1}. In the literature, various mechanisms have been proposed for making these choices; for example *policies* are used in [6], adversaries in [7], schedulers in [8], ... Here, we concentrate not on any such mechanism but rather the results of their application. In general, they reduce a nondeterministic structure, typically a pLTS, to a set of deterministic structures. To describe these deterministic structures, in Sect. 4.2, we adapted the notion of *resolution* defined in [9, 10] for probabilistic automata, to pLTSs.

Therefore, we have now seen at least two ways of associating sets of outcomes with the application of a test to the process. The first, in Sect. 6.4.1, uses extreme derivations in which nondeterministic choices are resolved dynamically as the derivation proceeds, while in the second, in Sect. 4.2, we associate with a test and a process a set of deterministic structures called resolutions. In this section, we show that the testing preorders obtained from these two approaches coincide.

First, let us see how an extreme derivation can be viewed as a method for dynamically generating a resolution.

Theorem 6.7 (Resolutions from Extreme Derivatives) *Suppose* $\Delta \Longrightarrow\!\!\!\succ \Delta'$ *in a pLTS* $\langle S, \Omega_\tau, \rightarrow\rangle$. *Then there is a resolution* $\langle R, \Theta, \rightarrow_R\rangle$ *of* Δ, *with resolving function* f, *such that* $\Theta \Longrightarrow\!\!\!\succ_R \Theta'$ *for some* Θ' *with* $\Delta' = Img_f(\Theta')$.

Proof Consider an extreme derivation of $\Delta \Longrightarrow\!\!\!\succ \Delta'$, as given in Definition 6.4 where all $\Delta_k^{\times}$ are assumed to be stable. To define the corresponding resolution $\langle R, \Theta, \longrightarrow_R\rangle$ we refer to Definition 4.3. First, let the set of states R be $S \times \mathbb{N}$ and the resolving function $f : R \rightarrow S$ be given by $f(s,k) = s$. To complete the description, we must define the partial functions $\xrightarrow{\alpha}$, for $\alpha = \omega$ and $\alpha = \tau$. These are always defined so that if $(s,k) \xrightarrow{\alpha} \Gamma$ then the only states in the support of Γ are of the form $(s', k+1)$. In the definition we use $\Theta^{\downarrow k}$, for any subdistribution Θ over S, to be the subdistribution over R given by

$$\Theta^{\downarrow k}(t) = \begin{cases} \Theta(s) & \text{if } t = (s,k) \\ 0 & \text{otherwise} \end{cases}$$

Note that by definition

(a) $\mathrm{Img}_f(\Theta^{\downarrow k)} = \Theta$
(b) $\Delta_k^{\downarrow k} = \Delta_k^{\rightarrow\downarrow k} + \Delta_k^{\times\downarrow k}$

The definition of $\xrightarrow{\omega}_R$ is straightforward; its domain consists of states (s,k), where $s \in \lceil \Delta_k^{\times} \rceil$ and is defined by letting $(s,k) \xrightarrow{\omega} \Delta_s^{\downarrow k+1}$ for some arbitrarily chosen $s \xrightarrow{\omega} \Delta_s$.

The definition of $\xrightarrow{\tau}_R$ is more complicated and is determined by the moves $\Delta_k^{\rightarrow} \xrightarrow{\tau} \Delta_{k+1}$. For a given k this move means that

$$\Delta_k^{\rightarrow} = \sum_{i\in I} p_i \cdot \overline{s_i}, \qquad \Delta_{k+1} = \sum_{i\in I} p_i \cdot \Gamma_i, \qquad s_i \xrightarrow{\tau} \Gamma_i$$

So for each k we let

$$(s,k) \xrightarrow{\tau}_R \sum_{s_i=s} p_i \cdot \Gamma_i^{\downarrow k+1}$$

This definition ensures

(c) $(\Delta_k^{\rightarrow})^{\downarrow k} \xrightarrow{\tau}_R (\Delta_{k+1})^{\downarrow k+1}$;
(d) $(\Delta_k^{\times})^{\downarrow k}$ is stable.

This completes our definition of the deterministic pLTS underlying the required resolution; it remains to find subdistributions Θ, Θ' over R such that $\Delta = \mathrm{Img}_f(\Theta)$, $\Delta' = \mathrm{Img}_f(\Theta')$ and $\Theta \Longrightarrow\!\!\succ \Theta'$.

Because of (b) (c) and (d), we have the following extreme derivation, which by part (ii) of Lemma 6.7 is the unique one from $\Delta_0^{\downarrow 0}$:

$$\begin{array}{rcll}
\Delta^{\downarrow 0} & = & (\Delta_0^{\rightarrow})^{\downarrow 0} & + \; (\Delta_0^{\times})^{\downarrow 0} \\
(\Delta_0^{\rightarrow})^{\downarrow 0} & \xrightarrow{\tau}_R & (\Delta_1^{\rightarrow})^{\downarrow 1} & + \; (\Delta_1^{\times})^{\downarrow 1} \\
\vdots & & & \quad\vdots \\
(\Delta_k^{\rightarrow})^{\downarrow k} & \xrightarrow{\tau}_R & (\Delta_{k+1}^{\rightarrow})^{\downarrow k+1} & + \; (\Delta_{k+1}^{\times})^{\downarrow k+1} \\
 & & & \quad\vdots \\
\hline
 & & & \Theta' = \sum_{k=0}^{\infty} (\Delta_k^{\times})^{\downarrow k}
\end{array}$$

Letting Θ be $\Delta^{\downarrow 0}$, we see that (a) above ensures $\Delta = \mathrm{Img}_f(\Theta)$; the same note and the linearity of f applied to distributions also gives $\Delta' = \mathrm{Img}_f(\Theta')$. □

The converse is somewhat simpler.

Proposition 6.5 (Extreme Derivatives from Resolutions) *Suppose $\langle R, \Theta, \rightarrow_R\rangle$ is a resolution of a subdistribution Δ in a pLTS $\langle S, \Omega_\tau, \rightarrow\rangle$ with resolving function f. Then $\Theta \Longrightarrow\!\!\succ_R \Theta'$ implies $\Delta \Longrightarrow\!\!\succ \mathit{Img}_f(\Theta')$.*

Proof Consider any derivation of $\Theta \Longrightarrow\!\!\succ_R \Theta'$ along the lines of Definition 6.4. By systematically applying the function f to the component subdistributions in this derivation we get a derivation $\mathrm{Img}_f(\Theta) \Longrightarrow \mathrm{Img}_f(\Theta')$, that is, $\Delta \Longrightarrow \mathrm{Img}_f(\Theta')$. To show that $\mathrm{Img}_f(\Theta')$ is actually an extreme derivative it suffices to show that s is stable for every $s \in \lceil \mathrm{Img}_f(\Theta') \rceil$. But if $s \in \lceil \mathrm{Img}_f(\Theta') \rceil$, then by definition there is some $t \in \lceil \Theta' \rceil$ such that $s = f(t)$. Since $\Theta \Longrightarrow\!\!\succ_R \Theta'$ the state t must be stable. The stability of s now follows from requirement (iii) of Definition 4.3. □

Our next step is to relate the outcomes extracted from extreme derivatives to those extracted from the corresponding resolutions. By Lemma 4.3, we know that the function $\mathcal{C} : (R \to [0,1]^\Omega) \to (R \to [0,1]^\Omega)$ defined in (4.1) for a deterministic pLTS is continuous. Then its least fixed point $\mathbb{V} : R \to [0,1]^\Omega$ is also continuous and can be captured by a chain of approximants. The functions $\mathbb{V}^n$, $n \geq 0$, are defined by induction on n:

$$\mathbb{V}^0(r)(\omega) = 0$$
$$\mathbb{V}^{n+1} = \mathcal{C}(\mathbb{V}^n)$$

Then $\mathbb{V} = \bigsqcup_{n \geq 0} \mathbb{V}^n$. This is used in the following result.

Lemma 6.8 *Let Δ be a subdistribution in an ω-respecting deterministic pLTS. If $\Delta \Longrightarrow\!\!\succ \Delta'$ then $\mathbb{V}(\Delta) = \mathbb{V}(\Delta')$.*

Proof Since the pLTS is ω-respecting, we know that $s \xrightarrow{\tau} \Delta$ implies $s \stackrel{\omega}{\not\longrightarrow}$ for any ω. Therefore, from the definition of the function $\mathcal{C}$ we have that $s \xrightarrow{\tau} \Delta$ implies $\mathbb{V}^{n+1}(s) = \mathbb{V}^n(\Delta)$, whence by lifting and linearity we get,

$$\text{If } \Delta \xrightarrow{\tau} \Delta' \text{ then } \mathbb{V}^{n+1}(\Delta) = \mathbb{V}^n(\Delta') \text{ for all } n \geq 0. \tag{6.13}$$

Now, suppose $\Delta \Longrightarrow\!\!\succ \Delta'$. Referring to Definition 6.4 and carrying out a straightforward induction based on (6.13), we have

$$\mathbb{V}^{n+1}(\Delta) = \mathbb{V}^0(\Delta_{n+1}) + \sum_{k=0}^{n} \mathbb{V}^{n-k+1}(\Delta_k^\times) \tag{6.14}$$

for all $n \geq 0$. This can be simplified further by noting

(i) $\mathbb{V}^0(\Delta)(\omega) = 0$, for every Δ
(ii) $\mathbb{V}^{m+1}(\Delta) = \mathbb{V}(\Delta)$ for every $m \geq 0$, provided Δ is stable.

Applying these remarks to (6.14) above, since all $\Delta_k^\times$ are stable, we obtain

$$\mathbb{V}^{n+1}(\Delta) \;=\; \sum_{k=0}^{n} \mathbb{V}(\Delta_k^\times) \tag{6.15}$$

We conclude by reasoning as follows

$$\begin{aligned} \mathbb{V}(\Delta) &= \bigsqcup_{n \geq 0} \mathbb{V}^{n+1}(\Delta) \\ &= \bigsqcup_{n \geq 0} \sum_{k=0}^{n} \mathbb{V}(\Delta_k^\times) && \text{from (6.15) above} \\ &= \bigsqcup_{n \geq 0} \mathbb{V}\left(\sum_{k=0}^{n} \Delta_k^\times\right) && \text{by linearity of } \mathbb{V} \end{aligned}$$

$$= \mathbb{V}\left(\bigsqcup_{n\geq 0}\sum_{k=0}^{n}\Delta_k^{\times}\right) \qquad \text{by continuity of } \mathbb{V}$$

$$= \mathbb{V}\left(\sum_{k=0}^{\infty}\Delta_k^{\times}\right)$$

$$= \mathbb{V}(\Delta') \qquad \square$$

We are now ready to compare the two methods for calculating the set of outcomes associated with a subdistribution:

- using extreme derivatives and the reward function \$ from Definition 6.9; and
- using resolutions and the evaluation function $\mathbb{V}$ from Sect. 4.2.

Corollary 6.2 *In an ω-respecting pLTS $\langle S, \Omega_\tau, \rightarrow\rangle$, the following statements hold.*

(a) If $\Delta \Longrightarrow\!\!\succ \Delta'$ then there is a resolution $\langle R, \Theta, \rightarrow_R\rangle$ of Δ such that $\mathbb{V}(\Theta) = \$(\Delta')$.
(b) For any resolution $\langle R, \Theta, \rightarrow_R\rangle$ of Δ, there exists an extreme derivative Δ' such that $\Delta \Longrightarrow\!\!\succ \Delta'$ and $\mathbb{V}(\Theta) = \$(\Delta')$.

Proof Suppose $\Delta \Longrightarrow\!\!\succ \Delta'$. By Theorem 6.7, there is a resolution $\langle R, \Theta, \rightarrow_R\rangle$ of Δ with resolving function f and subdistribution Θ such that $\Theta \Longrightarrow\!\!\succ \Theta'$ and, moreover, $\Delta' = \mathrm{Img}_f(\Theta')$. By Lemma 6.8, we have $\mathbb{V}(\Theta) = \mathbb{V}(\Theta')$.

Since Θ' and Δ' are extreme derivatives, all the states in their supports are stable. Therefore, a simple calculation, using the fact that $\Delta' = \mathrm{Img}_f(\Theta')$, will show that $\mathbb{V}(\Theta') = \$(\Delta')$, from which the required $\mathbb{V}(\Theta) = \$(\Delta')$ follows.

To prove part (b), suppose that $\langle R, \Theta, \rightarrow_R\rangle$ is a resolution of Δ with resolving function f, so that $\Delta = \mathrm{Img}_f(\Theta)$. We know from Lemma 6.7 that there exists a (unique) subdistribution Θ' such that $\Theta \Longrightarrow\!\!\succ \Theta'$. By Proposition 6.5, we have that $\Delta = \mathrm{Img}_f(\Theta) \Longrightarrow\!\!\succ \mathrm{Img}_f(\Theta')$. The same arguments as in the other direction show that $\mathbb{V}(\Theta) = \$(\mathrm{Img}_f(\Theta'))$. $\square$

Corollary 6.3 *For any process P and test T, we have $\mathcal{A}^d(T, P) = \mathcal{A}(T, P)$.* $\square$

6.5 Generating Weak Derivatives in a Finitary pLTS

Now let us restrict our attention to a finitary pLTS whose state space is assumed to be $S = \{s_1, \ldots, s_n\}$. Here by definition the sets $\{\Theta \mid s \xrightarrow{\alpha} \Theta\}$ are finite, for every state s and label α. This of course is no longer true for the weak arrows; the sets $\{\Theta \mid \overline{s} \stackrel{\alpha}{\Longrightarrow} \Theta\}$ are in general not finite, because of the infinitary nature of the weak derivative relation $\Longrightarrow$. The purpose of this section is to show that, nevertheless, they can be finitely represented, at least for finitary pLTSs.

This is explained in Sect. 6.5.1 and the ramifications are then explored in the following subsection. These include a very useful topological property of these sets of derivatives; they are *closed* in the sense (from analysis) of containing all its limit points where, in turn, limit depends on a Euclidean-style metric defining the distance between two subdistributions in a straightforward way. Another consequence is that we can find in any derivation that partially diverges (by no matter how small an amount) a point at which the divergence is *distilled* into a state that wholly diverges; we call this *distillation of divergence.*

6.5.1 Finite Generability

A subdistribution over the finite state space S can now be viewed as a point in $\mathbb{R}^n$, and therefore a set of subdistributions, such as the set of weak derivatives $\{\Delta \mid \overline{s} \Longrightarrow \Delta\}$ corresponds to a subset of $\mathbb{R}^n$. We endow $\mathbb{R}^n$ with the standard Euclidean metric and proceed to establish useful topological properties of such sets of subdistributions. Recall that a set $X \subseteq \mathbb{R}^n$ is (Cauchy) *closed* if for every Cauchy sequence $\{x_n \mid n \geq 0\}$ with limit x, if $x_n \in X$ for every $n \geq 0$ then x is also in X.

Lemma 6.9 *If X is a finite subset of $\mathbb{R}^n$ then $\updownarrow X$ is closed.*

Proof Straightforward. □

In Definition 4.8, we gave a definition of extreme policies for pLTSs of the form $\langle S, \Omega_\tau, \rightarrow\rangle$ and showed how they determine resolutions. Here, we generalise these to *derivative policies* and show that these generalised policies can also be used to generate arbitrary weak derivatives of subdistributions over S.

Definition 6.12 A (static) *derivative policy* for a pLTS $\langle S, \mathsf{Act}_\tau, \rightarrow\rangle$, is a partial function $\mathsf{dp} : S \rightharpoonup \mathcal{D}(S)$ with the property that $\mathsf{dp}(s) = \Delta$ implies $s \xrightarrow{\tau} \Delta$. If dp is undefined at s, we write $\mathsf{dp}(s)\uparrow$. Otherwise, we write $\mathsf{dp}(s)\downarrow$.

A derivative policy dp, as its name suggests, can be used to guide the derivation of a weak derivative. Suppose $\overline{s} \Longrightarrow \Delta$, using a derivation as given in Definition 6.4. Then, we write $\overline{s} \Longrightarrow_{\mathsf{dp}} \Delta$ whenever, for all $k \geq 0$,

(a) $\Delta_k^{\rightarrow}(s) = \begin{cases} \Delta_k(s) & \text{if } \mathsf{dp}(s)\downarrow \\ 0 & \text{otherwise} \end{cases}$

(b) $\Delta_{(k+1)} = \sum_{s \in \lceil \Delta_k^{\rightarrow} \rceil} \Delta_k^{\rightarrow}(s) \cdot \mathsf{dp}(s)$

Intuitively, these conditions mean that the derivation of Δ from s is guided at each stage by the policy dp:

- Condition (a) implies that the division of Δ_k into $\Delta_k^{\rightarrow}$, the subdistribution that will continue marching, and $\Delta_k^{\times}$, the subdistribution that will stop, is determined by the domain of the derivative policy dp.
- Condition (b) ensures that the derivation of the next stage Δ_{k+1} from $\Delta_k^{\rightarrow}$ is determined by the action of the function dp on the support of $\Delta_k^{\rightarrow}$.

Lemma 6.10 *Let* dp *be a derivative policy in a pLTS. Then*

(a) If $\overline{s} \Longrightarrow_{\mathsf{dp}} \Delta$ *and* $\overline{s} \Longrightarrow_{\mathsf{dp}} \Theta$ *then* $\Delta = \Theta$.
(b) For every state s *there exists some* Δ *such that* $\overline{s} \Longrightarrow_{\mathsf{dp}} \Delta$.

Proof To prove part (a) consider the derivation of $\overline{s} \Longrightarrow \Delta$ and $\overline{s} \Longrightarrow \Theta$ as in Definition 6.4, via the subdistributions Δ_k, $\Delta_k^{\rightarrow}$, $\Delta_k^{\times}$ and Θ_k, $\Theta_k^{\rightarrow}$, $\Theta_k^{\times}$, respectively. Because both derivations are guided by the same derivative policy dp it is easy to show by induction on k that

$$\Delta_k = \Theta_k \qquad \Delta_k^{\rightarrow} = \Theta_k^{\rightarrow} \qquad \Delta_k^{\times} = \Theta_k^{\times}$$

from which $\Delta = \Theta$ follows immediately.

To prove (b) we use dp to generate subdistributions Δ_k, $\Delta_k^{\rightarrow}$, $\Delta_k^{\times}$ for each $k \geq 0$ satisfying the constraints of Definition 6.4 and simultaneously those in Definition 6.12 above. The result will then follow by letting Δ be $\sum_{k\geq 0} \Delta_k^{\times}$. □

The net effect of this lemma is that a derivative policy dp determines a *total* function from states to derivations. Let $\mathbf{Der}_{\mathsf{dp}} : S \rightarrow \mathcal{D}(S)$ be defined by letting $\mathbf{Der}_{\mathsf{dp}}(s)$ be the unique Δ such that $\overline{s} \Longrightarrow_{\mathsf{dp}} \Delta$.

It should be clear that the use of derivative policies limits considerably the scope for deriving weak derivations. Each particular policy can only derive one weak derivative, and moreover in finitary pLTS there are only a finite number of derivative policies. Nevertheless, we will show that this limitation is more apparent than real. In Sect. 4.6.1, we saw how the more restrictive extreme policies ep could in fact, realise the maximum value attainable by any resolution of a finitely branching pLTS. Here, we generalise this result by replacing resolutions with arbitrary reward functions.

Definition 6.13 (Rewards and Payoffs) Let S be a set of states. A *reward function* is a function $\mathbf{r} : S \rightarrow [-1, 1]$. With $S = \{s_1, \ldots, s_n\}$, we often consider a reward function as the n-dimensional vector $\langle \mathbf{r}(s_1), \ldots, \mathbf{r}(s_n) \rangle$. In this way, we can use the notion $\mathbf{r} \cdot \Delta$ to stand for the inner product of two vectors.

Given such a reward function, we define the *payoff function* $\mathbb{P}^{\mathbf{r}}_{\max} : S \rightarrow \mathbb{R}$ by

$$\mathbb{P}^{\mathbf{r}}_{\max}(s) \;=\; \sup\{\, \mathbf{r} \cdot \Delta \mid \overline{s} \Longrightarrow \Delta \}.$$

A priori these payoff functions for a given state s are determined by the set of weak derivatives of s. However, the main result of this section is that they can, in fact, always be realised by derivative policies.

Theorem 6.8 (Realising Payoffs) *In a finitary pLTS, for every reward function* $\mathbf{r}$ *there exists some derivative policy* dp *such that* $\mathbb{P}^{\mathbf{r}}_{\max}(s) \;=\; \mathbf{r} \cdot \mathbf{Der}_{\mathsf{dp}}(s)$.

Proof As with Theorem 4.3, there is a temptation to give a constructive proof here, defining the effect of the required derivative policy dp at state s by considering the application of the reward function $\mathbf{r}$ to both s and all of its derivatives. However, this is not possible, as the example below explains.

Instead the proof is nonconstructive, requiring *discounted* policies. The overall structure of the proof is similar to that of Theorem 4.3, but the use of (discounted)

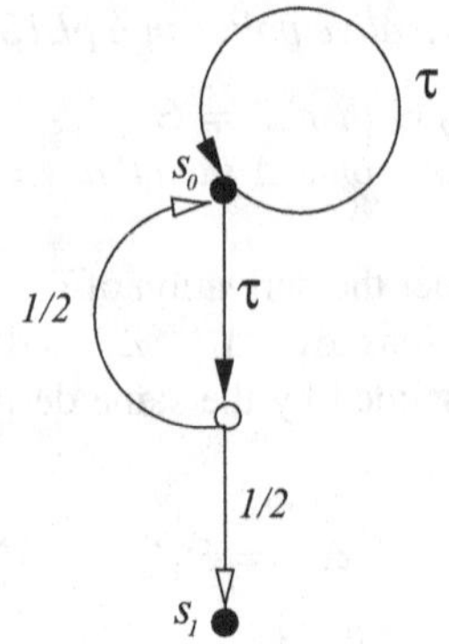

Fig. 6.3 Max-seeking policies

derivative policies rather than extreme policies makes the details considerably different. Consequently, the proof is spelled out in some detail in Sect. 6.5.2, cumulating in Theorem 6.10. □

Example 6.10 Let us say that a derivative policy dp is max-seeking with respect to a reward function $\mathbf{r}$ if for all $s \in S$ the following requirements are met.

1. If $\mathsf{dp}(s) \uparrow$ then $\mathbf{r}(s) \geq \mathbb{P}^{\mathbf{r}}_{\max}(\Delta_1)$ for all $s \xrightarrow{\tau} \Delta_1$.
2. If $\mathsf{dp}(s) = \Delta$ then
   a) $\mathbb{P}^{\mathbf{r}}_{\max}(\Delta) \geq \mathbf{r}(s)$ and
   b) $\mathbb{P}^{\mathbf{r}}_{\max}(\Delta) \geq \mathbb{P}^{\mathbf{r}}_{\max}(\Delta_1)$ for all $s \xrightarrow{\tau} \Delta_1$.

What a max-seeking policy does is to evaluate $\mathbb{P}^{\mathbf{r}}_{\max}$ in advance, for a given reward function $\mathbf{r}$, and then label each state s with the payoff value $\mathbb{P}^{\mathbf{r}}_{\max}(s)$. The policy at any state s is then to compare $\mathbf{r}(s)$ with the expected label values $\mathbb{P}^{\mathbf{r}}_{\max}(\Delta')$ (i.e. $\mathrm{Exp}_{\Delta'}(\mathbb{P}^{\mathbf{r}}_{\max})$) for each outgoing transition $s \xrightarrow{\tau} \Delta'$ and then to select the greatest among all those values. Note that for the policy to be well defined, we require that the pLTS under consideration is finitely branching.

In case that seems obvious, we now consider the pLTS in Fig. 6.3 and let us apply the above definition of max-seeking policies to the reward function given by $\mathbf{r}(s_0) = 0$, $\mathbf{r}(s_1) = 1$. For both states a payoff of 1 is attainable eventually, thus $\mathbb{P}^{\mathbf{r}}_{\max}(s_0) = \mathbb{P}^{\mathbf{r}}_{\max}(s_1) = 1$, because we have $s_0 \Longrightarrow \overline{s_1}$ and $s_1 \Longrightarrow \overline{s_1}$. Hence, both states will be $\mathbb{P}^{\mathbf{r}}_{\max}$-labelled with 1. At state s_0 the policy then makes a choice among three options: (1) to stay unmoved, yielding immediate payoff $\mathbf{r}(s_0) = 0$; (2) to take the transition $s_0 \xrightarrow{\tau} \overline{s_0}$; (3) to take the transition $s_0 \xrightarrow{\tau} \overline{s_0}\,{}_{1/2}\!\oplus \overline{s_1}$. Clearly one of the latter two is chosen — but which? If it is the second, then indeed the maximum payoff 1 can be achieved. If it is the first, then in fact the overall payoff will be 0 because of divergence, so the policy would fail to attain the maximum payoff 1.

However, for properly discounted max-seeking policies, we show in Proposition 6.6 that they always attain the maximum payoffs.

With Theorem 6.8 at hand, we are in the position to prove the main result of this section, which says that in a finitary pLTS the set of weak derivatives from any state

s, $\{\Delta \mid \overline{s} \Longrightarrow \Delta\}$, is generable by the convex closure of a finite set. The proof makes use of the separation theorem, Theorem 2.7.

Theorem 6.9 (Finite Generability) *Let* $P = \{\mathsf{dp}_1, \ldots, \mathsf{dp}_k\}$ *be the finite set of derivative policies in a finitary pLTS. For any state s and subdistribution* Δ *in the pLTS,* $\overline{s} \Longrightarrow \Delta$ *if and only if* $\Delta \in \updownarrow\{\mathbf{Der}_{\mathsf{dp}_i}(s) \mid 1 \leq i \leq k\}$.

Proof ($\Rightarrow$) For convenience, let X denote the set $\updownarrow\{\mathbf{Der}_{\mathsf{dp}_i}(s) \mid 1 \leq i \leq k\}$. Suppose, for a contradiction, that $\overline{s} \Longrightarrow \Delta$ for some Δ not in X. Recall that we are assuming that the underlying state space is $S = \{x_1, \ldots x_n\}$ so that X is a subset of $\mathbb{R}^n$. It is trivially bounded by $[-1, 1]^n$, and by definition it is convex; by Lemma 6.9 it follows that X is (Cauchy) closed.

By the separation theorem, Δ can be separated from X by a hyperplane H. What this means is that there is some function $\mathbf{r}_H : S \to \mathbb{R}$ and constant $c \in \mathbb{R}$ such that either

(a) $\mathbf{r}_H \cdot \Theta < c$ for all $\Theta \in X$ and $\mathbf{r}_H \cdot \Delta > c$ or,
(b) $\mathbf{r}_H \cdot \Theta > c$ for all $\Theta \in X$ and $\mathbf{r}_H \cdot \Delta < c$

In fact, from case (b) we can obtain case (a) by negating both the constant c and the components of the function $\mathbf{r}_H$; so we can assume (a) to be true. Moreover, by scaling with respect to the largest $\mathbf{r}_H(s_i)$, $1 \leq i \leq n$, we can assume that $\mathbf{r}_H$ is actually a reward function.

In particular (a) means that $\mathbf{r}_H \cdot \mathbf{Der}_{\mathsf{dp}_i}(s) < c$, and therefore that

$$\mathbf{r}_H \cdot \mathbf{Der}_{\mathsf{dp}_i}(s) \; < \; \mathbf{r}_H \cdot \Delta$$

for each of derivative polices dp_i. But this contradicts Theorem 6.8 which claims that there must be some $1 \leq i \leq n$ such that $\mathbf{r}_H \cdot \mathbf{Der}_{\mathsf{dp}_i}(s) = \mathbb{P}^{\mathbf{r}_H}_{\max}(s) \geq \mathbf{r}_H \cdot \Delta$.

($\Leftarrow$) Note that by definition $\overline{s} \Longrightarrow \mathbf{Der}_{\mathsf{dp}_i}(s)$ for every derivative policy dp_i with $1 \leq i \leq k$. By Proposition 6.3 and Remark 6.1, the relation $\Longrightarrow$ is convex. It follows that if $\Delta \in \updownarrow\{\mathbf{Der}_{\mathsf{dp}_1}(s), \ldots, \mathbf{Der}_{\mathsf{dp}_k}\}$, then $\overline{s} \Longrightarrow \Delta$. □

Extreme policies, as given in Definition 4.8, are particular kinds of derivative policies, designed for pLTSs of the form $\langle R, \Omega_\tau, \to_R\rangle$. The significant constraint on extreme policies is that for any state s if $s \xrightarrow{\tau}$ then $\mathsf{ep}(s)$ must be defined. As a consequence, in the computation determined by ep if a state can contribute to the computation at any stage it must contribute.

Lemma 6.11 *Let* ep *be any extreme policy. Then* $\overline{s} \Longrightarrow_{\mathsf{ep}} \Delta$ *implies* $\overline{s} \Longrightarrow\!\!\!\!\succ \Delta$.

Proof Consider the derivation of Δ as in Definition 6.4, and determined by the extreme policy ep. Since $\Delta = \sum_{k \geq 0} \Delta^\times_k$ it is sufficient to show that each $\Delta^\times_k$ is stable, that is $s \xrightarrow{\tau}$ implies $s \notin \lceil \Delta^\times_k \rceil$.

Since ep is an extreme policy, Definition 4.8 ensures that $\mathsf{ep}(s)$ is defined. From the definition of a computation, Definition 6.4, we know $\Delta_k = \Delta^\rightarrow_k + \Delta^\times_k$ and since the computation is guided by the policy ep we have that $\Delta^\rightarrow_k(s) = \Delta_k(s)$. An immediate consequence is that $\Delta^\times_k(s) = 0$. □

As a consequence, the finite generability result, Theorem 6.9, specialises to extreme derivatives.

Corollary 6.4 *Let* $\{\mathsf{ep}_1, \ldots, \mathsf{ep}_k\}$ *be the finite set of extreme policies of a finitary* ω*-respecting pLTS* $\langle S, \Omega_\tau, \rightarrow_R\rangle$. *For any state* s *and subdistribution* Δ *in the pLTS,* $\overline{s} \Longrightarrow\!\!\!\!\succ \Delta$ *if and only if* $\Delta \in \updownarrow\{\ \mathbf{Der}_{\mathsf{ep}_i}(s) \mid 1 \leq i \leq k\}$.

Proof One direction follows immediately from the previous lemma. Conversely, suppose $\overline{s} \Longrightarrow\!\!\!\!\succ \Delta$. By Theorem 6.9, $\Delta = \sum_{1\leq i \leq n} p_i \cdot \mathbf{Der}_{\mathsf{dp}_i}(s)$ for some finite collection of derivative policies dp_i, where we can assume that each $p_i \geq 0$. Because Δ is stable, that is $s \stackrel{\tau}{\not\longrightarrow}$ for every $s \in \lceil \Delta \rceil$, we show that each derivative policy dp_i can be transformed into an extreme policy ep_i such that $\mathbf{Der}_{\mathsf{ep}_i}(s) = \mathbf{Der}_{\mathsf{dp}_i}(s)$, from which the result will follow.

First, note it is sufficient to define ep_i on the set of states t accessible from s via the policy dp_i; on the remaining states in S ep_i can be defined arbitrarily, so as to satisfy the requirements of Definition 4.8. So consider the derivation of $\mathbf{Der}_{\mathsf{dp}_i}(s)$ as in Definition 6.4, determined by dp_i and suppose $t \in \lceil \Delta_k \rceil$ for some $k \geq 0$. There are three cases:

(i) Suppose $t \stackrel{\tau}{\longrightarrow}$. Since Δ is stable we know $t \notin \lceil \Delta_k^\times \rceil$, and therefore by definition $\mathsf{dp}_i(t)$ is defined. So in this case we let $\mathsf{ep}_i(t)$ be the same as $\mathsf{dp}_i(t)$.
(ii) Suppose $t \stackrel{\omega}{\longrightarrow}$, in which case, since the pLTS is ω-respecting, we know $t \stackrel{\tau}{\not\longrightarrow}$, and therefore $\mathsf{dp}_i(t)$ is not defined. Here we choose $\mathsf{ep}_i(t)$ arbitrarily so as to satisfy $t \stackrel{\omega}{\longrightarrow} \mathsf{ep}_i(t)$.
(iii) Otherwise we leave $\mathsf{ep}_i(t)$ undefined.

By definition, ep_i is an extreme policy since it satisfies conditions (a) and (b) in Definition 4.8 and by construction $\mathbf{Der}_{\mathsf{ep}_i}(s) = \mathbf{Der}_{\mathsf{dp}_i}(s)$. □

This corollary gives a useful method for calculating the set of extreme derivatives of a given state, and therefore of the result of applying a test to a process.

Example 6.11 Consider again Fig. 4.4, discussed in Example 6.9, where we have the ω-respecting pLTS obtained by applying the test $a.\omega$ to the process Q_2. There are only two extreme policies for this pLTS, denoted by ep_0 and ep_1. They differ only for the state s_1, with $\mathsf{ep}_0(s_1) = \Theta_0$ and $\mathsf{ep}_1(s_1) = \Theta_1$. The discussion in Example 6.9 explained how

$$\mathbf{Der}_{\mathsf{ep}_0}(s_1) = \overline{\omega.\mathbf{0}} \qquad \mathbf{Der}_{\mathsf{ep}_1}(s_1) = \frac{1}{2}\overline{s_3} + \frac{1}{2}\overline{\omega.\mathbf{0}}$$

By Corollary 6.4, we know that every possible extreme derivative of $[\![[\![T|_{\mathsf{Act}} Q_2]\!]]\!]$ takes the form

$$q \cdot \overline{\omega.\mathbf{0}} + (1-q)\cdot \left(\frac{1}{2}\overline{s_3} + \frac{1}{2}\overline{\omega.\mathbf{0}}\right)$$

for some $0 \leq q \leq 1$. Since $\$(\overline{\omega}) = \vec{\omega}$ and $\$(\frac{1}{2}\overline{s_3} + \frac{1}{2}\overline{\omega}) = \frac{1}{2}\ \vec{\omega}$ it follows that

$$\mathcal{A}^{\mathrm{d}}(T, Q_2) = \left\{ q\ \vec{\omega} \mid q \in \left[\frac{1}{2}, 1\right]\right\}.$$

6.5.2 Realising Payoffs

In this section, we expose some less obvious properties of derivations, relating to their behaviour at infinity. One important property is that if we associate each state with a reward, which is a value in $[-1, 1]$, then the maximum payoff realisable by following all possible weak derivations can in fact be achieved by some static derivative policy, as stated in Theorem 6.8. The property depends on our working within *finitary* pLTSs, that is, ones in which the state space is finite and the (unlifted) transition relation is finite-branching. We first need to formalise some concepts such as discounted weak derivation and discounted payoff.

Definition 6.14 (Discounted Weak Derivation) The *discounted weak derivation* $\Delta \Longrightarrow_\delta \Delta'$ for discount factor δ $(0 \leq \delta \leq 1)$ is obtained from a weak derivation by discounting each τ transition by δ. That is, there is a collection of $\Delta_k^\rightarrow$ and $\Delta_k^\times$ satisfying

$$\begin{array}{rcl} \Delta & = & \Delta_0^\rightarrow + \Delta_0^\times \\ \Delta_0^\rightarrow & \xrightarrow{\tau} & \Delta_1^\rightarrow + \Delta_1^\times \\ & \vdots & \\ \Delta_k^\rightarrow & \xrightarrow{\tau} & \Delta_{k+1}^\rightarrow + \Delta_{k+1}^\times \\ & \vdots & \end{array}$$

such that $\Delta' = \sum_{k=0}^{\infty} \delta^k \Delta_k^\times$.

It is trivial that the relation $\Longrightarrow_1$ coincides with $\Longrightarrow$ given in Definition 6.4.

Definition 6.15 (Discounted Payoff) Given a pLTS with state space S, a discount δ, and reward function $\mathbf{r}$, we define the *discounted payoff function* $\mathbb{P}^{\delta,\mathbf{r}}_{\max} : S \rightarrow \mathbb{R}$ by

$$\mathbb{P}^{\delta,\mathbf{r}}_{\max}(s) = \sup\{\mathbf{r} \cdot \Delta' \mid \overline{s} \Longrightarrow_\delta \Delta'\}$$

and we will generalise it to be of type $\mathcal{D}_{sub}(S) \rightarrow \mathbb{R}$ by letting

$$\mathbb{P}^{\delta,\mathbf{r}}_{\max}(\Delta) = \sum_{s \in \lceil \Delta \rceil} \Delta(s) \cdot \mathbb{P}^{\delta,\mathbf{r}}_{\max}(s).$$

Definition 6.16 (Max-seeking Policy) Given a pLTS, discount δ and reward function $\mathbf{r}$, we say a static derivative policy dp is *max-seeking* with respect to δ and $\mathbf{r}$ if for all s the following requirements are met.

1. If $\mathsf{dp}(s)\uparrow$, then $\mathbf{r}(s) \geq \delta \cdot \mathbb{P}^{\delta,\mathbf{r}}_{\max}(\Delta_1)$ for all $s \xrightarrow{\tau} \Delta_1$.
2. If $\mathsf{dp}(s) = \Delta$ then
   a) $\delta \cdot \mathbb{P}^{\delta,\mathbf{r}}_{\max}(\Delta) \geq \mathbf{r}(s)$ and
   b) $\mathbb{P}^{\delta,\mathbf{r}}_{\max}(\Delta) \geq \mathbb{P}^{\delta,\mathbf{r}}_{\max}(\Delta_1)$ for all $s \xrightarrow{\tau} \Delta_1$.

Lemma 6.12 *Given a finitely branching pLTS, discount δ and reward function $\mathbf{r}$, there always exists a max-seeking policy.*

Proof Given a pLTS, discount δ and reward function $\mathbf{r}$, the discounted payoff $\mathbb{P}^{\delta,\mathbf{r}}_{\max}(s)$ can be calculated for each state s. Then we can define a derivative policy dp in the following way. For any state s, if $\mathbf{r}(s) \geq \delta \cdot \mathbb{P}^{\delta,\mathbf{r}}_{\max}(\Delta_1)$ for all $s \xrightarrow{\tau} \Delta_1$, then we set dp undefined at s. Otherwise, we choose a transition $s \xrightarrow{\tau} \Delta$ among the finite number of outgoing transitions from s such that $\mathbb{P}^{\delta,\mathbf{r}}_{\max}(\Delta) \geq \mathbb{P}^{\delta,\mathbf{r}}_{\max}(\Delta_1)$ for all other transitions $s \xrightarrow{\tau} \Delta_1$, and we set $\mathsf{dp}(s) = \Delta$.

Given a pLTS, discount δ, reward function $\mathbf{r}$, and derivative policy dp, we define the function $F^{\delta,\mathsf{dp},\mathbf{r}} : (S \to \mathbb{R}) \to (S \to \mathbb{R})$ by letting

$$F^{\delta,\mathsf{dp},\mathbf{r}}(f)(s) = \begin{cases} \mathbf{r}(s) & \text{if } \mathsf{dp}(s)\uparrow \\ \delta \cdot f(\Delta) & \text{if } \mathsf{dp}(s) = \Delta \end{cases} \tag{6.16}$$

where $f(\Delta) = \sum_{s \in \lceil \Delta \rceil} \Delta(s) \cdot f(s)$.

Lemma 6.13 *Given a pLTS, discount $\delta < 1$, reward function $\mathbf{r}$, and derivative policy dp, the function $F^{\delta,\mathsf{dp},\mathbf{r}}$ has a unique fixed point.*

Proof We first show that the function $F^{\delta,\mathsf{dp},\mathbf{r}}$ is a contraction mapping. Let f, g be any two functions of type $S \to \mathbb{R}$.

$$\begin{aligned}
& |F^{\delta,\mathsf{dp},\mathbf{r}}(f) - F^{\delta,\mathsf{dp},\mathbf{r}}(g)| \\
= {} & \sup\{|F^{\delta,\mathsf{dp},\mathbf{r}}(f)(s) - F^{\delta,\mathsf{dp},\mathbf{r}}(g)(s)| \mid s \in S\} \\
= {} & \sup\{|F^{\delta,\mathsf{dp},\mathbf{r}}(f)(s) - F^{\delta,\mathsf{dp},\mathbf{r}}(g)(s)| \mid s \in S \text{ and } \mathsf{dp}(s)\downarrow\} \\
= {} & \delta \cdot \sup\{|f(\Delta) - g(\Delta)| \mid s \in S \text{ and } \mathsf{dp}(s) = \Delta \text{ for some } \Delta\} \\
\leq {} & \delta \cdot \sup\{|f(s') - g(s')| \mid s' \in S\} \\
= {} & \delta \cdot |f - g| \\
< {} & |f - g|
\end{aligned}$$

By the Banach unique fixed-point theorem (Theorem 2.8), the function $F^{\delta,\mathsf{dp},\mathbf{r}}$ has a unique fixed point. □

Lemma 6.14 *Given a pLTS, discount δ, reward function $\mathbf{r}$, and max-seeking policy dp, the function $\mathbb{P}^{\delta,\mathbf{r}}_{\max}$ is a fixed point of $F^{\delta,\mathsf{dp},\mathbf{r}}$.*

Proof We need to show that $F^{\delta,\mathsf{dp},\mathbf{r}}(\mathbb{P}^{\delta,\mathbf{r}}_{\max})(s) = \mathbb{P}^{\delta,\mathbf{r}}_{\max}(s)$ holds for any state s. We distinguish two cases.

1. If $\mathsf{dp}(s)\uparrow$, then $F^{\delta,\mathsf{dp},\mathbf{r}}(\mathbb{P}^{\delta,\mathbf{r}}_{\max})(s) = \mathbf{r}(s) = \mathbb{P}^{\delta,\mathbf{r}}_{\max}(s)$ as expected.
2. If $\mathsf{dp}(s) = \Delta$, then the arguments are more involved. First we observe that if $\overline{s} \Longrightarrow_\delta \Delta'$ then by Definition 6.14 there exist some $\Delta_0^{\rightarrow}, \Delta_0^{\times}, \Delta_1, \Delta''$ such that $\overline{s} = \Delta_0^{\rightarrow} + \Delta_0^{\times}$, $\Delta_0^{\rightarrow} \xrightarrow{\tau} \Delta_1$, $\Delta_1 \Longrightarrow_\delta \Delta''$ and $\Delta' = \Delta_0^{\times} + \delta \cdot \Delta''$. So we can do the following calculation.

$$
\begin{aligned}
&\mathbb{P}^{\delta,\mathbf{r}}_{\max}(s)\\
&= \sup\{\mathbf{r}\cdot\Delta' \mid \overline{s} \Longrightarrow_\delta \Delta'\}\\
&= \sup\{\mathbf{r}\cdot(\Delta_0^\times + \delta\cdot\Delta'') \mid \overline{s} = \Delta_0^\rightarrow + \Delta_0^\times, \Delta_0^\rightarrow \xrightarrow{\tau} \Delta_1, \text{ and } \Delta_1 \Longrightarrow_\delta \Delta''\\
&\qquad\qquad \text{for some } \Delta_0^\rightarrow, \Delta_0^\times, \Delta_1, \Delta''\}\\
&= \sup\{\mathbf{r}\cdot\Delta_0^\times + \delta\cdot\mathbf{r}\cdot\Delta'' \mid \overline{s} = \Delta_0^\rightarrow + \Delta_0^\times, \Delta_0^\rightarrow \xrightarrow{\tau} \Delta_1, \text{ and } \Delta_1 \Longrightarrow_\delta \Delta''\\
&\qquad\qquad \text{for some } \Delta_0^\rightarrow, \Delta_0^\times, \Delta_1, \Delta''\}\\
&= \sup\{\mathbf{r}\cdot\Delta_0^\times + \delta\cdot\sup\{\mathbf{r}\cdot\Delta'' \mid \Delta_1 \Longrightarrow_\delta \Delta''\} \mid \overline{s} = \Delta_0^\rightarrow + \Delta_0^\times \text{ and } \Delta_0^\rightarrow \xrightarrow{\tau} \Delta_1\\
&\qquad\qquad \text{for some } \Delta_0^\rightarrow, \Delta_0^\times, \Delta_1\}\\
&= \sup\{\mathbf{r}\cdot\Delta_0^\times + \delta\cdot\mathbb{P}^{\delta,\mathbf{r}}_{\max}(\Delta_1) \mid \overline{s} = \Delta_0^\rightarrow + \Delta_0^\times \text{ and } \Delta_0^\rightarrow \xrightarrow{\tau} \Delta_1\\
&\qquad\qquad \text{for some } \Delta_0^\rightarrow, \Delta_0^\times, \Delta_1\}\\
&= \sup\{(1-p)\cdot\mathbf{r}(s) + p\delta\cdot\mathbb{P}^{\delta,\mathbf{r}}_{\max}(\Delta_1) \mid p\in[0,1] \text{ and } \overline{s} \xrightarrow{\tau} \Delta_1 \text{ for some } \Delta_1\}\\
&\qquad\qquad [\overline{s} \text{ can be split into } p\overline{s} + (1-p)\overline{s} \text{ only}]\\
&= \sup\{(1-p)\cdot\mathbf{r}(s) + p\delta\cdot\mathbb{P}^{\delta,\mathbf{r}}_{\max}(\Delta_1) \mid p\in[0,1] \text{ and } s \xrightarrow{\tau} \Delta_1\\
&\qquad\qquad \text{for some } \Delta_1\}\\
&= \sup\{(1-p)\cdot\mathbf{r}(s) + p\delta\cdot\sup\{\mathbb{P}^{\delta,\mathbf{r}}_{\max}(\Delta_1) \mid s \xrightarrow{\tau} \Delta_1\} \mid p\in[0,1]\}\\
&= \mathit{max}(\mathbf{r}(s),\ \delta\cdot\sup\{\mathbb{P}^{\delta,\mathbf{r}}_{\max}(\Delta_1) \mid s \xrightarrow{\tau} \Delta_1\})\\
&= \delta\cdot\mathbb{P}^{\delta,\mathbf{r}}_{\max}(\Delta) \qquad [\text{as } \mathsf{dp} \text{ is max-seeking}]\\
&= F^{\delta,\mathsf{dp},\mathbf{r}}(\mathbb{P}^{\delta,\mathbf{r}}_{\max})(s)
\end{aligned}
$$

□

Definition 6.17 Let Δ be a subdistribution and dp a static derivative policy. We define a collection of subdistributions Δ_k as follows.

$$
\begin{aligned}
\Delta_0 &= \Delta\\
\Delta_{k+1} &= \sum\{\Delta_k(s)\cdot\mathsf{dp}(s) \mid s\in\lceil\Delta_k\rceil \text{ and } \mathsf{dp}(s)\downarrow\} \qquad \text{for all } k\geq 0.
\end{aligned}
$$

Then $\Delta_k^\times$ is obtained from Δ_k by letting

$$
\Delta_k^\times(s) = \begin{cases} 0 & \text{if } \mathsf{dp}(s)\downarrow\\ \Delta_k(s) & \text{otherwise}\end{cases}
$$

for all $k \geq 0$. Then we write $\Delta \Longrightarrow_{\delta,\mathsf{dp}} \Delta'$ for the discounted weak derivation that determines a unique subdistribution Δ' with $\Delta' = \sum_{k=0}^\infty \delta^k \Delta_k^\times$.

In other words, if $\Delta \Longrightarrow_{\delta,\mathsf{dp}} \Delta'$ then Δ comes from the discounted weak derivation $\Delta \Longrightarrow_\delta \Delta'$ that is constructed by following the derivative policy dp when choosing τ transitions from each state. In the special case when the discount factor $\delta = 1$, we see that $\Longrightarrow_{1,\mathsf{dp}}$ becomes $\Longrightarrow_{\mathsf{dp}}$ as defined in page 176.

Definition 6.18 (Policy-Following Payoff) Given a discount δ, reward function $\mathbf{r}$, and derivative policy dp, the *policy-following payoff function* $\mathbb{P}^{\delta,\mathsf{dp},\mathbf{r}} : S \to \mathbb{R}$ is defined by

$$\mathbb{P}^{\delta,\mathsf{dp},\mathbf{r}}(s) \;=\; \mathbf{r} \cdot \Delta'$$

where Δ' is determined by the discounted weak derivation $\overline{s} \Longrightarrow_{\delta,\mathsf{dp}} \Delta'$.

Lemma 6.15 *For any discount δ, reward function $\mathbf{r}$, and derivative policy dp, the function $\mathbb{P}^{\delta,\mathsf{dp},r}$ is a fixed point of $F^{\delta,\mathsf{dp},\mathbf{r}}$.*

Proof We need to show that $F^{\delta,\mathsf{dp},\mathbf{r}}(\mathbb{P}^{\delta,\mathsf{dp},\mathbf{r}})(s) = \mathbb{P}^{\delta,\mathsf{dp},\mathbf{r}}(s)$ holds for any state s. There are two cases.

1. If $\mathsf{dp}(s)\uparrow$, then $\overline{s} \Longrightarrow_{\delta,\mathsf{dp}} \Delta'$ implies $\Delta' = \overline{s}$. Therefore,

$$\mathbb{P}^{\delta,\mathsf{dp},\mathbf{r}}(s) \;=\; \mathbf{r}(s) \;=\; F^{\delta,\mathsf{dp},\mathbf{r}}(\mathbb{P}^{\delta,\mathsf{dp},\mathbf{r}})(s)$$

   as required.
2. Suppose $\mathsf{dp}(s) = \Delta_1$. If $\overline{s} \Longrightarrow_{\delta,\mathsf{dp}} \Delta'$ then $s \xrightarrow{\tau} \Delta_1$, $\Delta_1 \Longrightarrow_{\delta,\mathsf{dp}} \Delta''$ and $\Delta' = \delta\Delta''$ for some subdistribution Δ''. Therefore,

$$\begin{aligned}
&\mathbb{P}^{\delta,\mathsf{dp},\mathbf{r}}(s)\\
&= \mathbf{r} \cdot \Delta'\\
&= \mathbf{r} \cdot \delta\Delta''\\
&= \delta \cdot \mathbf{r} \cdot \Delta''\\
&= \delta \cdot \mathbb{P}^{\delta,\mathsf{dp},\mathbf{r}}(\Delta_1)\\
&= F^{\delta,\mathsf{dp},\mathbf{r}}(\mathbb{P}^{\delta,\mathsf{dp},\mathbf{r}})(s)
\end{aligned}$$

□

Proposition 6.6 *Let $\delta \in [0, 1)$ be a discount and $\mathbf{r}$ a reward function. If dp is a max-seeking policy with respect to δ and $\mathbf{r}$, then $\mathbb{P}^{\delta,r}_{\max} = \mathbb{P}^{\delta,\mathsf{dp},r}$.*

Proof By Lemma 6.13, the function $F^{\delta,\mathsf{dp},\mathbf{r}}$ has a unique fixed point. By Lemmas 6.14 and 6.15, both $\mathbb{P}^{\delta,\mathbf{r}}_{\max}$ and $\mathbb{P}^{\delta,\mathsf{dp},\mathbf{r}}$ are fixed points of the same function $F^{\delta,\mathsf{dp},\mathbf{r}}$, which means that $\mathbb{P}^{\delta,\mathbf{r}}_{\max}$ and $\mathbb{P}^{\delta,\mathsf{dp},\mathbf{r}}$ coincide with each other. □

Lemma 6.16 *Suppose $\overline{s} \Longrightarrow \Delta'$ with $\Delta' = \sum_{i=0}^{\infty} \Delta_i^{\times}$ for some properly related $\Delta_i^{\times}$. Let $\{\delta_j\}_{j=0}^{\infty}$ be a nondecreasing sequence of discount factors converging to 1. Then for any reward function $\mathbf{r}$ it holds that*

$$\mathbf{r} \cdot \Delta' \;=\; \lim_{j\to\infty} \sum_{i=0}^{\infty} \delta_j^i (\mathbf{r} \cdot \Delta_i^{\times}).$$

Proof Let $f : \mathbb{N} \times \mathbb{N} \to \mathbb{R}$ be the function defined by $f(i, j) = \delta_j^i(\mathbf{r} \cdot \Delta_i^{\times})$. We check that f satisfies the four conditions in Proposition 4.3.

1. f satisfies condition **C1**. For all $i, j_1, j_2 \in \mathbb{N}$, if $j_1 \leq j_2$ then $\delta^i_{j_1} \leq \delta^i_{j_2}$. It follows that

$$|f(i, j_1)| = |\delta^i_{j_1}(\mathbf{r} \cdot \Delta^\times_i)| \leq |\delta^i_{j_2}(\mathbf{r} \cdot \Delta^\times_i)| = |f(i, j_2)|.$$

2. f satisfies condition **C2**. For any $i \in \mathbb{N}$, we have

$$\lim_{j\to\infty} |f(i, j)| = \lim_{j\to\infty} |\delta^i_j(\mathbf{r} \cdot \Delta^\times_i)| = |\mathbf{r} \cdot \Delta^\times_i|. \tag{6.17}$$

3. f satisfies condition **C3**. For any $n \in \mathbb{N}$, the partial sum $S_n = \sum_{i=0}^n \lim_{j\to\infty} |f(i, j)|$ is bounded because

$$\sum_{i=0}^{n} \lim_{j\to\infty} |f(i, j)| = \sum_{i=0}^{n} |\mathbf{r} \cdot \Delta^\times_i| \leq \sum_{i=0}^{\infty} |\mathbf{r} \cdot \Delta^\times_i| \leq \sum_{i=0}^{\infty} |\Delta^\times_i| = |\Delta'|$$

   where the first equality is justified by (6.17).
4. f satisfies condition **C4**. For any $i, j_1, j_2 \in \mathbb{N}$ with $j_1 \leq j_2$, suppose we have $f(i, j_1) = \delta^i_{j_1}(\mathbf{r} \cdot \Delta^\times_i) > 0$. Then $\mathbf{r} \cdot \Delta^\times_i > 0$ and it follows immediately that $f(i, j_2) = \delta^i_{j_2}(\mathbf{r} \cdot \Delta^\times_i) > 0$.

Therefore, we can use Proposition 2.2 to do the following inference.

$$\begin{aligned}
&\lim_{j\to\infty} \textstyle\sum_{i=0}^\infty \delta^i_j(\mathbf{r} \cdot \Delta^\times_i) \\
&= \textstyle\sum_{i=0}^\infty \lim_{j\to\infty} \delta^i_j(\mathbf{r} \cdot \Delta^\times_i) \\
&= \textstyle\sum_{i=0}^\infty \mathbf{r} \cdot \Delta^\times_i \\
&= \mathbf{r} \cdot \textstyle\sum_{i=0}^\infty \Delta^\times_i \\
&= \mathbf{r} \cdot \Delta'
\end{aligned}$$

□

Corollary 6.5 *Let $\{\delta_j\}_{j=0}^\infty$ be a nondecreasing sequence of discount factors converging to 1. For any derivative policy* dp *and reward function* $\mathbf{r}$*, it holds that* $\mathbb{P}^{1,\mathsf{dp},\mathbf{r}} = \lim_{j\to\infty} \mathbb{P}^{\delta_j,\mathsf{dp},\mathbf{r}}$.

Proof We need to show that $\mathbb{P}^{1,\mathsf{dp},\mathbf{r}}(s) = \lim_{j\to\infty} \mathbb{P}^{\delta_j,\mathsf{dp},\mathbf{r}}(s)$, for any state s. Note that for any discount δ_j, each state s enables a unique discounted weak derivation $\overline{s} \Longrightarrow_{\delta_j,\mathsf{dp}} \Delta^j$ such that $\Delta^j = \sum_{i=0}^\infty \delta^i_j \Delta^\times_i$ for some properly related subdistributions $\Delta^\times_i$. Let $\Delta' = \sum_{i=0}^\infty \Delta^\times_i$. We have $\overline{s} \Longrightarrow_{1,\mathsf{dp}} \Delta'$. Then we can infer that

$$\begin{aligned}
&\lim_{j\to\infty} \mathbb{P}^{\delta_j,\mathsf{dp},\mathbf{r}}(s) \\
&= \lim_{j\to\infty} \mathbf{r} \cdot \Delta^j \\
&= \lim_{j\to\infty} \mathbf{r} \cdot \textstyle\sum_{i=0}^\infty \delta^i_j \Delta^\times_i \\
&= \lim_{j\to\infty} \textstyle\sum_{i=0}^\infty \delta^i_j(\mathbf{r} \cdot \Delta^\times_i) \\
&= \mathbf{r} \cdot \Delta' \qquad \text{by Lemma 6.16} \\
&= \mathbb{P}^{1,\mathsf{dp},\mathbf{r}}(s)
\end{aligned}$$

□

Theorem 6.10 *In a finitary pLTS, for any reward function* $\mathbf{r}$ *there exists a derivative policy* dp *such that* $\mathbb{P}^{1,r}_{\max} = \mathbb{P}^{1,\mathsf{dp},r}$.

Proof Let $\mathbf{r}$ be a reward function. By Proposition 6.6, for every discount factor $d < 1$ there exists a max-seeking derivative policy dp with respect to δ and $\mathbf{r}$ such that

$$\mathbb{P}^{\delta,\mathbf{r}}_{\max} = \mathbb{P}^{\delta,\mathsf{dp},\mathbf{r}}. \tag{6.18}$$

Since the pLTS is finitary, there are finitely many different static derivative policies. There must exist a derivative policy dp such that (6.18) holds for infinitely many discount factors. In other words, for every nondecreasing sequence $\{\delta_n\}_{n=0}^{\infty}$ converging to 1, there exists a subsequence $\{\delta_{n_j}\}_{j=0}^{\infty}$ and a derivative policy $\mathsf{dp}^\star$ such that

$$\mathbb{P}^{\delta_{n_j},\mathbf{r}}_{\max} = \mathbb{P}^{\delta_{n_j},\mathsf{dp}^\star,\mathbf{r}} \qquad \text{for all } j \geq 0. \tag{6.19}$$

For any state s, we infer as follows.

$$\begin{aligned}
&\mathbb{P}^{1,\mathbf{r}}_{\max}(s)\\
&= \sup\{\mathbf{r}\cdot\Delta' \mid \overline{s} \Longrightarrow \Delta'\}\\
&= \sup\{\lim_{j\to\infty}\textstyle\sum_{i=0}^{\infty}\delta_{n_j}^i(\mathbf{r}\cdot\Delta_i^{\times}) \mid \overline{s} \Longrightarrow \Delta' \text{ with } \Delta' = \sum_{i=0}^{\infty}\Delta_i^{\times}\} \quad \text{by Lemma 6.16}\\
&= \lim_{j\to\infty}\sup\{\textstyle\sum_{i=0}^{\infty}\delta_{n_j}^i(\mathbf{r}\cdot\Delta_i^{\times}) \mid \overline{s} \Longrightarrow \Delta' \text{ with } \Delta' = \sum_{i=0}^{\infty}\Delta_i^{\times}\}\\
&= \lim_{j\to\infty}\sup\{\mathbf{r}\cdot\textstyle\sum_{i=0}^{\infty}\delta_{n_j}^i\Delta_i^{\times} \mid \overline{s} \Longrightarrow \Delta' \text{ with } \Delta' = \sum_{i=0}^{\infty}\Delta_i^{\times}\}\\
&= \lim_{j\to\infty}\sup\{\mathbf{r}\cdot\Delta'' \mid \overline{s} \Longrightarrow_{\delta_{n_j}} \Delta''\}\\
&= \lim_{j\to\infty}\mathbb{P}^{\delta_{n_j},\mathbf{r}}_{\max}(s)\\
&= \lim_{j\to\infty}\mathbb{P}^{\delta_{n_j},\mathsf{dp}^\star,\mathbf{r}}(s) \qquad \text{by (6.19)}\\
&= \mathbb{P}^{1,\mathsf{dp}^\star,\mathbf{r}}(s) \qquad \text{by Corollary 6.5}
\end{aligned}$$

□

6.5.3 Consequences

In this section, we outline two major consequences of Theorem 6.9, which informally means that the set of weak derivatives from a given state is the convex-closure of a finite set. The first is straightforward and is explained in the following two results.

Lemma 6.17 *(Closure of* $\Longrightarrow$*) For any state s in a finitary pLTS the set of derivatives* $\{\Delta \mid \overline{s} \Longrightarrow \Delta\}$ *is closed and convex.*

Proof Let $\mathsf{dp}_1, \ldots, \mathsf{dp}_n$ $(n \geq 1)$ be all the derivative policies in the finitary pLTS. Consider two sets $C = \updownarrow\{\,\mathbf{Der}_{\mathsf{dp}_i}(s) \mid 1 \leq i \leq n\}$ and $D = \{\Delta' \mid \overline{s} \Longrightarrow \Delta'\}$. By Theorem 6.9 D coincides with C, the convex closure of a finite set. By Lemma 6.9, it is also Cauchy closed. □

The restriction here to finitary pLTSs is essential, as the following examples demonstrate.

Example 6.12 Consider the finite state but infinitely branching pLTS containing three states s_1, s_2, s_3 and the countable set of transitions given by

$$s_1 \xrightarrow{\tau} (\overline{s_2}\ {}_{\frac{1}{2^n}}\!\oplus \overline{s_3}) \qquad n \geq 1$$

For convenience, let Δ_n denote the distribution $(\overline{s_2}\ {}_{\frac{1}{2^n}}\!\oplus \overline{s_3})$. Then $\{\Delta_n \mid n \geq 1\}$ is a Cauchy sequence with limit $\overline{s_3}$. Trivially the set $\{\Delta \mid \overline{s_1} \Longrightarrow \Delta\}$ contains every Δ_n, but it does not contain the limit of the sequence, thus it is not closed.

Example 6.13 By adapting Example 6.12, we obtain the following pLTS that is finitely branching but has infinitely many states. Let t_1 and t_2 be two distinct states. Moreover, for each $n \geq 1$, there is a state s_n with two outgoing transitions: $s_n \xrightarrow{\tau} \overline{s_{n+1}}$ and $s_n \xrightarrow{\tau} \overline{t_1}\ {}_{\frac{1}{2^n}}\!\oplus \overline{t_2}$. Let Δ_n denote the distribution $\overline{t_1}\ {}_{\frac{1}{2^n}}\!\oplus \overline{t_2}$. Then $\{\Delta_n \mid n \geq 1\}$ is a Cauchy sequence with limit $\overline{t_2}$. The set $\{\Delta \mid \overline{s_1} \Longrightarrow \Delta\}$ is not closed because it contains each Δ_n but not the limit $\overline{t_2}$.

Corollary 6.6 *(Closure of $\stackrel{a}{\Longrightarrow}$) For any state s in a finitary pLTS the set*

$$\{\Delta \mid \overline{s} \stackrel{a}{\Longrightarrow} \Delta\}$$

is closed and convex.

Proof We first introduce a preliminary concept. We say a subset $D \subseteq \mathcal{D}_{sub}(S)$ is *finitely generable* whenever there is some finite set $F \subseteq \mathcal{D}_{sub}(S)$ such that $D = \updownarrow F$. A relation $\mathcal{R} \subseteq X \times \mathcal{D}_{sub}(S)$ is *finitely generable* if for every x in X the set $x \cdot \mathcal{R}$ is finitely generable. We observe that

(i) If a set is finitely generable, then it is closed and convex.
(ii) If $\mathcal{R}_1, \mathcal{R}_2 \subseteq \mathcal{D}_{sub}(S) \times \mathcal{D}_{sub}(S)$ are finitely generable then so is their composition $\mathcal{R}_1 \cdot \mathcal{R}_2$.

The first property is a direct consequence of the definition of finite generability. To prove the second property, we let $\mathcal{B}^i_\Phi$ be a finite set of subdistributions such that $\Phi \cdot \mathcal{R}_i = \updownarrow \mathcal{B}^i_\Phi$ for $i = 1, 2$. Then one can check that

$$\Delta \cdot (\mathcal{R}_1 \cdot \mathcal{R}_2) = \updownarrow \bigcup \{\mathcal{B}^2_\Theta \mid \Theta \in \mathcal{B}^1_\Delta\}$$

which implies that finite generability is preserved under composition of relations.

Notice that the relation $\stackrel{a}{\Longrightarrow}$ is a composition of three stages: $\Longrightarrow \cdot \xrightarrow{a} \cdot \Longrightarrow$. In the proof of Lemma 6.17, we have shown that $\Longrightarrow$ is finitely generable. In a finitary pLTS, the relation $\xrightarrow{a}$ is also finitely generable. It follows from property (ii) that $\stackrel{a}{\Longrightarrow}$ is finitely generable. By property (i) we have that $\stackrel{a}{\Longrightarrow}$ is closed and convex. □

Corollary 6.7 *In a finitary pLTS, the relation $\stackrel{a}{\Longrightarrow}$ is the lifting of the closed and convex relation $\Longrightarrow_S \xrightarrow{a} \Longrightarrow$, where $s \Longrightarrow_S \Delta$ means $\overline{s} \Longrightarrow \Delta$.*

Proof The relation $\Longrightarrow_S \xrightarrow{a} \Longrightarrow$ is $\stackrel{a}{\Longrightarrow}$ restricted to point distributions. We have shown that $\stackrel{a}{\Longrightarrow}$ is closed and convex in Corollary 6.6. Therefore, $\Longrightarrow_S \xrightarrow{a} \Longrightarrow$ is closed and convex. Its lifting coincides with $\stackrel{a}{\Longrightarrow}$, which can be shown by some arguments analogous to those in the proof of Proposition 6.3. □

The second consequence of Theorem 6.9 concerns the manner in which divergent computations arise in pLTSs. Consider again the infinite state pLTS given in Example 6.5. There is no state s that wholly diverges, that is satisfying $s \Longrightarrow \varepsilon$, yet there are many partially divergent computations. In fact for every $k \geq 2$ we have $s_k \Longrightarrow \frac{1}{k}[\![a.\mathbf{0}]\!]$. This cannot arise in a finitary pLTS; if there is any partial derivation in a finitary pLTS, $\Delta \Longrightarrow \Delta'$ with $|\Delta|>|\Delta'|$, then there is some state in the pLTS that wholly diverges.

We say a pLTS is *convergent* if $\overline{s} \Longrightarrow \varepsilon$ for no state $s \in S$.

Lemma 6.18 *Let Δ be a subdistribution in a* finite-state, convergent *and* deterministic *pLTS. If $\Delta \Longrightarrow \Delta'$ then $|\Delta| = |\Delta'|$.*

Proof Since the pLTS is convergent, then $\overline{s} \Longrightarrow \varepsilon$ for no state $s \in S$. In other words, each τ sequence from a state s is finite and ends with a distribution that cannot enable a τ transition. In a deterministic pLTS, each state has at most one outgoing transition. So from each s there is a unique τ sequence with length $n_s \geq 0$.

$$\overline{s} \xrightarrow{\tau} \Delta_1 \xrightarrow{\tau} \Delta_2 \xrightarrow{\tau} \cdots \xrightarrow{\tau} \Delta_{n_s} \not\xrightarrow{\tau}$$

Let p_s be $\Delta_{n_s}(s')$, where s' is any state in the support of Δ_{n_s}. We set

$$n = \max\{n_s \mid s \in S\}$$
$$p = \min\{p_s \mid s \in S\}$$

where n and p are well defined as S is assumed to be a finite set. Now let $\Delta \Longrightarrow \Delta'$ be any weak derivation constructed by a collection of $\Delta_k^{\rightarrow}, \Delta_k^{\times}$ such that

$$\begin{aligned} \Delta &= \Delta_0^{\rightarrow} + \Delta_0^{\times} \\ \Delta_0^{\rightarrow} &\xrightarrow{\tau} \Delta_1^{\rightarrow} + \Delta_1^{\times} \\ &\vdots \\ \Delta_k^{\rightarrow} &\xrightarrow{\tau} \Delta_{k+1}^{\rightarrow} + \Delta_{k+1}^{\times} \\ &\vdots \end{aligned}$$

with $\Delta' = \sum_{k=0}^{\infty} \Delta_k^{\times}$. From each $\Delta_{kn+i}^{\rightarrow}$ with $k, i \in \mathbb{N}$, the block of n steps of τ transition leads to $\Delta_{(k+1)n+i}^{\rightarrow}$ such that $|\Delta_{(k+1)n+i}^{\rightarrow}| \leq |\Delta_{kn+i}^{\rightarrow}|(1-p)$. It follows that

$$\begin{aligned} \textstyle\sum_{j=0}^{\infty} |\Delta_j^{\rightarrow}| &= \textstyle\sum_{i=0}^{n-1} \sum_{k=0}^{\infty} |\Delta_{kn+i}^{\rightarrow}| \\ &\leq \textstyle\sum_{i=0}^{n-1} \sum_{k=0}^{\infty} |\Delta_i^{\rightarrow}|(1-p)^k \\ &= \textstyle\sum_{i=0}^{n-1} |\Delta_i^{\rightarrow}| \frac{1}{p} \\ &\leq |\Delta_0^{\rightarrow}| \frac{n}{p} \end{aligned}$$

Therefore, we have that $\lim_{k\to\infty} \Delta_k^{\rightarrow} = 0$, which in turn means that $|\Delta'| = |\Delta|$. □

Corollary 6.8 (Zero-one Law - Deterministic Case) *If for some static derivative policy* dp *over a finite-state pLTS there is for some* s *a derivation* $\overline{s} \Longrightarrow_{\mathsf{dp}} \Delta'$ *with* $|\Delta'| < 1$ *then in fact for some (possibly different) state* s_ε *we have* $s_\varepsilon \Longrightarrow_{\mathsf{dp}} \varepsilon$.

Proof Suppose that for no state s do we have $\overline{s} \Longrightarrow_{\mathsf{dp}} \varepsilon$. Then the pLTS induced by dp is convergent. Since it is obviously finite-state and deterministic, we apply Lemma 6.18 and obtain $|\Delta'| = |\overline{s}| = 1$, contradicting the assumption that $|\Delta'| < 1$. Therefore, there must exist some state s_ε that wholly diverges. □

Although it is possible to have processes that diverge with some probability strictly between zero and one, in a finitary system it is possible to "distill" that divergence in the sense that in many cases we can limit our analyses to processes that either wholly diverge (can do so with probability one) or wholly converge (can diverge only with probability zero). This property is based on the zero-one law for finite-state probabilistic systems, and we present the aspects of it that we need here.

Lemma 6.19 (Distillation of Divergence - Deterministic Case) *For any state* s *and static derivative policy* dp *over a finite-state pLTS, if there is a derivation* $\overline{s} \Longrightarrow_{\mathsf{dp}} \Delta'$ *then there is a probability* p *and full distributions* $\Delta'_1, \Delta'_\varepsilon$ *such that* $\overline{s} \Longrightarrow (\Delta'_1 {}_p\!\oplus \Delta'_\varepsilon)$ *and* $\Delta' = p \cdot \Delta'_1$ *and* $\Delta'_\varepsilon \Longrightarrow \varepsilon$.

Proof We modify dp so as to obtain a static policy dp' by setting $\mathsf{dp}'(t) = \mathsf{dp}(t)$ except when $\overline{t} \Longrightarrow_{\mathsf{dp}} \varepsilon$, in which case we set $\mathsf{dp}'(t) \uparrow$. The new policy determines a unique weak derivation $\Delta \Longrightarrow_{\mathsf{dp}'} \Delta''$ for some subdistribution Δ'', and induces a sub-pLTS from the pLTS induced by dp. Note that the sub-pLTS is deterministic and convergent. By Lemma 6.18, we know that $|\Delta''| = |\overline{s}| = 1$. We split Δ'' up into $\Delta''_1 + \Delta''_\varepsilon$ so that each state in $\lceil \Delta''_\varepsilon \rceil$ is wholly divergent under policy dp and Δ''_1 is supported by all other states. From Δ''_ε the policy dp determines the weak derivation $\Delta''_\varepsilon \Longrightarrow_{\mathsf{dp}} \varepsilon$. Combining the two weak derivations we have $\overline{s} \Longrightarrow_{\mathsf{dp}'} \Delta''_1 + \Delta''_\varepsilon \Longrightarrow_{\mathsf{dp}} \Delta''_1$. As we only divide the original weak derivation into two stages and do not change the τ transition from each state, the final subdistribution will not change, thus $\Delta''_1 = \Delta'$. Finally, we determine p, Δ'_1 and Δ'_ε by letting $p = |\Delta'|$, $\Delta'_1 = \frac{1}{p}\Delta'$ and $\Delta'_\varepsilon = \frac{1}{1-p}\Delta''_\varepsilon$. □

Theorem 6.11 (Distillation of Divergence—General Case) *For any* s, Δ' *in a finitary pLTS with* $\overline{s} \Longrightarrow \Delta'$ *there is a probability* p *and full distributions* $\Delta'_1, \Delta'_\varepsilon$ *such that* $\overline{s} \Longrightarrow (\Delta'_1 {}_p\!\oplus \Delta'_\varepsilon)$ *and* $\Delta' = p \cdot \Delta'_1$ *and* $\Delta'_\varepsilon \Longrightarrow \varepsilon$.

Proof Let $\{\mathsf{dp}_i \mid i \in I\}$ (I is a finite index set) be all the static derivative policies in the finitary pLTS. Each policy determines a weak derivation $\overline{s} \Longrightarrow_{\mathsf{dp}_i} \Delta'_i$. From Theorem 6.9, we know that if $\overline{s} \Longrightarrow \Delta'$ then $\Delta' = \sum_{i\in I} p_i \Delta'_i$ for some p_i with $\sum_{i\in I} p_i = 1$. By Lemma 6.19, for each $i \in I$, there is a probability q_i and full distributions $\Delta'_{i,1}, \Delta'_{i,\varepsilon}$ such that $\overline{s} \Longrightarrow (\Delta'_{i,1} {}_{q_i}\!\oplus \Delta'_{i,\varepsilon})$, $\Delta'_i = q_i \cdot \Delta'_{i,1}$, and $\Delta'_{i,\varepsilon} \Longrightarrow \varepsilon$. Finally, we determine p, Δ'_1 and Δ'_ε by letting $p = |\sum_{i\in I} p_i q_i \cdot \Delta'_{i,1}|$, $\Delta'_1 = \frac{1}{p}\Delta'$ and $\Delta'_\varepsilon = \frac{1}{1-p}\sum_{i\in I} p_i(1-q_i)\Delta'_{i,\varepsilon}$. They satisfy our requirements just by noting that $\overline{s} \Longrightarrow \sum_{i\in I} p_i(\Delta'_{i,1} {}_{q_i}\!\oplus \Delta'_{i,\varepsilon}) = \Delta'_1 {}_p\!\oplus \Delta'_\varepsilon$ □

The requirement on the pLTS to be finitary is essential for this distillation of divergence, as we explain in the following examples.

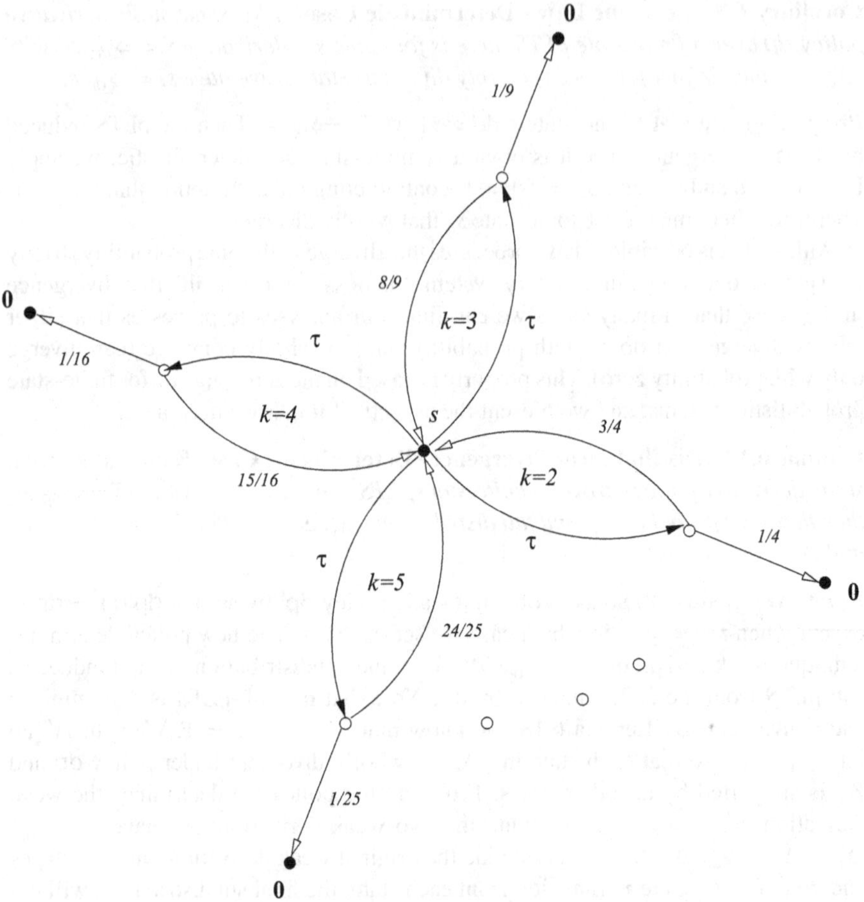

Fig. 6.4 Infinitely branching flower. There are two states s and $\mathbf{0}$. To diverge from s with probability $1 - 1/k$, start at "petal" k and take successive τ-loops anti-clockwise from there. Yet, although divergence with arbitrarily high probability is present, complete probability-1 divergence is nowhere possible. Either infinite states or infinite branching is necessary for this anomaly

Example 6.14 (Revisiting Example 6.5) The pLTS in Example 6.5 is an infinite state system over states s_k for all $k \geq 2$, where the probability of convergence is $1/k$ from any state s_k, thus a situation where distillation of divergence fails because all the states partially diverge, yet there is no single state that wholly diverges.

Example 6.15 Consider the finite state but infinitely branching pLTS described in Fig. 6.4; this consists of two states s and $\mathbf{0}$ together with a k-indexed set of transitions

$$s \xrightarrow{\tau}_k ([\![\mathbf{0}]\!]_{1/k^2} \oplus \overline{s}) \quad \text{for } k \geq 2, \tag{6.20}$$

This pLTS is obtained from the infinite state pLTS described in Example 6.5 by identifying all of the states s_i and replacing the state $a.\mathbf{0}$ with $\mathbf{0}$.

As we have seen, by taking transitions $s \xrightarrow{\tau}_k \cdot \xrightarrow{\tau}_{k+1} \cdot \xrightarrow{\tau}_{k+2} \cdots$ we have $\overline{s} \Longrightarrow \frac{1}{k} \cdot \overline{\mathbf{0}}$ for any $k \geq 2$; but crucially $\overline{s} \not\Longrightarrow \varepsilon$. Since trivially $\overline{\mathbf{0}} \not\Longrightarrow \varepsilon$ there is no full distribution Δ such that $\Delta \Longrightarrow \varepsilon$.

Now to contradict the distillation of divergence for this pLTS, note that $\overline{s} \Longrightarrow \frac{1}{2} \cdot \overline{\mathbf{0}}$, but this derivation cannot be factored in the required manner to $\overline{s} \Longrightarrow (\Delta'_1 {}_p\oplus \Delta'_\varepsilon)$, because no possible full distribution Δ'_ε can exist satisfying $\Delta'_\varepsilon \Longrightarrow \varepsilon$.

Corollary 6.8 and Lemma 6.19 are not affected by infinite branching, because they are restricted to the deterministic case (i.e. the case of no branching at all). What fails is the combination of a number of deterministic distillations to make a nondeterministic one, in Theorem 6.11: it depends on Theorem 6.9, which in turn requires finite branching.

Corollary 6.9 (Zero-One Law - General Case) *If in a finitary pLTS we have* Δ, Δ' *with* $\Delta \Longrightarrow \Delta'$ *and* $|\Delta| > |\Delta'|$ *then there is some state* s' *reachable with non-zero probability from* Δ *such that* $\overline{s'} \Longrightarrow \varepsilon$. *That is, the pLTS based on* Δ *must have a wholly diverging state somewhere.*

Proof Assume at first that $|\Delta|=1$; then the result is immediate from Theorem 6.11 since any $s' \in \lceil \Delta'_\varepsilon \rceil$ will do. The general result is obtained by dividing the given derivation by $|\Delta|$. □

6.6 The Failure Simulation Preorder

We have already defined a failure simulation preorder in Definition 5.5, which looks natural for finite processes. However, for general processes divergence often exists, which makes it subtle to formulate a good notion of failure simulation preorder.

This section is divided in four: the first subsection presents a definition of the *failure simulation preorder* in an arbitrary pLTS by taking divergence into account, together with some explanatory examples. It gives two equivalent characterisations of this preorder: a coinductive one as a largest relation between subdistributions satisfying certain transfer properties, and one that is obtained through lifting and an additional closure property from a relation between states and subdistributions that we call *failure similarity*. It also investigates some elementary properties of the failure simulation preorder and of failure similarity. In the second subsection, we restrict attention to finitary processes, and on this realm characterise the failure simulation preorder in terms of a *simple failure similarity*. All further results on the failure simulation preorder, in particular precongruence for the operators of rpCSP and soundness and completeness with respect to the must testing preorder, are in terms of this characterisation, and hence pertain to finitary processes only. The third subsection establishes monotonicity of the operators of rpCSP with respect to the failure simulation preorder. In other words, we show that the failure simulation preorder is a precongruence with respect to these operators. The last subsection is devoted to showing soundness with respect to must testing. Completeness is the subject of Sect. 6.7.

6.6.1 Two Equivalent Definitions and Their Rationale

Let Δ and its variants be subdistributions in a pLTS $\langle S, \mathsf{Act}_\tau, \rightarrow\rangle$. For $a \in \mathsf{Act}$, write $\Delta \stackrel{a}{\Longrightarrow} \Delta'$ whenever $\Delta \Longrightarrow \Delta^{\text{pre}} \stackrel{a}{\longrightarrow} \Delta^{\text{post}} \Longrightarrow \Delta'$. Extend this to Act_τ by allowing as a special case that $\stackrel{\tau}{\Longrightarrow}$ is simply $\Longrightarrow$, that is, including identity (rather than requiring at least one $\stackrel{\tau}{\longrightarrow}$). For example, referring to Example 6.4 we have $[\![Q_1]\!] \stackrel{a}{\Longrightarrow} [\![\mathbf{0}]\!]$, while in Example 6.5 we have $[\![s_2]\!] \stackrel{a}{\Longrightarrow} \frac{1}{2}[\![\mathbf{0}]\!]$ and $[\![s_2]\!] \Longrightarrow \stackrel{A}{\not\longrightarrow}$ for any set A not containing a, because $s_2 \Longrightarrow \frac{1}{2}[\![a.\mathbf{0}]\!]$.

Definition 6.19 (Failure Simulation Preorder) Define $\sqsubseteq_{FS}$ to be the largest relation in $\mathcal{D}_{sub}(S) \times \mathcal{D}_{sub}(S)$ such that if $\Theta \sqsubseteq_{FS} \Delta$ then

1. whenever $\Delta \stackrel{\alpha}{\Longrightarrow} (\sum_i p_i \Delta'_i)$, for $\alpha \in \mathsf{Act}_\tau$ and certain p_i with $\sum_i p_i \leq 1$, then there are $\Theta'_i \in \mathcal{D}_{sub}(S)$ with $\Theta \stackrel{\alpha}{\Longrightarrow} (\sum_i p_i \Theta'_i)$ and $\Theta'_i \sqsubseteq_{FS} \Delta'_i$ for each i, and
2. whenever $\Delta \Longrightarrow \stackrel{A}{\not\longrightarrow}$ then also $\Theta \Longrightarrow \stackrel{A}{\not\longrightarrow}$.

Sometimes we write $\Delta \sqsupseteq_{FS} \Theta$ for $\Theta \sqsubseteq_{FS} \Delta$.

In the first case of the above definition, the summation is allowed to be empty, which has the following useful consequence.

Lemma 6.20 *If $\Theta \sqsubseteq_{FS} \Delta$ and Δ diverges, then also Θ diverges.*

Proof Divergence of Δ means that $\Delta \Longrightarrow \varepsilon$, whence with $\Theta \sqsubseteq_{FS} \Delta$ we can take the empty summation in Definition 6.19 to conclude that also $\Theta \Longrightarrow \varepsilon$. □

Although the regularity of Definition 6.19 is appealing—for example it is trivial to see that $\sqsubseteq_{FS}$ is reflexive and transitive, as it should be—in practice, for specific processes, it is easier to work with a characterisation of the failure simulation preorder in terms of a relation between *states* and distributions.

Definition 6.20 (Failure Similarity) Define $\triangleleft^e_{FS}$ to be the largest binary relation in $S \times \mathcal{D}_{sub}(S)$ such that if $s \triangleleft^e_{FS} \Theta$ then

1. whenever $\overline{s} \stackrel{\alpha}{\Longrightarrow} \Delta'$, for $\alpha \in \mathsf{Act}_\tau$, then there is a $\Theta' \in \mathcal{D}_{sub}(S)$ with $\Theta \stackrel{\alpha}{\Longrightarrow} \Theta'$ and $\Delta' \; (\triangleleft^e_{FS})^\dagger \; \Theta'$, and
2. whenever $\overline{s} \Longrightarrow \stackrel{A}{\not\longrightarrow}$ then $\Theta \Longrightarrow \stackrel{A}{\not\longrightarrow}$.

Any relation $\mathcal{R} \subseteq S \times \mathcal{D}_{sub}(S)$ that satisfies the two clauses above is called a *failure simulation*.

This is very close to Definition 5.4. The main difference is the use of the weak transition $\overline{s} \stackrel{\alpha}{\Longrightarrow} \Delta'$. In particular, if $\overline{s} \stackrel{\tau}{\Longrightarrow} \varepsilon$, then Θ has to respond similarly by the transition $\Theta \stackrel{\tau}{\Longrightarrow} \varepsilon$.

Obviously, for any failure simulation $\mathcal{R}$ we have $\mathcal{R} \subseteq \triangleleft^e_{FS}$. The following two lemmas show that the lifted failure similarity relation $(\triangleleft^e_{FS})^\dagger \subseteq \mathcal{D}_{sub}(S) \times \mathcal{D}_{sub}(S)$ has simulating properties analogous to 1 and 2 above.

Lemma 6.21 *Suppose $\Delta \; (\triangleleft^e_{FS})^\dagger \; \Theta$ and $\Delta \stackrel{\alpha}{\Longrightarrow} \Delta'$ for $\alpha \in \mathsf{Act}_\tau$. Then $\Theta \stackrel{\alpha}{\Longrightarrow} \Theta'$ for some Θ' such that $\Delta' \; (\triangleleft^e_{FS})^\dagger \; \Theta'$.*

Proof By Lemma 6.1 $\Delta\ (\lhd^e_{FS})^\dagger\ \Theta$ implies that $\Delta = \sum_{i\in I} p_i \cdot \overline{s_i}$, $s_i \lhd^e_{FS} \Theta_i$, as well as $\Theta = \sum_{i\in I} p_i \cdot \Theta_i$. By Corollary 6.7 and Proposition 6.2 we know from $\Delta \overset{\alpha}{\Longrightarrow} \Delta'$ that $\overline{s_i} \overset{\alpha}{\Longrightarrow} \Delta'_i$ for $\Delta'_i \in \mathcal{D}_{sub}(S)$ such that $\Delta' = \sum_{i\in I} p_i \cdot \Delta'_i$. For each $i \in I$ we infer from $s_i \lhd^e_{FS} \Theta_i$ and $\overline{s_i} \overset{\alpha}{\Longrightarrow} \Delta'_i$ that there is a $\Theta'_i \in \mathcal{D}_{sub}(S)$ with $\Theta_i \overset{\alpha}{\Longrightarrow} \Theta'_i$ and $\Delta'_i\ (\lhd^e_{FS})^\dagger\ \Theta'$. Let $\Theta' := \sum_{i\in I} p_i \cdot \Theta'_i$. Then Definition 6.2(2) and Theorem 6.5(i) yield $\Delta'\ (\lhd^e_{FS})^\dagger\ \Theta'$ and $\Theta \overset{\alpha}{\Longrightarrow} \Theta'$. □

Lemma 6.22 *Suppose* $\Delta\ (\lhd^e_{FS})^\dagger\ \Theta$ *and* $\Delta \Longrightarrow\overset{A}{\not\longrightarrow}$. *Then* $\Theta \Longrightarrow\overset{A}{\not\longrightarrow}$.

Proof Suppose $\Delta\ (\lhd^e_{FS})^\dagger\ \Theta$ and $\Delta \Longrightarrow \Delta' \overset{A}{\not\longrightarrow}$. By Lemma 6.21 there exists some subdistribution Θ' such that $\Theta \Longrightarrow \Theta'$ and $\Delta'\ (\lhd^e_{FS})^\dagger\ \Theta'$. From Lemma 6.1 we know that $\Delta' = \sum_{i\in I} p_i \cdot \overline{s_i}$, $s_i \lhd^e_{FS} \Theta_i$, $\Theta' = \sum_{i\in I} p_i \cdot \Theta_i$, with $s_i \in \lceil \Delta' \rceil$ for all $i \in I$. Since $\Delta' \overset{A}{\not\longrightarrow}$, we have that $s_i \overset{A}{\not\longrightarrow}$ for all $i \in I$. It follows from $s_i \lhd^e_{FS} \Theta_i$ that $\Theta_i \Longrightarrow \Theta'_i \overset{A}{\not\longrightarrow}$. By Theorem 6.5(i), we obtain that $\sum_{i\in I} p_i \cdot \Theta_i \Longrightarrow \sum_{i\in I} p_i \cdot \Theta'_i \overset{A}{\not\longrightarrow}$. By the transitivity of $\Longrightarrow$ we have that $\Theta \Longrightarrow\overset{A}{\not\longrightarrow}$. □

The next result shows how the failure simulation preorder can alternatively be defined in terms of failure similarity. This is consistent with Definition 5.5 for finite processes.

Proposition 6.7 *For* $\Delta, \Theta \in \mathcal{D}_{sub}(S)$ *we have* $\Theta \sqsubseteq_{FS} \Delta$ *just when there is a* Θ^{match} *with* $\Theta \Longrightarrow \Theta^{match}$ *and* $\Delta\ (\lhd^e_{FS})^\dagger\ \Theta^{match}$.

Proof Let $\lhd'_{FS} \subseteq S \times \mathcal{D}_{sub}(S)$ be the relation given by $s \lhd'_{FS} \Theta$ iff $\Theta \sqsubseteq_{FS} \overline{s}$. Then $\lhd'_{FS}$ is a failure simulation; hence $\lhd'_{FS} \subseteq \lhd^e_{FS}$. Now suppose $\Theta \sqsubseteq_{FS} \Delta$. Let $\Delta := \sum_i p_i \cdot \overline{s_i}$. Then there are Θ_i with $\Theta \Longrightarrow \sum_i p_i \cdot \Theta_i$ and $\Theta_i \sqsubseteq_{FS} \overline{s_i}$ for each i, whence $s_i \lhd'_{FS} \Theta_i$, and thus $s_i \lhd^e_{FS} \Theta_i$. Take $\Theta^{\text{match}} := \sum_i p_i \cdot \Theta_i$. Definition 6.2 yields $\Delta\ (\lhd^e_{FS})^\dagger\ \Theta^{\text{match}}$.

For the other direction, it suffices to show that the relation $(\lhd^e_{FS})^\dagger \cdot \Longleftarrow$ satisfies the two clauses of Definition 6.19, yielding $(\lhd^e_{FS})^\dagger \cdot \Longleftarrow\ \subseteq\ \sqsupseteq_{FS}$. Here we write $\Longleftarrow$ for the inverse of $\Longrightarrow$. So suppose, for given $\Delta, \Theta \in \mathcal{D}_{sub}(S)$, there is a Θ^{match} with $\Theta \Longrightarrow \Theta^{\text{match}}$ and $\Delta\ (\lhd^e_{FS})^\dagger\ \Theta^{\text{match}}$.

Suppose $\Delta \overset{\alpha}{\Longrightarrow} \sum_{i\in I} p_i \cdot \Delta'_i$ for some $\alpha \in \mathsf{Act}_\tau$. By Lemma 6.21, there is some Θ' such that $\Theta^{\text{match}} \overset{\alpha}{\Longrightarrow} \Theta'$ and $(\sum_{i\in I} p_i \cdot \Delta'_i)\ (\lhd^e_{FS})^\dagger\ \Theta'$. From Proposition 6.2, we know that $\Theta' = \sum_{i\in I} p_i \cdot \Theta'_i$ for subdistributions Θ'_i such that $\Delta'_i\ (\lhd^e_{FS})^\dagger\ \Theta'_i$ for $i \in I$. Thus $\Theta \overset{\alpha}{\Longrightarrow} \sum_i p_i \cdot \Theta'_i$ by the transitivity of $\Longrightarrow$ (Theorem 6.6) and $\Delta'_i((\lhd^e_{FS})^\dagger \cdot \Longleftarrow)\Theta'_i$ for each $i \in I$ by the reflexivity of $\Longleftarrow$.

Suppose $\Delta \Longrightarrow\overset{A}{\not\longrightarrow}$. By Lemma 6.22 we have $\Theta^{\text{match}} \Longrightarrow\overset{A}{\not\longrightarrow}$. It follows that $\Theta \Longrightarrow\overset{A}{\not\longrightarrow}$ by the transitivity of $\Longrightarrow$. □

Note the appearance of the "anterior step" $\Theta \Longrightarrow \Theta^{\text{match}}$ in Proposition 6.7 immediately above. For the same reason explained in Example 5.14, defining $\sqsupseteq_{FS}$ simply to be $(\lhd^e_{FS})^\dagger$ (i.e. without anterior step) would not have been suitable.

Remark 6.3 For $s \in S$ and $\Theta \in \mathcal{D}_{sub}(S)$, we have $s \lhd^e_{FS} \Theta$ iff $\Theta \sqsubseteq_{FS} \overline{s}$; here no anterior step is needed. One direction of this statement has been obtained in the beginning of the proof of Proposition 6.7; for the other note that $s \lhd^e_{FS} \Theta$ implies

$\overline{s}\ (\lhd^e_{FS})^\dagger\ \Theta$ by Definition 6.2(1) which implies $\Theta \sqsubseteq_{FS} \overline{s}$ by Proposition 6.7 and the reflexivity of $\Longrightarrow$.

Example 5.14 also shows that $\sqsubseteq_{FS}$ cannot be obtained as the lifting of any relation; it lacks the decomposition property of Proposition 6.2. Nevertheless, $\sqsubseteq_{FS}$ enjoys the property of linearity, as occurs in Definition 6.2:

Lemma 6.23 *If $\Theta_i \sqsubseteq_{FS} \Delta_i$ for $i \in I$ then $\sum_{i\in I} p_i \cdot \Theta_i \sqsubseteq_{FS} \sum_{i\in I} p_i \cdot \Delta_i$ for any $p_i \in [0,1]$ ($i \in I$) with $\sum_{i\in I} p_i \leq 1$.*

Proof This follows immediately from the linearity of $(\lhd^e_{FS})^\dagger$ and $\Longrightarrow$ (cf. Theorem 6.5(i)), using Proposition 6.7. □

Example 6.16 (Divergence) From Example 6.3 we know that $[\![\mathsf{rec}\, x.x]\!] \Longrightarrow \varepsilon$. This, together with (6.1) in Sect. 6.3.1, and the fact that $\varepsilon \stackrel{A}{\not\longrightarrow}$ for any set of actions A, ensures that $s \lhd^e_{FS} [\![\mathsf{rec}\, x.x]\!]$ for any s, hence $\Theta\ (\lhd^e_{FS})^\dagger\ [\![\mathsf{rec}\, x.x]\!]$ for any Θ, and thus that $[\![\mathsf{rec}\, x.x]\!] \sqsubseteq_{FS} \Theta$. Indeed similar reasoning applies to any Δ with

$$\Delta = \Delta_0 \xrightarrow{\tau} \Delta_1 \xrightarrow{\tau} \cdots \xrightarrow{\tau} \cdots$$

because—as explained right before Example 6.3—this also ensures that $\Delta \Longrightarrow \varepsilon$. In particular, we have $\varepsilon \Longrightarrow \varepsilon$ and hence $[\![\mathsf{rec}\, x.x]\!] \simeq_{FS} \varepsilon$.

Yet $\mathbf{0} \not\sqsubseteq_{FS} [\![\mathsf{rec}\, x.x]\!]$, because the move $[\![\mathsf{rec}\, x.x]\!] \Longrightarrow \varepsilon$ cannot be matched by a corresponding move from $[\![\mathbf{0}]\!]$—see Lemma 6.20.

Example 6.16 shows again that the anterior move in Proposition 6.7 is necessary; although $[\![\mathsf{rec}\, x.x]\!] \sqsubseteq_{FS} \varepsilon$, we do not have $\varepsilon\ (\lhd^e_{FS})^\dagger\ [\![\mathsf{rec}\, x.x]\!]$, since by Lemma 6.2 any Θ with $\varepsilon\ (\lhd^e_{FS})^\dagger\ \Theta$ must have $|\Theta| = 0$.

Example 6.17 Referring to the process Q_1 of Example 6.4, with Proposition 6.7, we easily see that $Q_1 \sqsubseteq_{FS} a.\mathbf{0}$ because we have $a.\mathbf{0} \lhd^e_{FS} [\![Q_1]\!]$. Note that the move $[\![Q_1]\!] \Longrightarrow [\![a.\mathbf{0}]\!]$ is crucial, since it enables us to match the move $[\![a.\mathbf{0}]\!] \xrightarrow{a} [\![\mathbf{0}]\!]$ with $[\![Q_1]\!] \Longrightarrow [\![a.\mathbf{0}]\!] \xrightarrow{a} [\![\mathbf{0}]\!]$. It also enables us to match refusals; if $[\![a.\mathbf{0}]\!] \stackrel{A}{\not\longrightarrow}$ then A cannot contain the action a, and therefore also $[\![Q_1]\!] \Longrightarrow \stackrel{A}{\not\longrightarrow}$.

The converse, that $a.\mathbf{0} \sqsubseteq_{FS} Q_1$, is also true because it is straightforward to verify that the relation

$$\{(Q_1, [\![a.\mathbf{0}]\!]), (\tau.Q_1, [\![a.\mathbf{0}]\!]), (a.\mathbf{0}, [\![a.\mathbf{0}]\!]), (\mathbf{0}, [\![\mathbf{0}]\!])\}$$

is a failure simulation and thus is a subset of $\lhd^e_{FS}$. We, therefore, have $Q_1 \simeq_{FS} a.\mathbf{0}$.

Example 6.18 Let P be the process $a.\mathbf{0}\,{}_{\frac{1}{2}}\!\oplus \mathsf{rec}\, x.x$ and consider the state s_2 introduced in Example 6.5. First, note that $[\![P]\!]\ (\lhd^e_{FS})^\dagger\ \frac{1}{2} \cdot [\![a.\mathbf{0}]\!]$, since $\mathsf{rec}\, x.x \lhd^e_{FS} \varepsilon$. Then, because $s_2 \Longrightarrow \frac{1}{2} \cdot [\![a.\mathbf{0}]\!]$ we have $s_2 \sqsubseteq_{FS} [\![P]\!]$. The converse, that $[\![P]\!] \sqsubseteq_{FS} s_2$ holds, is true because $s_2 \lhd^e_{FS} [\![P]\!]$ follows from the fact that the relation

$$\{(s_k, [\![a.\mathbf{0}]\!]_{1/k}\!\oplus [\![\mathsf{rec}\, x.x]\!]) \mid k \geq 2\} \cup \{(a.\mathbf{0}, [\![a.\mathbf{0}]\!]), (\mathbf{0}, [\![\mathbf{0}]\!])\}$$

is a failure simulation that contains the pair $(s_2, [\![P]\!])$.

Our final examples pursue the consequences of the fact that the empty distribution ε is behaviourally indistinguishable from divergent processes like $[\![\mathsf{rec}\, x.x]\!]$.

Example 6.19 (Subdistributions formally unnecessary) For any subdistribution Δ, let Δ^e denote the (full) distribution defined by

$$\Delta^e := \Delta \;+\; (1 - |\Delta|) \cdot \overline{[\![\mathsf{rec}\, x.x]\!]}\;.$$

Intuitively, it is obtained from Δ by padding the missing support with the divergent state $[\![\mathsf{rec}\, x.x]\!]$.

Then $\Delta \simeq_{FS} \Delta^e$. This follows because $\Delta^e \Longrightarrow \Delta$, which is sufficient to establish $\Delta^e \sqsubseteq_{FS} \Delta$; but also $\Delta^e\ (\lhd^e_{FS})^\dagger\ \Delta$ because $[\![\mathsf{rec}\, x.x]\!] \lhd^e_{FS} \varepsilon$, and that implies the converse $\Delta \sqsubseteq_{FS} \Delta^e$. The equivalence shows that formally we have no need for subdistributions and that our technical development could be carried out using (full) distributions only.

But abandoning subdistributions comes at a cost; the definition of weak transition, Definition 6.4, would be much more complex if expressed with full distributions, as would syntactic manipulations such as those used in the proof of Theorem 6.6.

More significant, however is that, diverging processes have a special character in failure simulation semantics. Placing them at the bottom of the $\sqsubseteq_{FS}$ preorder—as we do—requires that they failure-simulate every processes, thus allowing all visible actions and all refusals and so behaving in a sense "chaotically"; yet applying the operational semantics of Fig. 6.2 to $\mathsf{rec}\, x.x$ literally would suggest exactly the opposite, since $\mathsf{rec}\, x.x$ allows no visible actions (all its derivatives enable only τ) and no refusals (all its derivatives have τ enabled). The case analyses that discrepancy would require are entirely escaped by allowing subdistributions, as the chaotic behaviour of the diverging ε follows naturally from the definitions, as we saw in Example 6.16.

We conclude with an example involving divergence and subdistributions.

Example 6.20 For $0 \le c \le 1$ let P_c be the process $\mathbf{0}\,{}_c\!\oplus\, \mathsf{rec}\, x.x$. We show that $[\![P_c]\!] \sqsubseteq_{FS} [\![P_{c'}]\!]$ just when $c \le c'$. (Refusals can be ignored, since P_c refuses every set of actions, for all c.)

Suppose first that $c \le c'$, and split the two processes as follows:

$$\begin{aligned}
[\![P_c]\!] &= c \cdot [\![\mathbf{0}]\!] + (c'-c) \cdot [\![\mathsf{rec}\, x.x]\!] + (1-c') \cdot [\![\mathsf{rec}\, x.x]\!] \\
[\![P_{c'}]\!] &= c \cdot [\![\mathbf{0}]\!] + (c'-c) \cdot [\![\mathbf{0}]\!] \qquad\quad\;\, + (1-c') \cdot [\![\mathsf{rec}\, x.x]\!]\;.
\end{aligned}$$

Because $\mathbf{0} \lhd^e_{FS} [\![\mathsf{rec}\, x.x]\!]$ (the middle terms), we have immediately $[\![P_{c'}]\!]\ (\lhd^e_{FS})^\dagger\ [\![P_c]\!]$, whence $[\![P_c]\!] \sqsubseteq_{FS} [\![P_{c'}]\!]$.

For the other direction, note that $[\![P_{c'}]\!] \Longrightarrow c' \cdot [\![\mathbf{0}]\!]$. If $[\![P_c]\!] \sqsubseteq_{FS} [\![P_{c'}]\!]$, then from Definition 6.19 we would have to have $[\![P_c]\!] \Longrightarrow c' \cdot \Theta'$ for some subdistribution Θ', a derivative of weight not more than c'. But the smallest weight P_c can reach via $\Longrightarrow$ is just c, so that we must have in fact $c \le c'$.

We end this subsection with two properties of failure similarity that will be useful later on.

Proposition 6.8 *The relation $\lhd^e_{FS}$ is convex.*

Proof Suppose $s \lhd^e_{FS} \Theta_i$ and $p_i \in [0,1]$ for $i \in I$, with $\sum_{i\in I} p_i = 1$. We need to show that $s \lhd^e_{FS} \sum_{i\in I} p_i \cdot \Theta_i$.

If $\overline{s} \stackrel{\alpha}{\Longrightarrow} \Delta'$ then there exist Θ'_i for each $i \in I$ such that $\Theta_i \stackrel{\alpha}{\Longrightarrow} \Theta'_i$ and $\Delta' \, (\lhd^e_{FS})^\dagger \, \Theta'_i$. By Corollary 6.7 and Theorem 6.4, we obtain that $\sum_{i\in I} p_i \cdot \Theta_i \stackrel{\alpha}{\Longrightarrow} \sum_{i\in I} p_i \cdot \Theta'_i$ and $\Delta' \, (\lhd^e_{FS})^\dagger \, \sum_{i\in I} p_i \cdot \Theta'_i$.

If $\overline{s} \Longrightarrow \stackrel{A}{\not\longrightarrow}$ for some $A \subseteq \mathsf{Act}$, then $\Theta_i \Longrightarrow \Theta'_i \stackrel{A}{\not\longrightarrow}$ for all $i \in I$. By definition we have $\sum_{i\in I} p_i \cdot \Theta'_i \stackrel{A}{\not\longrightarrow}$. Theorem 6.5(i) yields $\sum_{i\in I} p_i \cdot \Theta_i \Longrightarrow \sum_{i\in I} p_i \cdot \Theta'_i$.

So we have checked that $s \lhd^e_{FS} \sum_{i\in I} p_i \cdot \Theta_i$. It follows that $\lhd^e_{FS}$ is convex. □

Proposition 6.9 *The relation $(\lhd^e_{FS})^\dagger \subseteq \mathcal{D}_{sub}(S) \times \mathcal{D}_{sub}(S)$ is reflexive and transitive.*

Proof Reflexivity is easy; it relies on the fact that $s \lhd^e_{FS} \overline{s}$ for every state s.

For transitivity, we first show that $\lhd^e_{FS} \cdot (\lhd^e_{FS})^\dagger$ is a failure simulation. Suppose that $s \lhd^e_{FS} \Theta \, (\lhd^e_{FS})^\dagger \, \Phi$. If $\overline{s} \stackrel{\alpha}{\Longrightarrow} \Delta'$ then there is a Θ' such that $\Theta \stackrel{\alpha}{\Longrightarrow} \Theta'$ and $\Delta' \, (\lhd^e_{FS})^\dagger \, \Theta'$. By Lemma 6.21, there exists a Φ' such that $\Phi \stackrel{\alpha}{\Longrightarrow} \Phi'$ and $\Theta' \, (\lhd^e_{FS})^\dagger \, \Phi'$. Hence, $\Delta' \, (\lhd^e_{FS})^\dagger \cdot (\lhd^e_{FS})^\dagger \, \Phi'$. By Lemma 6.3 we know that

$$(\lhd^e_{FS})^\dagger \cdot (\lhd^e_{FS})^\dagger = (\, \lhd^e_{FS} \cdot (\lhd^e_{FS}) \,)^\dagger \tag{6.21}$$

Therefore, we obtain $\Delta' \, (\, \lhd^e_{FS} \cdot (\lhd^e_{FS})^\dagger \,)^\dagger \, \Phi'$.

If $s \Longrightarrow \stackrel{A}{\not\longrightarrow}$ for some $A \subseteq \mathsf{Act}$, then $\Theta \Longrightarrow \stackrel{A}{\not\longrightarrow}$ and hence $\Phi \Longrightarrow \stackrel{A}{\not\longrightarrow}$ by applying Lemma 6.22.

So we established that $\lhd^e_{FS} \cdot (\lhd^e_{FS})^\dagger \subseteq \lhd^e_{FS}$. It now follows from the monotonicity of the lifting operation and (6.21) that $(\lhd^e_{FS})^\dagger \cdot (\lhd^e_{FS})^\dagger \subseteq (\lhd^e_{FS})^\dagger$. □

6.6.2 A Simple Failure Similarity for Finitary Processes

Here, we present a simpler characterisation of failure similarity, valid when considering finitary processes only. It is in terms of this characterisation that we will establish soundness and completeness of the failure simulation preorder with respect to the must testing preorder; consequently we have these results for finitary processes only.

Definition 6.21 (Simple Failure Similarity) Let $\mathbf{F}$ be the function on $S \times \mathcal{D}_{sub}(S)$ such that for any binary relation $\mathcal{R} \in S \times \mathcal{D}_{sub}(S)$, state s and subdistribution Θ, if $s \, \mathbf{F}(\mathcal{R}) \, \Theta$ then

1. whenever $\overline{s} \Longrightarrow \varepsilon$ then also $\Theta \Longrightarrow \varepsilon$, otherwise
2. whenever $s \stackrel{\alpha}{\longrightarrow} \Delta'$ for some $\alpha \in \mathsf{Act}_\tau$, then there is a Θ' with $\Theta \stackrel{\alpha}{\Longrightarrow} \Theta'$ and $\Delta' \, \mathcal{R}^\dagger \, \Theta'$ and
3. whenever $s \stackrel{A}{\not\longrightarrow}$ then $\Theta \Longrightarrow \stackrel{A}{\not\longrightarrow}$.

Let $\lhd^S_{FS}$ be the greatest fixed point of $\mathbf{F}$.

The above definition is obtained by factoring out divergence from Clause 1 in Definition 6.20 and conservatively extends Definition 5.4 to compare processes that might not be finite. We first note that the relation $\lhd^S_{FS}$ is not interesting for infinitary processes since its lifted form $(\lhd^S_{FS})^\dagger$ is not a transitive relation for those processes.

Example 6.21 Consider the process defined by the following two transitions: $t_0 \xrightarrow{\tau} (\overline{\mathbf{0}}_{1/2}\oplus \overline{t_1})$ and $t_1 \xrightarrow{\tau} \overline{t_1}$. We compare state t_0 with state s in Example 6.15 and have that $t_0 \lhd^S_{FS} \overline{s}$. The transition $t_0 \xrightarrow{\tau} (\overline{\mathbf{0}}_{1/2}\oplus \overline{t_1})$ can be matched by $s \Longrightarrow \frac{1}{2}\overline{\mathbf{0}}$ because $(\overline{\mathbf{0}}_{1/2}\oplus \overline{t_1})\ (\lhd^S_{FS})^\dagger\ \frac{1}{2}\overline{\mathbf{0}}$ by noticing that $t_1 \lhd^S_{FS} \varepsilon$.

It also holds that $s \lhd^S_{FS} \overline{\mathbf{0}}$ because the relation $\{(s,\overline{\mathbf{0}}),(\mathbf{0},\overline{\mathbf{0}})\}$ is a simple failure simulation. The transition $s \xrightarrow{\tau}_k (\overline{\mathbf{0}}_{\frac{1}{k^2}} \oplus \overline{s})$ for any $k \geq 2$ is matched by $\mathbf{0} \Longrightarrow \overline{\mathbf{0}}$.

However, we do not have $t_0 \lhd^S_{FS} \overline{\mathbf{0}}$. The only candidate to simulate the transition $t_0 \xrightarrow{\tau} (\overline{\mathbf{0}}_{1/2}\oplus \overline{t_1})$ is $\mathbf{0} \Longrightarrow \overline{\mathbf{0}}$, but we do not have $(\overline{\mathbf{0}}_{\frac{1}{2}} \oplus \overline{t_1})\ (\lhd^S_{FS})^\dagger\ \overline{\mathbf{0}}$ because the divergent state t_1 cannot be simulated by $\mathbf{0}$.

Therefore, we have $\overline{t_0}\ (\lhd^S_{FS})^\dagger\ \overline{s}\ (\lhd^S_{FS})^\dagger\ \overline{\mathbf{0}}$ but not $\overline{t_0}\ (\lhd^S_{FS})^\dagger\ \overline{\mathbf{0}}$, thus transitivity of the relation $(\lhd^S_{FS})^\dagger$ fails. Here the state s is not finitely branching. As a matter of fact, transitivity of $(\lhd^S_{FS})^\dagger$ also fails for finitely branching but infinite state processes.

Consider an infinite state pLTS consisting of a collection of states s_k for $k \geq 2$ such that

$$s_k \xrightarrow{\tau} \overline{\mathbf{0}}_{\frac{1}{k^2}} \oplus \overline{s_{k+1}}. \qquad (6.22)$$

This pLTS is obtained from that in Example 6.5 by replacing $a.\mathbf{0}$ with $\mathbf{0}$. One can check that $\overline{t_0}\ (\lhd^S_{FS})^\dagger\ \overline{s_2}\ (\lhd^S_{FS})^\dagger\ \overline{\mathbf{0}}$ but we already know that $\overline{t_0}\ (\lhd^S_{FS})^\dagger\ \overline{\mathbf{0}}$ does not hold. Again, we loose the transitivity of $(\lhd^S_{FS})^\dagger$.

If we restrict our attention to finitary processes, then $\lhd^S_{FS}$ provides a simpler characterisation of failure similarity.

Theorem 6.12 (Equivalence of Failure- and Simple Failure Similarity) *For finitary distributions $\Delta, \Theta \in \mathcal{D}_{sub}(S)$ in a pLTS $\langle S, \mathsf{Act}_\tau, \rightarrow\rangle$ we have $\Delta \lhd^S_{FS} \Theta$ if and only if $\Delta \lhd^e_{FS} \Theta$.*

Proof Because $s \xrightarrow{\alpha} \Delta'$ implies $\overline{s} \xrightarrow{\alpha} \Delta'$ and $s \not\xrightarrow{A}$ implies $\overline{s} \Longrightarrow\not\xrightarrow{A}$ it is trivial that $\lhd^e_{FS}$ satisfies the conditions of Definition 6.21, so that $\lhd^e_{FS} \subseteq \lhd^S_{FS}$.

For the other direction, we need to show that $\lhd^S_{FS}$ satisfies Clause 1 of Definition 6.20 with $\alpha = \tau$, that is

$$\text{if } s \lhd^S_{FS} \Theta \text{ and } \overline{s} \Longrightarrow \Delta' \text{ then there is some } \Theta' \in \mathcal{D}_{sub}(S) \text{ with } \Theta \Longrightarrow \Theta' \text{ and}$$
$$\Delta'\ (\lhd^S_{FS})^\dagger\ \Theta'.$$

Once we have this, the relation $\lhd^S_{FS}$ clearly satisfies both clauses of Definition 6.20, so that we have $\lhd^S_{FS} \subseteq \lhd^e_{FS}$.

So suppose that $s \lhd^S_{FS} \Theta$ and that $\overline{s} \Longrightarrow \Delta'$ where—for the moment—we assume $|\Delta'| = 1$. Referring to Definition 6.4, there must be Δ_k, $\Delta_k^\rightarrow$ and $\Delta_k^\times$ for $k \geq 0$ such that $\overline{s} = \Delta_0$, $\Delta_k = \Delta_k^{\rightarrow} + \Delta_k^{\times}$, $\Delta_k^{\rightarrow} \xrightarrow{\tau} \Delta_{k+1}$ and $\Delta' = \sum_{k=1}^{\infty} \Delta_k^{\times}$. Since it holds that $\Delta_0^{\times} + \Delta_0^{\rightarrow} = \overline{s}\ (\lhd_{FS}^{S})^{\dagger}\ \Theta$, using Proposition 6.2 we can define $\Theta =: \Theta_0^{\times} + \Theta_0^{\rightarrow}$ so that $\Delta_0^{\times}\ (\lhd_{FS}^{S})^{\dagger}\ \Theta_0^{\times}$ and $\Delta_0^{\rightarrow}\ (\lhd_{FS}^{S})^{\dagger}\ \Theta_0^{\rightarrow}$. Since $\Delta_0^{\rightarrow} \xrightarrow{\tau} \Delta_1$ and $\Delta_0^{\rightarrow}\ (\lhd_{FS}^{S})^{\dagger}\ \Theta_0^{\rightarrow}$ it follows that $\Theta_0^{\rightarrow} \Longrightarrow \Theta_1$ with $\Delta_1\ (\lhd_{FS}^{S})^{\dagger}\ \Theta_1$.

Repeating the above procedure gives us inductively a series $\Theta_k, \Theta_k^{\rightarrow}, \Theta_k^{\times}$ of subdistributions, for $k \geq 0$, such that $\Theta_0 = \Theta$, $\Delta_k\ (\lhd_{FS}^{S})^{\dagger}\ \Theta_k$, $\Theta_k = \Theta_k^{\rightarrow} + \Theta_k^{\times}$, $\Delta_k^{\times}\ (\lhd_{FS}^{S})^{\dagger}\ \Theta_k^{\times}$, $\Delta_k^{\rightarrow}\ (\lhd_{FS}^{S})^{\dagger}\ \Theta_k^{\rightarrow}$ and $\Theta_k^{\rightarrow} \xRightarrow{\tau} \Theta_k$. We define $\Theta' := \sum_i \Theta_i^{\times}$. By Additivity (Remark 6.2), we have $\Delta'\ (\lhd_{FS}^{S})^{\dagger}\ \Theta'$. It remains to be shown that $\Theta \Longrightarrow \Theta'$.

For that final step, since $(\Theta \Longrightarrow)$ is closed according to Lemma 6.17, we can establish $\Theta \Longrightarrow \Theta'$ by exhibiting a sequence Θ_i' with $\Theta \Longrightarrow \Theta_i'$ for each i and with the Θ_i' being arbitrarily close to Θ'. Induction establishes for each i that

$$\Theta \Longrightarrow \Theta_i' := \left(\Theta_i^{\rightarrow} + \sum_{k \leq i} \Theta_k^{\times}\right).$$

Since $|\Delta'| = 1$, we are guaranteed to have that $\lim_{i\to\infty} |\Delta_i^{\rightarrow}| = 0$, whence by Lemma 6.2, using that $\Delta_i^{\rightarrow}\ (\lhd_{FS}^{S})^{\dagger}\ \Theta_i^{\rightarrow}$, also $\lim_{i\to\infty} |\Theta_i^{\rightarrow}| = 0$. Thus, these Θ_i's form the sequence we needed.

That concludes the case for $|\Delta'| = 1$. If on the other hand $\Delta' = \varepsilon$, that is we have $|\Delta'| = 0$, then $\Theta \Longrightarrow \varepsilon$ follows immediately from $s \lhd_{FS}^{S} \Theta$, and $\varepsilon\ (\lhd_{FS}^{S})^{\dagger}\ \varepsilon$ trivially.

In the general case, if $s \Longrightarrow \Delta'$, then by Theorem 6.11 we have $s \Longrightarrow \Delta_1'\,{}_p\!\oplus \Delta_\varepsilon'$ for some probability p and distributions $\Delta_1', \Delta_\varepsilon'$, with $\Delta' = p \cdot \Delta_1'$ and $\Delta_\varepsilon' \Longrightarrow \varepsilon$. From the mass-1 case above, we have $\Theta \Longrightarrow \Theta_1'\,{}_p\!\oplus \Theta_\varepsilon'$ with $\Delta_1'\ (\lhd_{FS}^{S})^{\dagger}\ \Theta_1'$ and $\Delta_\varepsilon'\ (\lhd_{FS}^{S})^{\dagger}\ \Theta_\varepsilon'$; from the mass-0 case, we have $\Theta_\varepsilon' \Longrightarrow \varepsilon$ and hence $\Theta_1'\,{}_p\!\oplus \Theta_\varepsilon' \Longrightarrow p \cdot \Theta_1'$ by Theorem 6.5(i); thus transitivity yields $\Theta \Longrightarrow p \cdot \Theta_1'$, with $\Delta' = p \cdot \Delta_1'\ (\lhd_{FS}^{S})^{\dagger}\ p \cdot \Theta_1'$ as required, using Definition 6.2(2). □

The proof of Theorem 6.12 refers to Theorem 6.11 where the underlying pLTS is assumed to be finitary. As we would expect, Theorem 6.12 fails for infinitary pLTSs.

Example 6.22 We have seen in Example 6.21 that the state s from (6.20) is related to $\overline{\mathbf{0}}$ via the relation $\lhd_{FS}^{S}$. We now compare s with $\overline{\mathbf{0}}$ according to $\lhd_{FS}^{e}$. From state s, we have the weak transition $\overline{s} \Longrightarrow \overline{\mathbf{0}}_{1/2}\!\oplus \varepsilon$, which cannot be matched by any transition from $\mathbf{0}$, thus $s \not\lhd_{FS}^{e} \overline{\mathbf{0}}$. This means that Theorem 6.12 fails for infinitely branching processes.

If we replace state s by the state s_2 from (6.22), similar phenomenon happens. Therefore, Theorem 6.12 also fails for finitely branching but infinite-state processes.

6.6.3 *Precongruence*

The purpose of this section is to show that the semantic relation $\sqsubseteq_{FS}$ is preserved by the constructs of rpCSP. The proofs follow closely the corresponding proofs in

Sect. 5.5.2, but here there is a significant extra proof obligation; in order to relate two processes we have to demonstrate that if the first diverges then so does the second.

Here, in order to avoid such complications, we introduce yet another version of failure simulation; it modifies Definition 6.21 by checking divergence coinductively instead of using a predicate.

Definition 6.22 Define $\lhd_{FS}^{C}$ to be the largest relation in $S \times \mathcal{D}_{sub}(S)$ such that if $s \lhd_{FS}^{C} \Theta$ then

1. whenever $\overline{s} \Longrightarrow \varepsilon$, there are some Δ', Θ' such that $\overline{s} \Longrightarrow\xrightarrow{\tau}\Longrightarrow \Delta' \Longrightarrow \varepsilon$, $\Theta \Longrightarrow\xrightarrow{\tau}\Longrightarrow \Theta'$ and $\Delta' \; (\lhd_{FS}^{C})^{\dagger} \; \Theta'$; otherwise
2. whenever $s \xrightarrow{\alpha} \Delta'$, for some $\alpha \in \mathsf{Act}_\tau$, then there is a Θ' with $\Theta \stackrel{\alpha}{\Longrightarrow} \Theta'$ and $\Delta' \; (\lhd_{FS}^{C})^{\dagger} \; \Theta'$ and
3. whenever $s \stackrel{A}{\not\longrightarrow}$ then $\Theta \Longrightarrow\stackrel{A}{\not\longrightarrow}$.

Lemma 6.24 *The following statements about divergence are equivalent.*

(1) $\Delta \Longrightarrow \varepsilon$.
(2) There is an infinite sequence $\Delta \xrightarrow{\tau} \Delta_1 \xrightarrow{\tau} \Delta_2 \xrightarrow{\tau} \ldots$.
(3) There is an infinite sequence $\Delta \Longrightarrow\xrightarrow{\tau}\Longrightarrow \Delta_1 \Longrightarrow\xrightarrow{\tau}\Longrightarrow \Delta_2 \Longrightarrow\xrightarrow{\tau}\Longrightarrow \ldots$.

Proof By the definition of weak transition, it is immediate that (1) $\Leftrightarrow$ (2). Clearly, we have (2) $\Rightarrow$ (3). To show that (3) $\Rightarrow$ (2), we introduce another characterisation of divergence. Let Δ be a subdistribution in a pLTS L. A *pLTS induced by* Δ is a pLTS whose states and transitions are subsets of those in L and all states are reachable from Δ.

(4) There is a pLTS induced by Δ where all states have outgoing τ transitions.

It holds that (3) $\Rightarrow$ (4) because we can construct a pLTS whose states and transitions are just those used in deriving the infinite sequence in (3). For this pLTS, each state has an outgoing τ transition, which gives (4) $\Rightarrow$ (2). □

The next lemma shows the usefulness of the relation $\lhd_{FS}^{C}$ by checking divergence in a coinductive way.

Lemma 6.25 *Suppose* $\Delta \; (\lhd_{FS}^{C})^{\dagger} \; \Theta$ *and* $\Delta \Longrightarrow \varepsilon$. *Then there exist* Δ', Θ' *such that* $\Delta \Longrightarrow\xrightarrow{\tau}\Longrightarrow \Delta' \Longrightarrow \varepsilon$, $\Theta \Longrightarrow\xrightarrow{\tau}\Longrightarrow \Theta'$, *and* $\Delta' \; (\lhd_{FS}^{C})^{\dagger} \; \Theta'$.

Proof Suppose $\Delta \; (\lhd_{FS}^{C})^{\dagger} \; \Theta$ and $\Delta \Longrightarrow \varepsilon$. In analogy with Proposition 6.8, we can show that $\lhd_{FS}^{C}$ is convex. By Corollary 6.1, we can decompose Θ as $\sum_{s \in \lceil \Delta \rceil} \Delta(s) \cdot \Theta_s$ and $s \lhd_{FS}^{C} \Theta_s$ for each $s \in \lceil \Delta \rceil$. Now each s must also diverge. So there exist Δ'_s, Θ'_s such that $\overline{s} \Longrightarrow\xrightarrow{\tau}\Longrightarrow \Delta'_s \Longrightarrow \varepsilon$, $\Theta_s \Longrightarrow\xrightarrow{\tau}\Longrightarrow \Theta'_s$ and $\Delta'_s \; (\lhd_{FS}^{C})^{\dagger} \; \Theta'_s$ for each $s \in \lceil \Delta \rceil$. Let $\Delta' = \sum_{s \in \lceil \Delta \rceil} \Delta(s) \cdot \Delta'_s$ and $\Theta' = \sum_{s \in \lceil \Delta \rceil} \Delta(s) \cdot \Theta'_s$. By Definition 6.2 and Theorem 6.5(i), we have $\Delta' \; (\lhd_{FS}^{C})^{\dagger} \; \Theta'$, $\Delta \Longrightarrow\xrightarrow{\tau}\Longrightarrow \Delta'$, and $\Theta \Longrightarrow\xrightarrow{\tau}\Longrightarrow \Theta'$. We also have that $\Delta' \Longrightarrow \varepsilon$ because for each state s in Δ' it holds that $s \in \lceil \Delta'_s \rceil$ for some Δ'_s and $\Delta'_s \Longrightarrow \varepsilon$, which means $\overline{s} \Longrightarrow \varepsilon$. □

Lemma 6.26 *$\lhd^C_{FS}$ coincides with $\lhd^S_{FS}$.*

Proof We only need to check that the first clause in Definition 6.21 is equivalent to the first clause in Definition 6.22. For one direction, we consider the relation

$$\mathcal{R} := \{(s, \Theta) \mid \overline{s} \Longrightarrow \varepsilon \text{ and } \Theta \Longrightarrow \varepsilon\}$$

and show $\mathcal{R} \subseteq \lhd^C_{FS}$. Suppose $s\mathcal{R}\Theta$. By Lemma 6.24 there are two infinite sequences $s \xrightarrow{\tau} \Delta_1 \xrightarrow{\tau} \Delta_2 \xrightarrow{\tau} \ldots$ and $\Theta \xrightarrow{\tau} \Theta_1 \xrightarrow{\tau} \ldots$. Then we have both $\Delta_1 \Longrightarrow \varepsilon$ and $\Theta_1 \Longrightarrow \varepsilon$. Note that $\Delta_1 \Longrightarrow \varepsilon$ if and only if $\overline{t} \Longrightarrow \varepsilon$ for each $t \in \lceil \Delta_1 \rceil$. Therefore, $\Delta_1 \; \mathcal{R}^\dagger \; \Theta_1$ as we have $\Delta_1 = \sum_{t \in \lceil \Delta_1 \rceil} \Delta_1(t) \cdot \overline{t}$, $\Theta_1 = \sum_{t \in \lceil \Delta_1 \rceil} \Delta_1(t) \cdot \Theta_1$, and $t\mathcal{R}\Theta_1$. Here $|\Delta_1| = 1$ because Δ_1, like $\overline{s}$, is a distribution.

For the other direction, we show that $\Delta \; (\lhd^C_{FS})^\dagger \; \Theta$ and $\Delta \Longrightarrow \varepsilon$ imply $\Theta \Longrightarrow \varepsilon$. Then as a special case, we get $s \lhd^C_{FS} \Theta$ and $s \Longrightarrow \varepsilon$ imply $\Theta \Longrightarrow \varepsilon$. By repeated application of Lemma 6.25, we can obtain two infinite sequences

$$\Delta \Longrightarrow\xrightarrow{\tau}\Longrightarrow \Delta_1 \Longrightarrow\xrightarrow{\tau}\Longrightarrow \ldots \quad \text{and} \quad \Theta \Longrightarrow\xrightarrow{\tau}\Longrightarrow \Theta_1 \Longrightarrow\xrightarrow{\tau}\Longrightarrow \ldots$$

such that $\Delta_i \; (\lhd^C_{FS})^\dagger \; \Theta_i$ for all $i \geq 1$. By Lemma 6.24 this implies $\Theta \Longrightarrow \varepsilon$. □

The advantage of this new relation $\lhd^C_{FS}$ over $\lhd^S_{FS}$ is that in order to check $s \lhd^C_{FS} \Theta$ when s diverges, it is sufficient to find a single matching move $\Theta \Longrightarrow\xrightarrow{\tau}\Longrightarrow \Theta'$, rather than an infinite sequence of moves. However, to construct this matching move we cannot rely on clause 2 in Definition 6.22, as the move generated there might actually be empty, which we have seen in Example 6.2. Instead, we need a method for generating weak moves that contain at least one occurrence of a τ-action.

Definition 6.23 (Productive Moves) Let us write $s|_A t \xrightarrow{\tau}_p \Theta$ whenever we can infer $s|_A t \xrightarrow{\tau}_p \Theta$ from the last two rules in Fig. 6.2. In effect this means that t must contribute to the action.

These *productive* actions are extended to subdistributions in the standard manner, giving $\Delta \stackrel{\tau}{\Longrightarrow}_p \Theta$.

The following lemma is adapted from Lemma 5.8 in the last chapter, which still holds in our current setting.

Lemma 6.27

(1) If $\Phi \Longrightarrow \Phi'$ then $\Phi|_A\Delta \Longrightarrow \Phi'|_A\Delta$ and $\Delta|_A\Phi \Longrightarrow \Delta|_A\Phi'$.
(2) If $\Phi \xrightarrow{a} \Phi'$ and $a \notin A$ then $\Phi|_A\Delta \xrightarrow{a} \Phi'|_A\Delta$ and $\Delta|_A\Phi \xrightarrow{a} \Delta|_A\Phi'$.
(3) If $\Phi \xrightarrow{a} \Phi'$, $\Delta \xrightarrow{a} \Delta'$ and $a \in A$ then $\Delta|_A\Phi \xrightarrow{\tau} \Delta'|_A\Phi'$.
(4) $(\sum_{j\in J} p_j \cdot \Phi_j)|_A(\sum_{k\in K} q_k \cdot \Delta_k) = \sum_{j\in J}\sum_{k\in K}(p_j \cdot q_k)\cdot(\Phi_j|_A\Delta_k)$.
(5) Let $\mathcal{R}, \mathcal{R}' \subseteq S \times \mathcal{D}_{sub}(S)$ be two relations satisfying $u\mathcal{R}\Psi$ whenever $u = s|_A t$ and $\Psi = \Theta|_A t$ with $s\mathcal{R}'\Theta$ and $t \in S$. Then $\Delta \; \mathcal{R}'^\dagger \; \Theta$ and $\Phi \in \mathcal{D}_{sub}(S)$ implies $(\Delta|_A\Phi) \; \mathcal{R}^\dagger \; (\Theta|_A\Phi)$. □

Proposition 6.10 *Suppose* $\Delta \ (\lhd_{FS}^{C})^{\dagger}\ \Theta$ *and* $\Delta|_A t \xrightarrow{\tau}_p \Gamma$. *Then* $\Theta|_A t \Longrightarrow\xrightarrow{\tau}\Longrightarrow \Psi$ *for some* Ψ *such that* $\Gamma\ \mathcal{R}^{\dagger}\ \Psi$, *where* $\mathcal{R}$ *is the relation* $\{(s|_A t, \Theta|_A t) \mid s \lhd_{FS}^{C} \Theta\}$.

Proof We first show a simplified version of the result. Suppose that $s \lhd_{FS}^{C} \Theta$ and $s|_A t \xrightarrow{\tau}_p \Gamma$; we prove this entails $\Theta|_A t \Longrightarrow\xrightarrow{\tau}\Longrightarrow \Psi$ such that $\Gamma\ \mathcal{R}^{\dagger}\ \Psi$. There are only two possibilities for inferring the above productive move from $s|_A t$:

(i) $\Gamma = s|_A \Phi$ where $t \xrightarrow{\tau} \Phi$ and
(ii) $\Gamma = \Delta|_A \Phi$ where for some $a \in A$, $s \xrightarrow{a} \Delta$ and $t \xrightarrow{a} \Phi$.

In the first case, we have $\Theta|_A t \xrightarrow{\tau} \Theta|_A \Phi$ by using Lemma 6.27(2) and also that $(s|_A \Phi)\ \mathcal{R}^{\dagger}\ (\Theta|_A \Phi)$ by Lemma 6.27(5), whereas in the second case, $s \lhd_{FS}^{C} \Theta$ implies $\Theta \Longrightarrow\xrightarrow{a}\Longrightarrow \Theta'$ for some $\Theta' \in \mathcal{D}_{sub}(S)$ with $\Delta\ (\lhd_{FS}^{C})^{\dagger}\ \Theta'$, and we have $\Theta|_A t \Longrightarrow\xrightarrow{\tau}\Longrightarrow \Theta'|_A \Phi$ by Lemma 6.27(1) and (3), and $(\Delta|_A \Phi)\ \mathcal{R}^{\dagger}\ (\Theta'|_A \Phi)$ by Lemma 6.27(5).

The general case now follows using a standard decomposition/recomposition argument. Since $\Delta|_A t \xrightarrow{\tau}_p \Gamma$, Lemma 6.1 yields

$$\Delta = \sum_{i\in I} p_i \cdot \overline{s_i}, \qquad s_i|_A t \xrightarrow{\tau}_p \Gamma_i, \qquad \Gamma = \sum_{i\in I} p_i \cdot \Gamma_i,$$

for certain $s_i \in S$, $\Gamma_i \in \mathcal{D}_{sub}(S)$ and $\sum_{i\in I} p_i \le 1$. In analogy with Proposition 6.8, we can show that $\lhd_{FS}^{C}$ is convex. Hence, since $\Delta\ (\lhd_{FS}^{C})^{\dagger}\ \Theta$, Corollary 6.1 yields that $\Theta = \sum_{i\in I} p_i \cdot \Theta_i$ for some $\Theta_i \in \mathcal{D}_{sub}(S)$ such that $s_i \lhd_{FS}^{C} \Theta_i$ for $i \in I$. By the above argument, we have $\Theta_i|_A t \Longrightarrow\xrightarrow{\tau}\Longrightarrow \Psi_i$ for some $\Psi_i \in \mathcal{D}_{sub}(S)$ such that $\Gamma_i\ \mathcal{R}^{\dagger}\ \Psi_i$. The required Ψ can be taken to be $\sum_{i\in I} p_i \cdot \Psi_i$ as Definition 6.2(2) yields $\Gamma\ \mathcal{R}^{\dagger}\ \Psi$ and Theorem 6.5(i) and Definition 6.2(2) yield $\Theta|_A t \Longrightarrow\xrightarrow{\tau}\Longrightarrow \Psi$. □

Our next result shows that we can always factor out productive moves from an arbitrary action of a parallel process.

Lemma 6.28 *Suppose* $\Delta|_A t \xrightarrow{\tau} \Gamma$. *Then there exists subdistributions* $\Delta^{\rightarrow}$, $\Delta^{\times}$, Δ^{next}, $\Gamma^{\times}$ *(possibly empty) such that*

(i) $\Delta = \Delta^{\rightarrow} + \Delta^{\times}$
(ii) $\Delta^{\rightarrow} \xrightarrow{\tau} \Delta^{\text{next}}$
(iii) $\Delta^{\times}|_A t \xrightarrow{\tau}_p \Gamma^{\times}$
(iv) $\Gamma = \Delta^{\text{next}}|_A t + \Gamma^{\times}$

Proof By Lemma 6.1 $\Delta|_A t \xrightarrow{\tau} \Gamma$ implies that

$$\Delta = \sum_{i\in I} p_i \cdot \overline{s_i}, \qquad s_i|_A t \xrightarrow{\tau} \Gamma_i, \qquad \Gamma = \sum_{i\in I} p_i \cdot \Gamma_i,$$

for certain $s_i \in S$, $\Gamma_i \in \mathcal{D}_{sub}(S)$ and $\sum_{i\in I} p_i \le 1$. Let $J = \{i \in I \mid s_i|_A t \xrightarrow{\tau}_p \Gamma_i\}$. Note that for each $i \in (I - J)$ the subdistribution Γ_i has the form $\Gamma_i'|_A t$, where $s_i \xrightarrow{\tau} \Gamma_i'$. Now let

$$\Delta^{\rightarrow} = \sum_{i\in(I-J)} p_i \cdot \overline{s_i}, \quad \Delta^{\times} = \sum_{i\in J} p_i \cdot \overline{s_i}$$

$$\Delta^{\text{next}} = \sum_{i\in(I-J)} p_i \cdot \Gamma'_i, \qquad \Gamma^{\times} = \sum_{i\in J} p_i \cdot \Gamma_i$$

By construction (i) and (iv) are satisfied, and (ii) and (iii) follow by property (2) of Definition 6.2. □

Lemma 6.29 *If $\Delta|_A t \Longrightarrow \varepsilon$ then there is a $\Delta' \in \mathcal{D}_{sub}(S)$ such that $\Delta \Longrightarrow \Delta'$ and $\Delta'|_A t \xrightarrow{\tau}_p \Longrightarrow \varepsilon$.*

Proof Suppose $\Delta_0|_A t \Longrightarrow \varepsilon$. By Lemma 6.24 there is an infinite sequence

$$\Delta_0|_A t \xrightarrow{\tau} \Psi_1 \xrightarrow{\tau} \Psi_2 \xrightarrow{\tau} \ldots \tag{6.23}$$

By induction on $k \geq 0$, we find distributions Γ_{k+1}, $\Delta_k^{\rightarrow}$, $\Delta_k^{\times}$, Δ_{k+1}, $\Gamma_{k+1}^{\times}$ such that

(i) $\Delta_k|_A\, t \xrightarrow{\tau} \Gamma_{k+1}$
(ii) $\Gamma_{k+1} \leq \Psi_{k+1}$
(iii) $\Delta_k = \Delta_k^{\rightarrow} + \Delta_k^{\times}$
(iv) $\Delta_k^{\rightarrow} \xrightarrow{\tau} \Delta_{k+1}$
(v) $\Delta_k^{\times}|_A\, t \xrightarrow{\tau}_p \Gamma_{k+1}^{\times}$
(vi) $\Gamma_{k+1} = \Delta_{k+1}|_A\, t + \Gamma_{k+1}^{\times}$.

Induction Base Take $\Gamma_1 := \Psi_1$ and apply Lemma 6.28.
Induction Step Assume we already have Γ_k, Δ_k and $\Gamma_k^{\times}$. Since $\Delta_k|_A t \leq \Gamma_k \leq \Psi_k$ and $\Psi_k \xrightarrow{\tau} \Psi_{k+1}$, Proposition 6.2 gives us a Γ_{k+1} such that $\Delta_k|_A t \xrightarrow{\tau} \Gamma_{k+1}$ and $\Gamma_{k+1} \leq \Psi_{k+1}$. Now apply Lemma 6.28.

Let $\Delta' := \sum_{k=0}^{\infty} \Delta_k^{\times}$. By (iii) and (iv) above, we obtain a weak τ move $\Delta_0 \Longrightarrow \Delta'$. Since $\Delta'|_A t = \sum_{k=0}^{\infty} (\Delta_k^{\times}|_A t)$, by (v) and Definition 6.2 we have $\Delta'|_A t \xrightarrow{\tau}_p \sum_{k=1}^{\infty} \Gamma_k^{\times}$. Note that here it does not matter if $\Delta' = \varepsilon$. Since $\Gamma_k^{\times} \leq \Gamma_k \leq \Psi_k$ and $\Psi_k \Longrightarrow \varepsilon$, it follows from Theorem 6.5(ii) that $\Gamma_k^{\times} \Longrightarrow \varepsilon$. Hence by using Theorem 6.5(i), we obtain that $\sum_{k=1}^{\infty} \Gamma_k^{\times} \Longrightarrow \varepsilon$. □

We are now ready to prove the main result of this section, namely that $\sqsubseteq_{FS}$ is preserved by the parallel operator.

Proposition 6.11 *In a finitary pLTS, if $\Theta \sqsubseteq_{FS} \Delta$ then $\Theta|_A \Phi \sqsubseteq_{FS} \Delta|_A \Phi$.*

Proof We first construct the following relation

$$\mathcal{R} := \{(s|_A t, \Theta|_A t) \mid s \lhd_{FS}^{C} \Theta\}$$

and check that $\mathcal{R} \subseteq \lhd_{FS}^{C}$. As in the proof of Theorem 5.1, one can check that each strong transition from $s|_A t$ can be matched by a transition from $\Theta|_A t$, and the matching of failures can also be established. So we concentrate on the requirement involving divergence.

Suppose $s \lhd_{FS}^{C} \Theta$ and $\overline{s|_A t} \Longrightarrow \varepsilon$. We need to find some Γ, Ψ such that

(a) $\overline{s|_A t} \Longrightarrow \xrightarrow{\tau} \Longrightarrow \Gamma \Longrightarrow \varepsilon$,
(b) $\Theta|_A t \Longrightarrow \xrightarrow{\tau} \Longrightarrow \Psi$ and $\Gamma\ \mathcal{R}^\dagger\ \Psi$.

By Lemma 6.29, there are $\Delta', \Gamma \in \mathcal{D}_{sub}(S)$ with $\overline{s} \Longrightarrow \Delta'$ and $\Delta'|_A t \xrightarrow{\tau}_p \Gamma \Longrightarrow \varepsilon$. Since for finitary processes $\lhd^C_{FS}$ coincides with $\lhd^S_{FS}$ and $\lhd^e_{FS}$ by Lemma 6.26 and Theorem 6.12, there must be a $\Theta' \in \mathcal{D}_{sub}(S)$ such that $\Theta \Longrightarrow \Theta'$ and $\Delta'\ (\lhd^C_{FS})^\dagger\ \Theta'$. By Proposition 6.10, we have $\Theta'|_A t \Longrightarrow \xrightarrow{\tau} \Longrightarrow \Psi$ for some Ψ such that $\Gamma\ \mathcal{R}^\dagger\ \Psi$. Now, $s|_A t \Longrightarrow \Delta'|_A t \xrightarrow{\tau} \Gamma \Longrightarrow \varepsilon$ and $\Theta|_A t \Longrightarrow \Theta'|_A t \Longrightarrow \xrightarrow{\tau} \Longrightarrow \Psi$ with $\Gamma\ \mathcal{R}^\dagger\ \Psi$, which had to be shown.

Therefore, we have shown that $\mathcal{R} \subseteq \lhd^C_{FS}$. Now let us focus our attention on the statement of the proposition, which involves $\sqsubseteq_{FS}$.

Suppose $\Theta \sqsubseteq_{FS} \Delta$. By Proposition 6.7, this means that there is some Θ^{match} such that $\Theta \Longrightarrow \Theta^{\text{match}}$ and $\Delta\ (\lhd^e_{FS})^\dagger\ \Theta^{\text{match}}$. By Theorem 6.12 and Lemma 6.26, we have $\Delta\ (\lhd^C_{FS})^\dagger\ \Theta^{\text{match}}$. Then Lemma 6.27(5) yields $(\Delta|_A \Phi)\ \mathcal{R}^\dagger\ (\Theta^{\text{match}}|_A \Phi)$. Therefore, we have $(\Delta|_A \Phi)\ (\lhd^C_{FS})^\dagger\ (\Theta^{\text{match}}|_A \Phi)$, that is, $(\Delta|_A \Phi)\ (\lhd^e_{FS})^\dagger\ (\Theta^{\text{match}}|_A \Phi)$ by Lemma 6.26 and Theorem 6.12. By using Lemma 6.27(1), we also have that $(\Theta|_A \Phi) \Longrightarrow (\Theta^{\text{match}}|_A \Phi)$, which had to be established according to Proposition 6.7. □

In the proof of Proposition 6.11, we use the characterisation of $\lhd^e_{FS}$ as $\lhd^S_{FS}$, which assumes the pLTS to be finitary. In general, the relation $\lhd^S_{FS}$ is not closed under parallel composition.

Example 6.23 We use a modification of the infinite state pLTS's in Example 6.5 that as before has states s_k with $k \geq 2$, but we add an extra a-looping state s_a to give all together the system

$$\text{for } k \geq 2 \qquad s_k \xrightarrow{\tau} (\overline{s_a}\ {}_{\frac{1}{k^2}}\!\oplus \overline{s_{k+1}}) \qquad \text{and} \qquad s_a \xrightarrow{a} \overline{s_a}\ .$$

There is a failure simulation so that $s_k \lhd^S_{FS} (\overline{s_a}\ {}_{\frac{1}{k}}\!\oplus \overline{\mathbf{0}})$ because from state s_k the transition $s_k \xrightarrow{\tau} (\overline{s_a}\ {}_{\frac{1}{k^2}}\!\oplus \overline{s_{k+1}})$ can be matched by a transition to $(\overline{s_a}\ {}_{\frac{1}{k^2}}\!\oplus (\overline{s_a}\ {}_{\frac{1}{k+1}}\!\oplus \overline{\mathbf{0}}))$ that simplifies to just $(\overline{s_a}\ {}_{\frac{1}{k}}\!\oplus \overline{\mathbf{0}})$ again, that is, a sufficient simulating transition would be the identity instance of $\Longrightarrow$.

Now $s_2|_{\{a\}} s_a$ wholly diverges even though s_2 itself does not, and (recall from above) we have $s_2 \lhd^S_{FS} (\overline{s_a}\ {}_{\frac{1}{2}}\!\oplus \mathbf{0})$. Yet $(\overline{s_a}\ {}_{\frac{1}{2}}\!\oplus \mathbf{0})|_{\{a\}} s_a$ does not diverge, and therefore $s_2|_{\{a\}} s_a \not\lhd^S_{FS} (\overline{S_a}\ {}_{\frac{1}{2}}\!\oplus \mathbf{0})|_{\{a\}} S_a$.

Note that this counter-example does not go through if we use failure similarity $\lhd^e_{FS}$ instead of simple failure similarity $\lhd^S_{FS}$, since $s_2 \not\lhd^e_{FS} (\overline{s_a}\ {}_{\frac{1}{2}}\!\oplus \mathbf{0})$—the former has the transition $s_2 \Longrightarrow s_a\ {}_{\frac{1}{2}}\!\oplus \varepsilon$, which cannot be matched by $s_a\ {}_{\frac{1}{2}}\!\oplus \mathbf{0}$.

Proposition 6.12 (Precongruence) *In a finitary pLTS, if $P \sqsubseteq_{FS} Q$ then it holds that $\alpha.P \sqsubseteq_{FS} \alpha.Q$ for any $\alpha \in$ Act, and similarly if $P_1 \sqsubseteq_{FS} Q_1$ and $P_2 \sqsubseteq_{FS} Q_2$ then $P_1 \odot P_2 \sqsubseteq_{FS} Q_1 \odot Q_2$ for $\odot$ being any of the operators $\sqcap$, $\Box$, ${}_p\!\oplus$ and $|_A$.*

Proof The most difficult case is the closure of failure simulation under parallel composition, which is proved in Proposition 6.11. The other cases are simpler, thus omitted. □

Lemma 6.30 *In a finitary pLTS, if* $P \sqsubseteq_{FS} Q$ *then* $[P|_{\mathsf{Act}}T] \sqsubseteq_{FS} [Q|_{\mathsf{Act}}T]$ *for any test* T.

Proof We first construct the following relation

$$\mathcal{R} := \{(s|_{\mathsf{Act}}\, t, \Theta|_{\mathsf{Act}}\, t) \mid s \lhd^{C}_{FS} \Theta\}$$

where $s|_{\mathsf{Act}}\, t$ is a state in $[P|_{\mathsf{Act}}T]$ and $\Theta|_{\mathsf{Act}}\, t$ is a subdistribution in $[Q|_{\mathsf{Act}}T]$, and show that $\mathcal{R} \subseteq \lhd^{C}_{FS}$.

1. The matching of divergence between $s|_{\mathsf{Act}}\, t$ and $\Theta|_{\mathsf{Act}}\, t$ is almost the same as the proof of Proposition 6.11, besides that we need to check the requirements $t \stackrel{\omega}{\not\longrightarrow}$ and $\Gamma \stackrel{\omega}{\not\longrightarrow}$ are always met there.
2. We now consider the matching of transitions.
   - If $s|_{\mathsf{Act}}\, t \stackrel{\omega}{\longrightarrow}$ then this action is actually performed by t. Suppose $t \stackrel{\omega}{\longrightarrow} \Gamma$. Then $s|_{\mathsf{Act}}\, t \stackrel{\omega}{\longrightarrow} s|_{\mathsf{Act}}\Gamma$ and $\Theta|_{\mathsf{Act}}\, t \stackrel{\omega}{\longrightarrow} \Theta|_{\mathsf{Act}}\Gamma$. Obviously we have $(s|_{\mathsf{Act}}\Gamma, \Theta|_{\mathsf{Act}}\Gamma) \in \mathcal{R}^{\dagger}$.
   - If $s|_{\mathsf{Act}}\, t \stackrel{\tau}{\longrightarrow}$ then we must have $s|_{\mathsf{Act}}\, t \stackrel{\omega}{\not\longrightarrow}$, otherwise the τ transition would be a "scooting" transition and the pLTS is not ω-respecting. It follows that $t \stackrel{\omega}{\not\longrightarrow}$. There are three subcases.
     - $t \stackrel{\tau}{\longrightarrow} \Gamma$. So the transition $s|_{\mathsf{Act}}\, t \stackrel{\tau}{\longrightarrow} s|_{\mathsf{Act}}\Gamma$ can simply be matched by $\Theta|_{\mathsf{Act}}\, t \stackrel{\tau}{\longrightarrow} \Theta|_{\mathsf{Act}}\Gamma$.
     - $s \stackrel{\tau}{\longrightarrow} \Delta$. Since $s \lhd^{C}_{FS} \Theta$, there exists some Θ' such that $\Theta \Longrightarrow \Theta'$ and $\Delta\ (\lhd^{C}_{FS})^{\dagger}\ \Theta'$. Note that in this case $t \stackrel{\omega}{\not\longrightarrow}$. Hence $\Theta|_{\mathsf{Act}}\, t \Longrightarrow \Theta'|_{\mathsf{Act}}\, t$ which can match the transition $s|_{\mathsf{Act}}\, t \longrightarrow \Delta|_{\mathsf{Act}}\, t$ because we also have $(\Delta|_{\mathsf{Act}}\, t, \Theta'|_{\mathsf{Act}}\, t) \in \mathcal{R}^{\dagger}$.
     - $s \stackrel{a}{\longrightarrow} \Delta$ and $t \stackrel{a}{\longrightarrow} \Gamma$ for some action $a \in \mathsf{Act}$. Since $s \lhd^{C}_{FS} \Theta$, there exists some Θ' such that $\Theta \stackrel{a}{\Longrightarrow} \Theta'$ and $\Delta\ (\lhd^{C}_{FS})^{\dagger}\ \Theta'$. Note that in this case $t \stackrel{\omega}{\not\longrightarrow}$. It easily follows that $\Theta|_{\mathsf{Act}}\, t \Longrightarrow \Theta'|_{\mathsf{Act}}\,\Gamma$ which can match the transition $s|_{\mathsf{Act}}\, t \longrightarrow \Delta|_{\mathsf{Act}}\Gamma$ because $(\Delta|_{\mathsf{Act}}\,\Gamma, \Theta'|_{\mathsf{Act}}\,\Gamma) \in \mathcal{R}^{\dagger}$.
   - Suppose $s|_{\mathsf{Act}} \stackrel{A}{\not\longrightarrow}$ for any $A \subseteq \mathsf{Act} \cup \{\omega\}$. There are two possibilities.
     - If $s|_{\mathsf{Act}}\, t \stackrel{\omega}{\not\longrightarrow}$, then $t \stackrel{\omega}{\not\longrightarrow}$ and there are two subsets A_1, A_2 of A such that $s \stackrel{A_1}{\not\longrightarrow}, t \stackrel{A_2}{\not\longrightarrow}$ and $A = A_1 \cup A_2$. Since $s \lhd^{C}_{FS} \Theta$ there exists some Θ' such that $\Theta \Longrightarrow \Theta'$ and $\Theta' \stackrel{A_1}{\not\longrightarrow}$. Therefore, we have $\Theta|_{\mathsf{Act}}\, t \Longrightarrow \Theta'|_{\mathsf{Act}}\, t \stackrel{A}{\not\longrightarrow}$.
     - If $s|_{\mathsf{Act}}\, t \stackrel{\omega}{\longrightarrow}$ then $t \stackrel{\omega}{\longrightarrow}$ and $\omega \notin A$. Therefore, we have $\Theta|_{\mathsf{Act}}\, t \stackrel{\omega}{\longrightarrow}$ and $\Theta|_{\mathsf{Act}}\, t \stackrel{\tau}{\not\longrightarrow}$ because there is no "scooting" transition in $\Theta|_{\mathsf{Act}}\, t$. It follows that $\Theta|_{\mathsf{Act}}\, t \stackrel{A}{\not\longrightarrow}$.

Therefore, we have shown that $\mathcal{R} \subseteq \lhd^{C}_{FS}$, from which our expected result can be established by using similar arguments as in the last part of the proof of Proposition 6.11. □

6.6.4 Soundness

In this section, we prove that failure simulations are sound for showing that processes are related via the failure-based testing preorder. We assume initially that we are using only one success action ω, so that $|\Omega| = 1$.

Because we prune our pLTSs before extracting values from them, we will be concerned mainly with ω-respecting structures.

Definition 6.24 Let Δ be a subdistribution in a pLTS $\langle S, \{\omega, \tau\}, \rightarrow\rangle$. We write $\mathcal{V}(\Delta)$ for the set of testing outcomes $\{\$\Delta' \mid \Delta \Longrightarrow\!\!\succ \Delta'\}$.

Lemma 6.31 *Let Δ and Θ be subdistributions in an ω-respecting pLTS given by $\langle S, \{\tau, \omega\}, \rightarrow\rangle$. If Δ is stable and $\Delta\ (\lhd^e_{FS})^\dagger\ \Theta$, then $\mathcal{V}(\Theta) \leq_{\mathrm{Sm}} \mathcal{V}(\Delta)$.*

Proof We first show that if s is stable and $s \lhd^e_{FS} \Theta$ then $\mathcal{V}(\Theta) \leq_{\mathrm{Sm}} \mathcal{V}(s)$. Since s is stable, we have only two cases:

(i) $s \not\longrightarrow$. Here $\mathcal{V}(s) = \{0\}$ and since $s \lhd^e_{FS} \Theta$ we have $\Theta \Longrightarrow \Theta'$ with $\Theta' \not\longrightarrow$, whence in fact $\Theta \Longrightarrow\!\!\succ \Theta'$ and $\$\Theta' = 0$. Therefore $0 \in \mathcal{V}(\Theta)$ that means $\mathcal{V}(\Theta) \leq_{\mathrm{Sm}} \mathcal{V}(s)$.
(ii) $s \xrightarrow{\omega} \Delta'$ for some Δ'. Here $\mathcal{V}(s) = \{1\}$ and $\Theta \Longrightarrow \Theta' \xrightarrow{\omega}$ with $\$\Theta' = |\Theta'|$. Because the pLTS is ω-respecting, in fact $\Theta \Longrightarrow\!\!\succ \Theta'$ and so again we have $\mathcal{V}(\Theta) \leq_{\mathrm{Sm}} \mathcal{V}(s)$.

Now, for the general case, we suppose $\Delta\ (\lhd^e_{FS})^\dagger\ \Theta$. Use Proposition 6.2 to decompose Θ into $\sum_{s \in \lceil\Delta\rceil} \Delta(s) \cdot \Theta_s$ such that $s \lhd^e_{FS} \Theta_s$ for each $s \in \lceil\Delta\rceil$, and recall each such state s is stable. From above we have that $\mathcal{V}(\Theta_s) \leq_{\mathrm{Sm}} \mathcal{V}(s)$ for those s, and so $\mathcal{V}(\Theta) = \sum_{s \in \lceil\Delta\rceil} \Delta(s) \cdot \mathcal{V}(\Theta_s) \leq_{\mathrm{Sm}} \sum_{s \in \lceil\Delta\rceil} \Delta(s) \cdot \mathcal{V}(s) = \mathcal{V}(\Delta)$. □

Lemma 6.32 *Let Δ be a subdistribution in an ω-respecting pLTS $\langle S, \{\tau, \omega\}, \rightarrow\rangle$. If $\Delta \Longrightarrow \Delta'$ then $\mathcal{V}(\Delta') \subseteq \mathcal{V}(\Delta)$.*

Proof Note that if $\Delta' \Longrightarrow\!\!\succ \Delta''$ then $\Delta \Longrightarrow \Delta' \Longrightarrow\!\!\succ \Delta''$, so that every extreme derivative of Δ' is also an extreme derivative of Δ. □

Lemma 6.33 *Let Δ and Θ be subdistributions in an ω-respecting pLTS given by $\langle S, \{\tau, \omega\}, \rightarrow\rangle$. If $\Theta \sqsubseteq_{FS} \Delta$, then it holds that $\mathcal{V}(\Theta) \leq_{\mathrm{Sm}} \mathcal{V}(\Delta)$.*

Proof Let Δ and Θ be subdistributions in an ω-respecting pLTS $\langle S, \{\tau, \omega\}, \rightarrow\rangle$. We first claim that

$$\text{If } \Delta\ (\lhd^e_{FS})^\dagger\ \Theta, \text{ then } \mathcal{V}(\Theta) \leq_{\mathrm{Sm}} \mathcal{V}(\Delta).$$

We assume that $\Delta\ (\lhd^e_{FS})^\dagger\ \Theta$. For any $\Delta \Longrightarrow\!\!\succ \Delta'$ we have the matching transition $\Theta \Longrightarrow \Theta'$ such that $\Delta'\ (\lhd^e_{FS})^\dagger\ \Theta'$. It follows from Lemmas 6.31 and 6.32 that

$$\mathcal{V}(\Theta) \supseteq \mathcal{V}(\Theta') \leq_{\mathrm{Sm}} \mathcal{V}(\Delta').$$

Consequently, we obtain $\mathcal{V}(\Theta) \leq_{\mathrm{Sm}} \mathcal{V}(\Delta)$.

Now suppose $\Theta \sqsubseteq_{FS} \Delta$. By Proposition 6.7, there exists some Θ' with $\Theta \Longrightarrow \Theta'$ and $\Delta\ (\lhd^{e}_{FS})^{\dagger}\ \Theta'$. By the above claim and Lemma 6.32 we obtain

$$\mathcal{V}(\Theta) \supseteq \mathcal{V}(\Theta') \leq_{\mathrm{Sm}} \mathcal{V}(\Delta)',$$

thus $\mathcal{V}(\Theta) \leq_{\mathrm{Sm}} \mathcal{V}(\Delta)$. □

Theorem 6.13 *For any finitary processes P and Q, if $P \sqsubseteq_{FS} Q$ then $P \sqsubseteq_{pmust} Q$.*

Proof We reason as follows.

	$P \sqsubseteq_{FS} Q$	
implies	$[P\vert_{\mathsf{Act}} T] \sqsubseteq_{FS} [Q\vert_{\mathsf{Act}} T]$	Lemma 6.30, for any test T
implies	$\mathcal{V}([P\vert_{\mathsf{Act}} T]) \leq_{\mathrm{Sm}} \mathcal{V}([Q\vert_{\mathsf{Act}} T])$	$[\cdot]$ is ω-respecting; Lemma 6.33
iff	$\mathcal{A}^{\mathrm{d}}(T, P) \leq_{\mathrm{Sm}} \mathcal{A}^{\mathrm{d}}(T, Q)$	Definition 6.24
iff	$P \sqsubseteq_{\mathrm{pmust}} Q$.	Definition 6.11

□

In the proof of the soundness result above, we use Lemma 6.30, which holds for finitary processes only. For infinitary processes, a preorder induced by $\lhd^{S}_{FS}$ is not sound for must testing.

Example 6.24 We have seen in Example 6.21 that the state s from (6.20) is related to $\overline{\mathbf{0}}$ via the relation $\lhd^{S}_{FS}$. If we apply test $\tau.\omega$ to both s and $\mathbf{0}$, we obtain $\{q\ \vec{\omega} \mid q \in [\frac{1}{2}, 1]\}$ as an outcome set for the former and $\{\vec{\omega}\}$ for the latter. Although $\overline{s}\ (\lhd^{S}_{FS})^{\dagger}\ \overline{\mathbf{0}}$, we have $\mathcal{A}^{\mathrm{d}}(\tau.\omega, \mathbf{0}) \not\leq_{\mathrm{Sm}} \mathcal{A}^{\mathrm{d}}(\tau.\omega, s)$.

If we replace state s by the state s_2 from (6.22), similar phenomenon happens. Although $\overline{s_2}\ (\lhd^{S}_{FS})^{\dagger}\ \overline{\mathbf{0}}$, we have

$$\mathcal{A}^{\mathrm{d}}(\tau.\omega, \mathbf{0}) = \{\vec{\omega}\} \not\leq_{\mathrm{Sm}} \left\{ q\vec{\omega} \mid q \in \left[\frac{1}{2}, 1\right]\right\} = \mathcal{A}^{\mathrm{d}}(\tau.\omega, s_2)\ .$$

6.7 Failure Simulation is Complete for Must Testing

This section establishes the completeness of the failure simulation preorder with respect to the must testing preorder. It does so in three steps. First, we provide a characterisation of the preorder relation $\sqsubseteq_{FS}$ by an inductively defined relation. Secondly, using this, we develop a modal logic that can be used to characterise the failure simulation preorder on finitary pLTSs. Finally, we adapt the results of Sect. 5.7 to show that the modal formulae can in turn be characterised by tests; again this result depends on the underlying pLTS being finitary. From this, completeness follows.

6.7.1 Inductive Characterisation

The relation $\triangleleft_{FS}^{S}$ of Definition 6.21 is given coinductively; it is the largest fixed point of an equation $\mathcal{R} = \mathbf{F}(\mathcal{R})$. An alternative approach, therefore, is to use that $\mathbf{F}(-)$ to define $\triangleleft_{FS}^{S}$ as a limit of approximants:

Definition 6.25 For every $k \geq 0$ we define the relations $\triangleleft_{FS}^{k} \subseteq S \times \mathcal{D}_{sub}(S)$ as follows:

(i) $\triangleleft_{FS}^{0} := S \times \mathcal{D}_{sub}(S)$
(ii) $\triangleleft_{FS}^{k+1} := \mathbf{F}(\triangleleft_{FS}^{k})$

Finally, let $\triangleleft_{FS}^{\infty} := \bigcap_{k=0}^{\infty} \triangleleft_{FS}^{k}$.

A simple inductive argument ensures that $\triangleleft_{FS}^{S} \subseteq \triangleleft_{FS}^{k}$, for every $k \geq 0$, and therefore that $\triangleleft_{FS}^{S} \subseteq \triangleleft_{FS}^{\infty}$. The converse is, however, not true in general.

A (nonprobabilistic) example is well-known in the literature; it makes essential use of an infinite branching. Let P be the process $\mathsf{rec}\, x.a.x$ and s a state in a pLTS that starts by making an infinitary choice, namely for each $k \geq 1$ it has the option to perform a sequence of a actions with length k in succession and then deadlock. This can be described by the infinitary CSP expression $\Box_{k=1}^{\infty} a^k$. Then $[\![P]\!] \not\triangleleft_{FS}^{S} s$, because the move $[\![P]\!] \xrightarrow{a} [\![P]\!]$ cannot be matched by s. However, an easy inductive argument shows that $[\![P]\!] \triangleleft_{FS}^{k} a^k$ for every k, and therefore that $[\![P]\!] \triangleleft_{FS}^{\infty} s$.

Once we restrict our nonprobabilistic systems to be finitely branching, however, a simple counting argument will show that $\triangleleft_{FS}^{S}$ coincides with $\triangleleft_{FS}^{\infty}$; see [11, Theorem 2.1] for the argument applied to bisimulation equivalence. In the probabilistic case, we restrict to both finite-state *and* finitely branching systems, and the effect of that is captured by topological *compactness*. Finiteness is lost unavoidably when we remember that, for example, the process $a.\mathbf{0} \sqcap b.\mathbf{0}$ can move via $\Longrightarrow$ to a distribution $[\![a.\mathbf{0}]\!]_p\oplus[\![b.\mathbf{0}]\!]$ for any of the uncountably many probabilities $p \in [0, 1]$. Nevertheless, those uncountably many weak transitions can be generated by arbitrary interpolation of two transitions $[\![a.\mathbf{0} \sqcap b.\mathbf{0}]\!] \xrightarrow{\tau} [\![a.\mathbf{0}]\!]$ and $[\![a.\mathbf{0} \sqcap b.\mathbf{0}]\!] \xrightarrow{\tau} [\![b.\mathbf{0}]\!]$, and that is the key structural property that compactness captures.

Because compactness follows from closure and boundedness, we approach this topic via closure.

Note that the metric spaces $(\mathcal{D}_{sub}(S), d_1)$ with $d_1(\Delta, \Theta) = max_{s\in S}|\Delta(s) - \Theta(s)|$ and $(S \to \mathcal{D}_{sub}(S), d_2)$ with $d_2(f, g) = max_{s\in S} d_1(f(s), g(s))$ are complete. Let X be a subset of either $\mathcal{D}_{sub}(S)$ or $S \to \mathcal{D}_{sub}(S)$. Clearly, X is bounded. So if X is closed, it is also compact.

Definition 6.26 A relation $\mathcal{R} \subseteq S \times \mathcal{D}_{sub}(S)$ is said to be *closed* if for every $s \in S$ the set $s \cdot \mathcal{R} = \{\Delta \mid s\mathcal{R}\Delta\}$ is closed.

Two examples of closed relations are $\Longrightarrow$ and $\stackrel{a}{\Longrightarrow}$ for any action a, as shown by Lemma 6.17 and Corollary 6.6.

Our next step is to show that each of the relations $\triangleleft_{FS}^{k}$ are closed. This requires some results to be first established.

Lemma 6.34 *Let $\mathcal{R} \subseteq S \times \mathcal{D}_{sub}(S)$ be closed. Then* $\mathbf{Ch}(\mathcal{R})$ *is also closed.*

Proof Straightforward. □

Corollary 6.10 *Let $\mathcal{R} \subseteq S \times \mathcal{D}_{sub}(S)$ be closed and convex. Then $\mathcal{R}^{\dagger}$ is also closed.*

Proof For any $\Delta \in \mathcal{D}_{sub}(S)$, we know from Proposition 6.1 that

$$\Delta \cdot \mathcal{R}^{\dagger} = \{\text{Exp}_{\Delta}(f) \mid f \in \mathbf{Ch}(\mathcal{R})\}.$$

The function $\text{Exp}_{\Delta}(-)$ is continuous. By Lemma 6.34, the set of choice functions of $\mathcal{R}$ is closed and it is also bounded, thus being compact. Its image is also compact, thus being closed. □

Lemma 6.35 *Let $\mathcal{R} \subseteq S \times \mathcal{D}_{sub}(S)$ be closed and convex, and $C \subseteq \mathcal{D}_{sub}(S)$ be closed. Then the set $\{\Delta \mid \Delta \cdot \mathcal{R}^{\dagger} \cap C \neq \emptyset\}$ is also closed.*

Proof First define $\mathcal{E} : \mathcal{D}_{sub}(S) \times (S \to \mathcal{D}_{sub}(S)) \to \mathcal{D}_{sub}(S)$ by $\mathcal{E}(\Theta, f) = \text{Exp}_{\Theta}(f)$, which is obviously continuous. Then we know from the previous lemma that $\mathbf{Ch}(\mathcal{R})$ is closed. Finally let

$$Z = \pi_1(\mathcal{E}^{-1}(C) \cap (\mathcal{D}_{sub}(S) \times \mathbf{Ch}(\mathcal{R}))),$$

where π_1 is the projection onto the first component of a pair. We observe that the continuity of $\mathcal{E}$ ensures that the inverse image of the closed set C is closed. Furthermore, $\mathcal{E}^{-1}(C) \cap (\mathcal{D}_{sub}(S) \times \mathbf{Ch}(\mathcal{R}))$ is compact because it is both closed and bounded. Its image under the continuous function π_1 is also compact. It follows that Z is closed. But $Z = \{\Delta \mid \Delta \cdot \mathcal{R}^{\dagger} \cap C \neq \emptyset\}$ because

$$\begin{aligned}
\Delta \in Z \ &\text{iff } (\Delta, f) \in \mathcal{E}^{-1}(C) \text{ for some } f \in \mathbf{Ch}(\mathcal{R}) \\
&\text{iff } \mathcal{E}(\Delta, f) \in C \text{ for some } f \in \mathbf{Ch}(\mathcal{R}) \\
&\text{iff } \text{Exp}_{\Delta}(f) \in C \text{ for some } f \in \mathbf{Ch}(\mathcal{R}) \\
&\text{iff } \Delta \ \mathcal{R}^{\dagger} \ \Delta' \text{ for some } \Delta' \in C
\end{aligned}$$

The reasoning in the last line is an application of Proposition 6.1, which requires the convexity of $\mathcal{R}$. □

An immediate corollary of this last result is:

Corollary 6.11 *In a finitary pLTS the following sets are closed:*

(i) $\{\Delta \mid \Delta \longrightarrow \varepsilon\}$
(ii) $\{\Delta \mid \Delta \Longrightarrow \stackrel{A}{\not\longrightarrow}\}$

Proof By Lemma 6.17, we see that $\Longrightarrow$ is closed and convex. Therefore, we can apply the previous lemma with $C = \{\varepsilon\}$ to obtain the first result. To obtain the second we apply it with $C = \{\Theta \mid \Theta \stackrel{A}{\not\longrightarrow}\}$, which is easily seen to be closed. □

The result is also used in the proof of:

Proposition 6.13 *In a finitary pLTS, for every $k \geq 0$ the relation $\lhd_{FS}^{k}$ is closed and convex.*

Proof By induction on k. For $k = 0$ it is obvious. So let us assume that $\lhd_{FS}^{k}$ is closed and convex. We have to show that

$$s \cdot \lhd_{FS}^{(k+1)} \text{ is closed and convex, for every state } s \tag{6.24}$$

If $s \Longrightarrow \varepsilon$, then this follows from the corollary above, since in this case $s \cdot \lhd_{FS}^{(k+1)}$ coincides with $\{\Delta \mid \Delta \Longrightarrow \varepsilon\}$. So let us assume that this is not the case.

For every $A \subseteq \mathsf{Act}$ let $R_A = \{\Delta \mid \Delta \Longrightarrow \stackrel{A}{\not\longrightarrow}\}$, which we know by the corollary above to be closed and is obviously convex. Also for every Θ and α we let

$$G_{\Theta,\alpha} := \{\Delta \mid (\Delta \cdot \stackrel{\alpha}{\Longrightarrow}) \cap (\Theta \cdot (\lhd_{FS}^{k})^{\dagger}) \neq \emptyset\}.$$

By Corollary 6.7, the relation $\stackrel{\alpha}{\Longrightarrow}$ is lifted from a closed convex relation. By Corollary 6.10, the assumption that $\lhd_{FS}^{k}$ is closed and convex implies that $(\lhd_{FS}^{k})^{\dagger}$ is also closed. So we can appeal to Lemma 6.35 and conclude that each $G_{\Theta,\alpha}$ is closed. By Definition 6.2(2), it is also easy to see that $G_{\Theta,\alpha}$ is convex. But it follows that $s \cdot \lhd_{FS}^{(k+1)}$ is also closed and convex as it can be written as

$$\cap\{R_A \mid s \stackrel{A}{\not\longrightarrow}\} \;\cap\; \cap\{G_{\Theta,\alpha} \mid s \stackrel{\alpha}{\longrightarrow} \Theta\}$$

Before the main result of this section we need one more technical lemma. □

Lemma 6.36 *Let S be a finite set of states. Suppose $\mathcal{R}^k \subseteq S \times \mathcal{D}_{sub}(S)$ is a sequence of closed convex relations such that $\mathcal{R}^{(k+1)} \subseteq \mathcal{R}^k$. Then*

$$\cap_{k=0}^{\infty} (\mathcal{R}^k)^{\dagger} \subseteq (\cap_{k=0}^{\infty} \mathcal{R}^k)^{\dagger}.$$

Proof Let $\mathcal{R}^{\infty}$ denote $(\cap_{k=0}^{\infty} \mathcal{R}^k)$ and suppose $\Delta \; (\mathcal{R}^k)^{\dagger} \; \Theta$ for every $k \geq 0$. We have to show that $\Delta \; (\mathcal{R}^{\infty})^{\dagger} \; \Theta$.

Let $G = \{f : S \to \mathcal{D}_{sub}(S) \mid \Theta = \text{Exp}_{\Delta}(f)\}$,which is easily seen to be a closed set. For each k, we know from Lemma 6.34 that the set $\mathbf{Ch}(\mathcal{R}^k)$ is closed. Finally, consider the collection of closed sets $H^k = \mathbf{Ch}(\mathcal{R}^k) \cap G$; since $\Delta \; (\mathcal{R}^k)^{\dagger} \; \Theta$, Proposition 6.1 assures us that all of these are non-empty. Also $H^{(k+1)} \subseteq H^k$ and, therefore, by the finite-intersection property (Theorem 2.4) $\cap_{k=0}^{\infty} H^k$ is also nonempty.

Let f be an arbitrary element of this intersection. For any state $s \in dom(\mathcal{R}^{\infty})$ and for every $k \geq 0$, since $dom(\mathcal{R}^{\infty}) \subseteq dom(\mathcal{R}^k)$ we have $s \mathcal{R}^k f(s)$, that is $s \mathcal{R}^{\infty} f(s)$. So f is a choice function for $\mathcal{R}^{\infty}$, $f \in \mathbf{Ch}(\mathcal{R}^{\infty})$. From convexity and Proposition 6.1, it follows that $\Delta \; (\mathcal{R}^{\infty})^{\dagger} \; \text{Exp}_{\Delta}(f)$. But from the definition of the G we know that $\Theta = \text{Exp}_{\Delta}(f)$, and the required result follows. □

Theorem 6.14 *In a finitary pLTS, $s \lhd_{FS}^{S} \Theta$ if and only if $s \lhd_{FS}^{\infty} \Theta$.*

Proof Since $\lhd_{FS}^{S} \subseteq \lhd_{FS}^{\infty}$ it is sufficient to show the opposite inclusion, which by definition holds if $\lhd_{FS}^{\infty}$ is a failure simulation, viz. if $\lhd_{FS}^{\infty} \subseteq \mathbf{F}(\lhd_{FS}^{\infty})$. Suppose $s \lhd_{FS}^{\infty} \Theta$, which means that $s \lhd^k_{FS} \Theta$ for every $k \geq 0$. According to Definition 6.21, in order to show $s\ \mathbf{F}(\lhd^{\infty}_{FS})\ \Theta$, we have to establish three properties, the first and last of which are trivial (for they are independent on the argument of $\mathbf{F}$).

So suppose $s \xrightarrow{\alpha} \Delta'$. We have to show that $\Theta \xRightarrow{\alpha} \Theta'$ for some Θ' such that $\Delta'\ (\lhd^{\infty}_{FS})^{\dagger}\ \Theta'$.

For every $k \geq 0$, there exists some Θ'_k such that $\Theta \xRightarrow{\alpha} \Theta'_k$ and $\Delta'\ (\lhd^k_{FS})^{\dagger}\ \Theta'_k$. Now construct the sets

$$D^k = \{\Theta' \mid \Theta \xRightarrow{\alpha} \Theta' \text{ and } \Delta'\ (\lhd^k_{FS})^{\dagger}\ \Theta'\}.$$

From Lemma 6.17 and Proposition 6.13, we know that these are closed. They are also nonempty and $D^{k+1} \subseteq D^k$. So by the finite-intersection property the set $\bigcap_{k=0}^{\infty} D^k$ is nonempty. For any Θ' in it, we know $\Theta \xRightarrow{\alpha} \Theta'$ and $\Delta'\ (\lhd^k_{FS})^{\dagger}\ \Theta'$ for every $k \geq 0$. By Proposition 6.13, the relations $\lhd^k_{FS}$ are all closed and convex. Therefore, Lemma 6.36 may be applied to them, which enables us to conclude $\Delta'\ (\lhd^{\infty}_{FS})^{\dagger}\ \Theta'$. □

For Theorem 6.14 to hold, it is crucial that the pLTS is assumed to be finitary.

Example 6.25 Consider an infinitely branching pLTS with four states $s, t, u, v, \mathbf{0}$ and the transitions are

- $s \xrightarrow{a} \overline{\mathbf{0}}\ {}_{\frac{1}{2}}\!\oplus \overline{s}$
- $t \xrightarrow{a} \overline{\mathbf{0}},\ t \xrightarrow{a} \overline{t}$
- $u \xrightarrow{a} \overline{u}$
- $v \xrightarrow{\tau} \overline{u}\ {}_p\!\oplus \overline{t}$ for all $p \in (0, 1)$.

This is a finite-state but not finitely branching system, due to the infinite branch in v. We have that $s \lhd^k_{FS} \overline{v}$ for all $k \geq 0$ but we do not have $s \lhd^S_{FS} \overline{v}$.

We first observe that $s \lhd^S_{FS} \overline{v}$ does not hold because s will eventually deadlock with probability 1, whereas a fraction of v will go to u and never deadlock.

We now show that $s \lhd^k_{FS} \overline{v}$ for all $k \geq 0$. For any k we start the simulation by choosing the move $v \xrightarrow{\tau} (\overline{u}\ {}_{\frac{1}{2^k}}\!\oplus \overline{t})$. By induction on k we show that

$$s \lhd^k_{FS} (\overline{u}\ {}_{\frac{1}{2^k}}\!\oplus \overline{t}). \tag{6.25}$$

The base case $k = 0$ is trivial. So suppose we already have (6.25). We now show that $s \lhd^{(k+1)}_{FS} (\overline{u}\ {}_{\frac{1}{2^{k+1}}}\!\oplus \overline{t})$. Neither s nor t nor u can diverge or refuse $\{a\}$, so the only relevant move is the a-move. We know that s can do the move $s \xrightarrow{a} \overline{\mathbf{0}}\ {}_{\frac{1}{2}}\!\oplus \overline{s}$. This can be matched by $(\overline{u}\ {}_{\frac{1}{2^{k+1}}}\!\oplus \overline{t}) \xrightarrow{a} (\overline{\mathbf{0}}\ {}_{\frac{1}{2}}\!\oplus (\overline{u}\ {}_{\frac{1}{2^k}}\!\oplus \overline{t}))$.

Analogously to what we did for $\lhd^S_{FS}$, we also give an inductive characterisation of $\sqsubseteq_{FS}$: For every $k \geq 0$ let $\Theta \sqsubseteq^k_{FS} \Delta$ if there exists a transition $\Theta \Longrightarrow \Theta^{\mathrm{match}}$ such that $\Delta\ (\lhd^k_{FS})^{\dagger}\ \Theta^{\mathrm{match}}$, and let $\sqsubseteq^{\infty}_{FS}$ denote $\bigcap_{k=0}^{\infty} \sqsubseteq^k_{FS}$.

Corollary 6.12 *In a finitary pLTS,* $\Theta \sqsubseteq_{FS} \Delta$ *if and only if* $\Theta \sqsubseteq^{\infty}_{FS} \Delta$.

Proof Since $\lhd^S_{FS} \subseteq \lhd^k_{FS}$, for every $k \geq 0$, it is straightforward to prove one direction: $\Theta \sqsubseteq_{FS} \Delta$ implies $\Theta \sqsubseteq^{\infty}_{FS} \Delta$. For the converse, $\Theta \sqsubseteq^{\infty}_{FS} \Delta$ means that for every k

we have some Θ^k satisfying $\Theta \Longrightarrow \Theta^k$ and $\Delta\ (\lhd_{FS}^k)^\dagger\ \Theta^k$. By Proposition 6.7, we have to find some Θ^∞ such that $\Theta \Longrightarrow \Theta^\infty$ and $\Delta\ (\lhd_{FS}^k)^\dagger\ \Theta^\infty$. This can be done exactly as in the proof of Theorem 6.14. □

6.7.2 The Modal Logic

We add to the modal language $\mathcal{F}$ given in Sect. 5.6 a new constant **div**, representing the ability of a process to diverge. The extended language, still written as $\mathcal{F}$ in this chapter, has the set of modal formulae defined inductively as follows:

- **div**, $\top \in \mathcal{F}$,
- **ref**(X) $\in \mathcal{F}$ when $X \subseteq \mathsf{Act}$,
- $\langle a \rangle \varphi \in \mathcal{F}$ when $\varphi \in \mathcal{F}$ and $a \in \mathsf{Act}$,
- $\varphi_1 \wedge \varphi_2 \in \mathcal{F}$ when $\varphi_1, \varphi_2 \in \mathcal{F}$,
- $\varphi_1\, {}_p{\oplus}\, \varphi_2 \in \mathcal{F}$ when $\varphi_1, \varphi_2 \in \mathcal{F}$ and $p \in [0,1]$.

Relative to a given pLTS $\langle S, \mathsf{Act}_\tau, \rightarrow \rangle$ the *satisfaction relation* $\models \subseteq \mathcal{D}_{sub}(S) \times \mathcal{F}$ is given by:

- $\Delta \models \top$ for any $\Delta \in \mathcal{D}_{sub}(S)$,
- $\Delta \models$ **div** iff $\Delta \Longrightarrow \varepsilon$,
- $\Delta \models$ **ref**(X) iff $\Delta \Longrightarrow \stackrel{X}{\not\longrightarrow}$,
- $\Delta \models \langle a \rangle \varphi$ iff there is a Δ' with $\Delta \stackrel{a}{\Longrightarrow} \Delta'$ and $\Delta' \models \varphi$,
- $\Delta \models \varphi_1 \wedge \varphi_2$ iff $\Delta \models \varphi_1$ and $\Delta \models \varphi_2$,
- $\Delta \models \varphi_1\, {}_p{\oplus}\, \varphi_2$ iff there are $\Delta_1, \Delta_2 \in \mathcal{D}_{sub}(S)$ with $\Delta_1 \models \varphi_1$ and $\Delta_2 \models \varphi_2$, such that $\Delta \Longrightarrow p \cdot \Delta_1 + (1-p) \cdot \Delta_2$.

We write $\Theta \sqsubseteq^{\mathcal{F}} \Delta$ when $\Delta \models \varphi$ implies $\Theta \models \varphi$ for all $\varphi \in \mathcal{F}$—note the opposing directions. This is because the modal formulae express "bad" properties of our processes, ultimately divergence and refusal; thus $\Theta \sqsubseteq^{\mathcal{F}} \Delta$ means that any bad thing implementation Δ does must have been allowed by the specification Θ.

For rpCSP processes, we use $P \sqsubseteq^{\mathcal{F}} Q$ to abbreviate $[\![P]\!] \sqsubseteq^{\mathcal{F}} [\![Q]\!]$ in the pLTS given in Sect. 6.2.

The set of formulae used here is obtained from that in Sect. 5.6 by adding one operator, **div**. But the interpretation is quite different, as it uses the new silent move relation $\Longrightarrow$. As a result, our satisfaction relation no longer enjoys a natural, and expected, property. In the nonprobabilistic setting if a recursive CCS process P satisfies a modal formula from the Hennessy–Milner logic, then there is a recursion-free finite unwinding of P that also satisfies it. Intuitively, this reflects the fact that if a nonprobabilistic process does a bad thing, then at some (finite) point it must actually do it. But this is not true in our new, probabilistic setting: for example, the process Q_1 given in Example 6.8 can do an a and then refuse anything; but all finite unwindings of it achieve that with probability strictly less than one. That is, whereas $[\![Q_1]\!] \models \langle a \rangle \top$, no finite unwinding of Q_1 will satisfy $\langle a \rangle \top$.

Our first task is to show that the interpretation of the logic is consistent with the operational semantics of processes.

Theorem 6.15 *If $\Theta \sqsubseteq_{FS} \Delta$ then $\Theta \sqsubseteq^{\mathcal{F}} \Delta$.*

Proof We must show that if $\Theta \sqsubseteq_{FS} \Delta$ then whenever $\Delta \models \varphi$ we have $\Theta \models \varphi$. The proof proceeds by induction on φ:

- The case when $\varphi = \top$ is trivial.
- Suppose φ is **div**. Then $\Delta \models$ **div** means that $\Delta \Longrightarrow \varepsilon$ and we have to show $\Theta \Longrightarrow \varepsilon$, which is immediate from Lemma 6.20.
- Suppose φ is $\langle a\rangle\varphi_a$. In this case we have $\Delta \overset{a}{\Longrightarrow} \Delta'$ for some Δ' that satisfies $\Delta' \models \varphi_a$. The existence of a corresponding Θ' is immediate from Definition 6.19 Case 1 and the induction hypothesis.
- The case when φ is **ref**(*X*) follows by Definition 6.19 Clause 2, and the case $\varphi_1 \wedge \varphi_2$ by induction.
- When φ is $\varphi_1\,{}_p{\oplus}\,\varphi_2$ we appeal again to Definition 6.19 Case 1, using $\alpha := \tau$ to infer the existence of suitable Θ_1' and Θ_2'. □

We proceed to show that the converse to this theorem also holds, so that the failure simulation preorder $\sqsubseteq_{FS}$ coincides with the logical preorder $\sqsubseteq^{\mathcal{F}}$.

The idea is to mimic the development in Sect. 5.7, by designing *characteristic formulae* that capture the behaviour of states in a pLTS. But here the behaviour is not characterised relative to $\lhd_{FS}^{S}$, but rather to the sequence of approximating relations $\lhd_{FS}^{k}$.

Definition 6.27 In a finitary pLTS $\langle S, \mathsf{Act}_\tau, \rightarrow\rangle$, the kth *characteristic formulae* φ_s^k, φ_Δ^k of states $s \in S$ and subdistributions $\Delta \in \mathcal{D}_{sub}(S)$ are defined inductively as follows:

- $\varphi_s^0 = \top$ and $\varphi_\Delta^0 = \top$,
- $\varphi_s^{k+1} =$ **div**, provided $\overline{s} \Longrightarrow \varepsilon$,
- $\varphi_s^{k+1} = \mathbf{ref}(X) \wedge \bigwedge_{s \xrightarrow{a} \Delta} \langle a\rangle \varphi_\Delta^k$ where $X = \{a \in \mathsf{Act} \mid s \overset{a}{\not\longrightarrow}\}$, provided $s \overset{\tau}{\not\longrightarrow}$,
- $\varphi_s^{k+1} = \bigwedge_{s \xrightarrow{a} \Delta} \langle a\rangle \varphi_\Delta^k \wedge \bigwedge_{s \xrightarrow{\tau} \Delta} \varphi_\Delta^k$ otherwise,
- and $\varphi_\Delta^{k+1} = (\mathbf{div})\ {}_{1-|\Delta|}{\oplus}\ \left(\bigoplus_{s \in \lceil\Delta\rceil} \frac{\Delta(s)}{|\Delta|} \cdot \varphi_s^{k+1}\right)$.

Lemma 6.37 *For every $k \geq 0$, $s \in S$ and $\Delta \in \mathcal{D}_{sub}(S)$ we have $\overline{s} \models \varphi_s^k$ and $\Delta \models \varphi_\Delta^k$.*

Proof By induction on *k*, with the case when $k = 0$ being trivial. The inductive case of the first statement proceeds by an analysis of the possible moves from *s*, from which that of the second statement follows immediately. □

Lemma 6.38 *For $k \geq 0$,*

(i) $\Theta \models \varphi_s^k$ *implies* $s \lhd_{FS}^k \Theta$,
(ii) $\Theta \models \varphi_\Delta^k$ *implies* $\Theta \Longrightarrow \Theta^{match}$ *such that* $\Delta\ (\lhd_{FS}^k)^\dagger\ \Theta^{match}$,
(iii) $\Theta \models \varphi_\Delta^k$ *implies* $\Theta \sqsubseteq_{FS}^k \Delta$.

Proof For every k part (iii) follows trivially from (ii). We prove (i) and (ii) simultaneously, by induction on k, with the case $k = 0$ being trivial. The inductive case, for $k+1$, follows the argument in the proof of Lemma 5.12.

(i) First suppose $\overline{s} \Longrightarrow \varepsilon$. Then $\varphi_s^{k+1} = \mathbf{div}$ and therefore $\Theta \models \mathbf{div}$, which gives the required $\Theta \Longrightarrow \varepsilon$.
Now suppose $s \xrightarrow{\tau} \Delta$. Here, there are two cases; if in addition $\overline{s} \Longrightarrow \varepsilon$ we have already seen that $\Theta \Longrightarrow \varepsilon$ and this is the required matching move from Θ, since $\Delta\ (\lhd_{FS}^k)^\dagger\ \varepsilon$. So let us assume that $\overline{s} \not\Longrightarrow \varepsilon$. Then by the definition of φ_s^{k+1} we must have that $\Theta \models \varphi_\Delta^k$, and we obtain the required matching move from Θ from the inductive hypothesis: induction on part (ii) gives some Θ' such that $\Theta \Longrightarrow \Theta'$ and $\Delta\ (\lhd_{FS}^k)^\dagger\ \Theta'$.
The matching move for $s \xrightarrow{a} \Theta$ is obtained in a similar manner.
Finally suppose $s \stackrel{X}{\not\longrightarrow}$. Since this implies $s \stackrel{\tau}{\not\longrightarrow}$, by the definition of φ_s^{k+1} we must have that $\Theta \models \mathbf{ref}(X)$, which actually means that $\Theta \Longrightarrow\stackrel{X}{\not\longrightarrow}$.

(ii) Note that $\varphi_\Delta^{k+1} = (\mathbf{div})\ {}_{1-|\Delta|} \oplus\ (\bigoplus_{s\in\lceil\Delta\rceil} \frac{\Delta(s)}{|\Delta|} \cdot \varphi_s^{k+1})$ and therefore by definition $\Theta \Longrightarrow (1-|\Delta|) \cdot \Theta_{\mathbf{div}} + \sum_{s\in\lceil\Delta\rceil} \Delta(s) \cdot \Theta_s$ such that $\Theta_{\mathbf{div}} \models \mathbf{div}$ and $\Theta_s \models \varphi_s^{k+1}$. By definition, $\Theta_{\mathbf{div}} \Longrightarrow \varepsilon$, so by Theorem 6.5(i) and the reflexivity and transitivity of $\Longrightarrow$ we obtain $\Theta \Longrightarrow \sum_{s\in\lceil\Delta\rceil} \Delta(s) \cdot \Theta_s$.
By part (i) we have $s \lhd_{FS}^{k+1} \Theta_s$ for every s in $\lceil\Delta\rceil$, which in turn means that $\Delta\ (\lhd_{FS}^{k+1})^\dagger \sum_{s\in\lceil\Delta\rceil} \Delta(s) \cdot \Theta_s$.

□

Theorem 6.16 *In a finitary pLTS, $\Theta \sqsubseteq^{\mathcal{F}} \Delta$ if and only if $\Theta \sqsubseteq_{FS} \Delta$.*

Proof One direction follows immediately from Theorem 6.15. For the opposite direction, suppose $\Theta \sqsubseteq^{\mathcal{F}} \Delta$. By Lemma 6.37 we have $\Delta \models \varphi_\Delta^k$, and hence $\Theta \models \varphi_\Delta^k$, for all $k \geq 0$. By part (iii) of the previous lemma we thus know that $\Theta \sqsubseteq_{FS}^\infty \Delta$. That $\Theta \sqsubseteq_{FS} \Delta$ now follows from Corollary 6.12. □

6.7.3 Characteristic Tests for Formulae

The import of Theorem 6.16 is that we can obtain completeness of the failure simulation preorder with respect to the must-testing preorder by designing for each formula φ a test that in some sense characterises the property of a process of satisfying φ. This has been achieved for the pLTS generated by the recursion free fragment of rpCSP in Sect. 5.7. Here, we generalise this technique to the pLTS generated by the set of finitary rpCSP terms.

As in Sect. 5.7, the generation of these tests depends on crucial characteristics of the testing function $\mathcal{A}^{\mathrm{d}}(-,-)$, which are summarised in Lemmas 6.39 and 6.42 below, corresponding to Lemmas 5.9 and 5.10, respectively.

Lemma 6.39 *Let Δ be a rpCSP process, and T, T_i be tests.*

1. $o \in \mathcal{A}^{\mathrm{d}}(\omega, \Delta)$ *iff* $o = |\Delta| \cdot \vec{\omega}$.

2. $\vec{0} \in \mathcal{A}^d(\tau.\omega, \Delta)$ *iff* $\Delta \Longrightarrow \varepsilon$.
3. $\vec{0} \in \mathcal{A}^d(\Box_{a\in X} a.\omega, \Delta)$ *iff* $\Delta \Longrightarrow \stackrel{A}{\not\longrightarrow}$.
4. *Suppose the action* ω *does not occur in the test* T. *Then* $o \in \mathcal{A}^d(\tau.\omega \,\Box\, a.T, \Delta)$ *with* $o(\omega) = 0$ *iff there is a* $\Delta' \in \mathcal{D}_{sub}(\mathsf{sCSP})$ *with* $\Delta \stackrel{a}{\Longrightarrow} \Delta'$ *and* $o \in \mathcal{A}^d(T, \Delta')$.
5. $o \in \mathcal{A}^d(T_1 \,{}_p\!\oplus T_2, \Delta)$ *iff* $o = p \cdot o_1 + (1-p) \cdot o_2$ *for certain* $o_i \in \mathcal{A}^d(T_i, \Delta)$.
6. $o \in \mathcal{A}^d(T_1 \sqcap T_2, \Delta)$ *if there are a* $q \in [0,1]$ *and* $\Delta_1, \Delta_2 \in \mathcal{D}_{sub}(\mathsf{sCSP})$ *such that* $\Delta \Longrightarrow q \cdot \Delta_1 + (1-q) \cdot \Delta_2$ *and* $o = q \cdot o_1 + (1-q) \cdot o_2$ *for certain* $o_i \in \mathcal{A}^d(T_i, \Delta_i)$.

Proof 1. Since $\omega|_{\mathsf{Act}}\,\Delta \stackrel{\omega}{\longrightarrow}$, the states in the support of $[\omega|_{\mathsf{Act}}\,\Delta]$ have no other outgoing transitions than ω. Therefore, $[\omega|_{\mathsf{Act}}\,\Delta]$ is the unique extreme derivative of itself, and as $\$[\omega|_{\mathsf{Act}}\,\Delta] = |\Delta| \cdot \vec{\omega}$ we have $\mathcal{A}^{\mathrm{d}}(\omega, \Delta) = \{|\Delta| \cdot \vec{\omega}\}$.

2. ($\Leftarrow$) Assume $\Delta \Longrightarrow \varepsilon$. By Lemma 6.27(1), we have $\tau.\omega|_{\mathsf{Act}}\,\Delta \Longrightarrow \tau.\omega|_{\mathsf{Act}}\,\varepsilon$. All states involved in this derivation (that is, all states u in the support of the intermediate distributions $\Delta_i^{\rightarrow}$ and $\Delta_i^{\times}$ of Definition 6.4) must have the form $\tau.\omega|_{\mathsf{Act}} s$, and thus satisfy $u \stackrel{\omega}{\not\longrightarrow}$ for all $\omega \in \Omega$. Therefore, it follows that $[\tau.\omega|_{\mathsf{Act}}\,\Delta] \Longrightarrow [\tau.\omega|_{\mathsf{Act}}\,\varepsilon]$. Trivially, $[\tau.\omega|_{\mathsf{Act}}\,\varepsilon] = \varepsilon$ is stable, and hence an extreme derivative of $[\tau.\omega|_{\mathsf{Act}}\,\Delta]$. Moreover, $\$\varepsilon = \vec{0}$, so $\vec{0} \in \mathcal{A}^{\mathrm{d}}(\tau.\omega, \Delta)$.

   ($\Rightarrow$) Suppose $\vec{0} \in \mathcal{A}^{\mathrm{d}}(\tau.\omega, \Delta)$, that is, there is some extreme derivative Γ of $[\tau.\omega|_{\mathsf{Act}}\,\Delta]$ such that $\$\Gamma = \vec{0}$. Given the operational semantics of rpCSP, all states $u \in \lceil \Gamma \rceil$ must have one of the forms $u = [\tau.\omega|_{\mathsf{Act}}\,t]$ or $u = [\omega|_{\mathsf{Act}}\,t]$. As $\$\Gamma = \vec{0}$, the latter possibility cannot occur. It follows that all transitions contributing to the derivation $[\tau.\omega|_{\mathsf{Act}}\,\Delta] \Longrightarrow\!\!\succ \Gamma$ do not require any action from $\tau.\omega$, and, in fact, Γ has the form $[\tau.\omega|_{\mathsf{Act}}\,\Delta']$ for some distribution Δ' with $\Delta \Longrightarrow \Delta'$. As Γ must be stable, yet none of the states in its support are, it follows that $\lceil \Gamma \rceil = \emptyset$, that is, $\Delta' = \varepsilon$.

3. Let $T := \Box_{a\in X} a.\omega$.

   ($\Leftarrow$) Assume $\Delta \Longrightarrow \Delta' \stackrel{X}{\not\longrightarrow}$ for some Δ'. Then, $T|_{\mathsf{Act}}\,\Delta \Longrightarrow T|_{\mathsf{Act}}\Delta'$ by Lemma 6.27(1), and by the same argument as in the previous case, we have $[T|_{\mathsf{Act}}\,\Delta] \Longrightarrow [T|_{\mathsf{Act}}\Delta']$. All states in the support of $T|_{\mathsf{Act}}\Delta'$ are deadlocked. So $[T|_{\mathsf{Act}}\,\Delta] \Longrightarrow\!\!\succ [T|_{\mathsf{Act}}\Delta]$ and $\$(T|_{\mathsf{Act}}\Delta) = \vec{0}$. Thus, we have $\vec{0} \in \mathcal{A}^{\mathrm{d}}(T, \Delta)$.

   ($\Rightarrow$) Suppose $\vec{0} \in \mathcal{A}^{\mathrm{d}}(T, \Delta)$. By the very same reasoning as in Case 2 we find that $\Delta \Longrightarrow \Delta'$ for some Δ' such that $T|_{\mathsf{Act}}\,\Delta'$ is stable. This implies $\Delta' \stackrel{X}{\not\longrightarrow}$.

4. Let T be a test in which the success action ω does not occur, and let U be an abbreviation for $\tau.\omega \Box a.T$.

   ($\Leftarrow$) Assume there is a $\Delta' \in \mathcal{D}_{sub}(\mathsf{sCSP})$ with $\Delta \stackrel{a}{\Longrightarrow} \Delta'$ and $o \in \mathcal{A}^{\mathrm{d}}(T, \Delta')$. Without loss of generality we may assume that $\Delta \Longrightarrow \Delta^{\mathrm{pre}} \stackrel{a}{\longrightarrow} \Delta^{\mathrm{post}} \Longrightarrow \Delta'$. Using Lemma 6.27(1) and (3), and the same reasoning as in the previous cases, $[U|_{\mathsf{Act}}\,\Delta] \Longrightarrow [U|_{\mathsf{Act}}\,\Delta^{\mathrm{pre}}] \stackrel{\tau}{\longrightarrow} [T|_{\mathsf{Act}}\,\Delta^{\mathrm{post}}] \Longrightarrow [T|_{\mathsf{Act}}\,\Delta'] \Longrightarrow \Gamma$ for a stable subdistribution Γ with $\$\Gamma = o$. It follows that $o \in \mathcal{A}^{\mathrm{d}}(U, \Delta)$.

   ($\Rightarrow$) Suppose $o \in \mathcal{A}^{\mathrm{d}}(U, \Delta)$ with $o(\omega) = 0$. Then, there is a stable subdistribution Γ such that $[U|_{\mathsf{Act}}\,\Delta] \Longrightarrow \Gamma$ and $\$\Gamma = o$. Since $o(\omega) = 0$, there is no state in the support of Γ of the form $\omega|_{\mathsf{Act}}\, t$. Hence, there must be a $\Delta' \in \mathcal{D}_{sub}(\mathsf{sCSP})$ such that $\Delta \Longrightarrow\xrightarrow{a} \Delta'$ and $[T|_{\mathsf{Act}}\, \Delta'] \Longrightarrow \Gamma$. It follows that $o \in \mathcal{A}^{\mathrm{d}}(T, \Delta')$.

5. ($\Leftarrow$) Assume $o_i \in \mathcal{A}^{\mathrm{d}}(T_i, \Delta)$ for $i = 1, 2$. Then, $[T_i|_{\mathsf{Act}}\, \Delta] \Longrightarrow \Gamma_i$ for some stable Γ_i with $\$\Gamma_i = o_i$. By Theorem 6.5(i) we have

$$[(T_1\,{}_p{\oplus}T_2)|_{\mathsf{Act}}\, \Delta] = p \cdot [T_1|_{\mathsf{Act}}\, \Delta] + (1-p) \cdot [T_2|_{\mathsf{Act}}\, \Delta] \Longrightarrow p \cdot \Gamma_1 + (1-p) \cdot \Gamma_2,$$

and $p \cdot \Gamma_1 + (1-p) \cdot \Gamma_2$ is stable. Moreover,

$$\$(p \cdot \Gamma_1 + (1-p) \cdot \Gamma_2) = p \cdot o_1 + (1-p) \cdot o_2,$$

so $o \in \mathcal{A}^{\mathrm{d}}(T_1\,{}_p{\oplus}T_2, \Delta)$.

($\Rightarrow$) Suppose $o \in \mathcal{A}^{\mathrm{d}}(T_1\,{}_p{\oplus}T_2, \Delta)$. Then, there is a stable Γ with $\$\Gamma = o$ such that $[(T_1\,{}_p{\oplus}T_2)|_{\mathsf{Act}}\, \Delta] = p \cdot [T_1|_{\mathsf{Act}}\, \Delta] + (1-p) \cdot [T_2|_{\mathsf{Act}}\, \Delta] \Longrightarrow \Gamma$. By Theorem 6.5(ii), there are Γ_i for $i = 1, 2$, such that $[T_i|_{\mathsf{Act}}\, \Delta] \Longrightarrow \Gamma_i$ and also $\Gamma = p \cdot \Gamma_1 + (1-p) \cdot \Gamma_2$. As Γ_1 and Γ_2 are stable, we have $\$\Gamma_i \in \mathcal{A}^{\mathrm{d}}(T_i, \Delta)$ for $i = 1, 2$. Moreover, $o = \$\Gamma = p \cdot \$\Gamma_1 + (1-p) \cdot \$\Gamma_2$.

6. Suppose $q \in [0, 1]$ and $\Delta_1, \Delta_2 \in \mathcal{D}_{sub}(\mathsf{rpCSP})$ with $\Delta \Longrightarrow q \cdot \Delta_1 + (1-q) \cdot \Delta_2$ and $o_i \in \mathcal{A}^{\mathrm{d}}(T_i, \Delta_i)$. Then there are stable Γ_i with $[T_i|_{\mathsf{Act}}\, \Delta_i] \Longrightarrow \Gamma_i$ and $\$\Gamma_i = o_i$. Now

$$\begin{aligned}
[(T_1 \sqcap T_2)|_{\mathsf{Act}}\, \Delta] &\Longrightarrow q \cdot [(T_1 \sqcap T_2)|_{\mathsf{Act}}\, \Delta_1] + (1-q) \cdot [(T_1 \sqcap T_2)|_{\mathsf{Act}}\, \Delta_2] \\
&\xrightarrow{\tau} q \cdot [T_1|_{\mathsf{Act}}\, \Delta_1] + (1-q) \cdot [T_2|_{\mathsf{Act}}\, \Delta_2] \\
&\Longrightarrow q \cdot \Gamma_1 + (1-q) \cdot \Gamma_2
\end{aligned}$$

The latter subdistribution is stable and satisfies

$$\$(q \cdot \Gamma_1 + (1-q) \cdot \Gamma_2) = q \cdot o_1 + (1-q) \cdot o_2.$$

Hence $q \cdot o_1 + (1-q) \cdot o_2 \in \mathcal{A}^{\mathrm{d}}(T_1 \sqcap T_2, \Delta)$. □

We also have the converse to part (6) of this lemma by mimicking Lemma 5.10. For that purpose, we use two technical lemmas whose proofs are similar to those for Lemmas 6.28 and 6.29, respectively.

Lemma 6.40 *Suppose* $\Delta|_A(T_1 \sqcap T_2) \xrightarrow{\tau} \Gamma$. *Then there exist subdistributions* $\Delta^{\rightarrow}$, $\Delta_1^{\times}$, $\Delta_2^{\times}$, Δ^{next} *(possibly empty) such that*

(i) $\Delta = \Delta^{\rightarrow} + \Delta_1^{\times} + \Delta_2^{\times}$
(ii) $\Delta^{\rightarrow} \xrightarrow{\tau} \Delta^{\mathrm{next}}$
(iii) $\Gamma = \Delta^{\mathrm{next}}|_A(T_1 \sqcap T_2) + \Delta_1^{\times}|_A T_1 + \Delta_2^{\times}|_A T_2$

Proof By Lemma 6.1 $\Delta|_A(T_1 \sqcap T_2) \xrightarrow{\tau} \Gamma$ implies that

$$\Delta = \sum_{i \in I} p_i \cdot \overline{s_i}, \qquad s_i|_A(T_1 \sqcap T_2) \xrightarrow{\tau} \Gamma_i, \qquad \Gamma = \sum_{i \in I} p_i \cdot \Gamma_i,$$

for certain $s_i \in S$, $\Gamma_i \in \mathcal{D}_{sub}(\mathsf{sCSP})$ and $\sum_{i\in I} p_i \le 1$. Let $J_1 = \{i \in I \mid \Gamma_i = s_i|_A T_1\}$ and $J_2 = \{i \in I \mid \Gamma_i = s_i|_A T_2\}$. Note that for each $i \in (I - J_1 - J_2)$ we have Γ_i in the form $\Gamma_i'|_A(T_1 \sqcap T_2)$, where $s_i \xrightarrow{\tau} \Gamma_i'$. Now let

$$\Delta^{\rightarrow} = \sum_{i\in(I-J_1-J_2)} p_i \cdot \overline{s_i}, \qquad \Delta_k^{\times} = \sum_{i\in J_k} p_i \cdot \overline{s_i}, \qquad \Delta^{\text{next}} = \sum_{i\in(I-J_1-J_2)} p_i \cdot \Gamma_i' .$$

where $k = 1, 2$. By construction (i) and (iii) are satisfied, and (ii) follows by property (2) of Definition 6.2. □

Lemma 6.41 *If $\Delta|_A(T_1 \sqcap T_2) \Longrightarrow\!\!\!\!\!\!\succ \Psi$ then there are Φ_1 and Φ_2 such that*

(i) $\Delta \Longrightarrow \Phi_1 + \Phi_2$
(ii) $\Phi_1|_A T_1 + \Phi_2|_A T_2 \Longrightarrow\!\!\!\!\!\!\succ \Psi$

Proof Suppose $\Delta_0|_A(T_1 \sqcap T_2) \Longrightarrow\!\!\!\!\!\!\succ \Psi$. We know from Definition 6.4 that there is a collection of subdistributions $\Psi_k, \Psi_k^{\rightarrow}, \Psi_k^{\times}$, for $k \ge 0$, satisfying the properties

$$\begin{array}{lcl}
\Delta_0|_A(T_1 \sqcap T_2) & = & \Psi_0 = \Psi_0^{\rightarrow} + \Psi_0^{\times} \\
\Psi_0^{\rightarrow} & \xrightarrow{\tau} & \Psi_1 = \Psi_1^{\rightarrow} + \Psi_1^{\times} \\
\vdots & & \qquad\qquad \vdots \\
\Psi_k^{\rightarrow} & \xrightarrow{\tau} & \Psi_{k+1} = \Psi_{k+1}^{\rightarrow} + \Psi_{k+1}^{\times} \\
 & & \qquad\qquad \vdots \\
\hline
 & & \Psi = \sum_{k=0}^{\infty} \Psi_k^{\times}
\end{array}$$

and Ψ is stable.

Take $\Gamma_0 := \Psi_0$. By induction on $k \ge 0$, we find distributions Γ_{k+1}, $\Delta_k^{\rightarrow}$, $\Delta_{k1}^{\times}$, $\Delta_{k2}^{\times}$, Δ_{k+1} such that

(i) $\Delta_k|_A(T_1 \sqcap T_2) \xrightarrow{\tau} \Gamma_{k+1}$
(ii) $\Gamma_{k+1} \le \Psi_{k+1}$
(iii) $\Delta_k = \Delta_k^{\rightarrow} + \Delta_{k1}^{\times} + \Delta_{k2}^{\times}$
(iv) $\Delta_k^{\rightarrow} \xrightarrow{\tau} \Delta_{k+1}$
(v) $\Gamma_{k+1} = \Delta_{k+1}|_A(T_1 \sqcap T_2) + \Delta_{k1}^{\times}|_A T_1 + \Delta_{k2}^{\times}|_A T_2$

Induction Step: Assume we already have Γ_k and Δ_k. Note that

$$\Delta_k|_A(T_1 \sqcap T_2) \le \Gamma_k \le \Psi_k = \Psi_k^{\rightarrow} + \Psi_k^{\times}$$

and $T_1 \sqcap T_2$ can make a τ move. Since Ψ is stable, we know that there are two possibilities— either $\Psi_k^{\times} = \varepsilon$ or $\Psi_k^{\times} \stackrel{\tau}{\not\longrightarrow}$. In both cases it holds that

$$\Delta_k|_A(T_1 \sqcap T_2) \le \Psi_k^{\rightarrow} .$$

Proposition 6.2 gives a subdistribution $\Gamma_{k+1} \le \Psi_{k+1}$ such that there exists the transition $\Delta_k|_A(T_1 \sqcap T_2) \xrightarrow{\tau} \Gamma_{k+1}$. Now apply Lemma 6.40.

Let $\Phi_1 = \sum_{k=0}^{\infty} \Delta_{k1}^{\times}$ and $\Phi_2 = \sum_{k=0}^{\infty} \Delta_{k2}^{\times}$. By (iii) and (iv) above we obtain a weak τ move $\Delta \Longrightarrow \Phi_1 + \Phi_2$. For $k \geq 0$, let $\Gamma_k^{\rightarrow} := \Delta_k|_A(T_1 \sqcap T_2)$, let $\Gamma_0^{\times} := \varepsilon$ and let $\Gamma_{k+1}^{\times} := \Delta_{k1}^{\times}|_A T_1 + \Delta_{k2}^{\times}|_A T_2$. Moreover, $\Gamma := \Phi_1|_A T_1 + \Phi_2|_A T_2$. Now all conditions of Definition 6.5 are fulfilled, so $\Delta_0|_A(T_1 \sqcap T_2) \Longrightarrow \Gamma$ is an initial segment of $\Delta_0|_A(T_1 \sqcap T_2) \Longrightarrow \Psi$. By Proposition 6.4, we have $\Phi_1|_A T_1 + \Phi_2|_A T_2 \Longrightarrow\!\!\!\!\!\succ \Psi$. □

Lemma 6.42 *If $o \in \mathcal{A}^d(T_1 \sqcap T_2, \Delta)$ then there are a $q \in [0, 1]$ and subdistributions $\Delta_1, \Delta_2 \in \mathcal{D}_{sub}(sCSP)$ such that $\Delta \Longrightarrow q \cdot \Delta_1 + (1-q) \cdot \Delta_2$ and $o = q \cdot o_1 + (1-q) \cdot o_2$ for certain $o_i \in \mathcal{A}^d(T_i, \Delta_i)$.*

Proof If $o \in \mathcal{A}^d(T_1 \sqcap T_2, \Delta)$ then there is an extreme derivative of $[(T_1 \sqcap T_2)|_{\mathsf{Act}}\, \Delta]$, say Ψ, such that $\$\Psi = o$. By Lemma 6.41 there are $\Phi_{1,2}$ such that

(i) $\Delta \Longrightarrow \Phi_1 + \Phi_2$
(ii) and $[T_1|_{\mathsf{Act}}\, \Phi_1] + [T_2|_{\mathsf{Act}}\, \Phi_2] \Longrightarrow\!\!\!\!\!\succ \Psi$.

By Theorem 6.5(ii), there are some subdistributions Ψ_1 and Ψ_2 such that $\Psi = \Psi_1 + \Psi_2$ and $T_i|_{\mathsf{Act}}\, \Phi_i \Longrightarrow\!\!\!\!\!\succ \Psi_i$ for $i = 1, 2$. Let $o_i' = \$\Psi_i$. As Ψ_i is stable we obtain that $o_i' \in \mathcal{A}^d(T_i, \Phi_i)$. We also have $o = \$\Psi = \$\Psi_1 + \$\Psi_2 = o_1' + o_2'$.

We now distinguish two cases:

- If $\Psi_1 = \varepsilon$, then we take $\Delta_i = \Phi_i$, $o_i = o_i'$ for $i = 1, 2$ and $q = 0$. Symmetrically, if $\Psi_2 = \varepsilon$, then we take $\Delta_i = \Phi_i$, $o_i = o_i'$ for $i = 1, 2$ and $q = 1$.
- If $\Psi_1 \neq \varepsilon$ and $\Psi_2 \neq \varepsilon$, then we let $q = \frac{|\Phi_1|}{|\Phi_1 + \Phi_2|}$, $\Delta_1 = \frac{1}{q}\Phi_1$, $\Delta_2 = \frac{1}{1-q}\Phi_2$, $o_1 = \frac{1}{q}o_1'$ and $o_2 = \frac{1}{1-q}o_2'$.

It is easy to check that $q \cdot \Delta_1 + (1-q) \cdot \Delta_2 = \Phi_1 + \Phi_2$, $q \cdot o_1 + (1-q) \cdot o_2 = o_1' + o_2'$ and $o_i \in \mathcal{A}^d(T_i, \Delta_i)$ for $i = 1, 2$. □

Proposition 6.14 *For every formula $\varphi \in \mathcal{F}$ there exists a pair (T_φ, v_φ) with T_φ an Ω-test and $v_\varphi \in [0, 1]^\Omega$ such that*

$$\Delta \models \varphi \text{ if and only if } \exists o \in \mathcal{A}^d(T_\varphi, \Delta) : o \leq v_\varphi. \tag{6.26}$$

T_φ is called a characteristic test *of φ and v_φ its* target value.

Proof The proof is adapted from that of Lemma 5.13, from where we take the following remarks: As in vector-based testing Ω is assumed to be countable (cf. page 74) and Ω-tests are finite expressions, for every Ω-test there is an $\omega \in \Omega$ not occurring in it. Furthermore, if a pair (T_φ, v_φ) satisfies requirement (6.26), then any pair obtained from (T_φ, v_φ) by bijectively renaming the elements of Ω also satisfies that requirement. Hence, two given characteristic tests can be assumed to be *Ω-disjoint*, meaning that no $\omega \in \Omega$ occurs in both of them.

Our modal logic $\mathcal{F}$ is identical to that used in Sect. 5.7, with the addition of one extra constant **div**. So we need a new characteristic test and target value for this latter formula and reuse those from Sect. 5.7 for the rest of the language.

- Let $\varphi = \top$. Take $T_\varphi := \omega$ for some $\omega \in \Omega$, and $v_\varphi := \vec{\omega}$.
- Let $\varphi = \mathbf{div}$. Take $T_\varphi := \tau.\omega$ for some $\omega \in \Omega$, and $v_\varphi := \vec{0}$.

- Let $\varphi = \mathbf{ref}(X)$ with $X \subseteq \mathsf{Act}$. Take $T_\varphi := \Box_{a \in X} a.\omega$ for some $\omega \in \Omega$, and set $v_\varphi := \vec{0}$.
- Let $\varphi = \langle a \rangle \psi$. By induction, ψ has a characteristic test T_ψ with target value v_ψ. Take $T_\varphi := \tau.\omega \Box a.T_\psi$ where $\omega \in \Omega$ does not occur in T_ψ, and $v_\varphi := v_\psi$.
- Let $\varphi = \varphi_1 \wedge \varphi_2$. Choose an Ω-disjoint pair (T_i, v_i) of characteristic tests T_i with target values v_i, for $i = 1, 2$. Furthermore, let $p \in (0, 1]$ be chosen arbitrarily, and take $T_\varphi := T_1 {}_p\!\oplus T_2$ and $v_\varphi := p \cdot v_1 + (1 - p) \cdot v_2$.
- Let $\varphi = \varphi_1 {}_p\!\oplus \varphi_2$. Again choose an Ω-disjoint pair (T_i, v_i) of characteristic tests T_i with target values v_i, $i = 1, 2$, this time ensuring that there are two distinct success actions ω_1, ω_2 that do not occur in any of these tests. Let $T_i' := T_i \,{}_{\frac{1}{2}}\!\oplus \omega_i$ and $v_i' := \frac{1}{2} v_i + \frac{1}{2} \vec{\omega}_i$. Note that for $i = 1, 2$ we have that T_i' is also a characteristic test of φ_i with target value v_i'. Take $T_\varphi := T_1' \sqcap T_2'$ and $v_\varphi := p \cdot v_1' + (1 - p) \cdot v_2'$.

Note that $v_\varphi(\omega) = 0$ whenever $\omega \in \Omega$ does not occur in T_φ.

As in the proof of Lemma 5.13, we now check by induction on φ that (6.26) above holds; the proof relies on Lemmas 6.39 and 6.42.

- Let $\varphi = \top$. For all $\Delta \in \mathcal{D}_{sub}(\mathsf{sCSP})$ we have $\Delta \models \varphi$ and it always holds that $\exists o \in \mathcal{A}^{\mathrm{d}}(T_\varphi, \Delta) : o \leq v_\varphi$, using Lemma 6.39(1).
- Let $\varphi = \mathbf{div}$. Suppose $\Delta \models \varphi$. Then, we have that $\Delta \Longrightarrow \varepsilon$. By Lemma 6.39(2), $\vec{0} \in \mathcal{A}^{\mathrm{d}}(T_\varphi, \Delta)$.

  Now suppose $\exists o \in \mathcal{A}^{\mathrm{d}}(T_\varphi, \Delta) : o \leq v_\varphi$. This implies $o = \vec{0}$, so we apply Lemma 6.39(2) and obtain $\Delta \Longrightarrow \varepsilon$. Hence $\Delta \models \varphi$.
- Let $\varphi = \mathbf{ref}\ (X)$ with $X \subseteq \mathsf{Act}$. Suppose $\Delta \models \varphi$. Then $\Delta \Longrightarrow \xrightarrow{X}\!\!\!\!\!\not\;\;$. We can use Lemma 6.39(3) and obtain $\vec{0} \in \mathcal{A}^{\mathrm{d}}(T_\varphi, \Delta)$.

  Now suppose $\exists o \in \mathcal{A}^{\mathrm{d}}(T_\varphi, \Delta) : o \leq v_\varphi$. This implies $o = \vec{0}$, so $\Delta \Longrightarrow \xrightarrow{A}\!\!\!\!\!\not\;\;$ by Lemma 6.39(3). Hence $\Delta \models \varphi$.
- Let $\varphi = \langle a \rangle \psi$ with $a \in \mathsf{Act}$. Suppose $\Delta \models \varphi$. Then there is a Δ' with $\Delta \stackrel{a}{\Longrightarrow} \Delta'$ and $\Delta' \models \psi$. By induction, $\exists o \in \mathcal{A}^{\mathrm{d}}(T_\psi, \Delta') : o \leq v_\psi$. By Lemma 6.39(4), we get $o \in \mathcal{A}^{\mathrm{d}}(T_\varphi, \Delta)$.

  Now suppose $\exists o \in \mathcal{A}^{\mathrm{d}}(T_\varphi, \Delta) : o \leq v_\varphi$. This implies $o(\omega) = 0$, hence by Lemma 6.39(4) there is a Δ' with $\Delta \stackrel{a}{\Longrightarrow} \Delta'$ and $o \in \mathcal{A}^{\mathrm{d}}(T_\psi, \Delta')$. By induction, $\Delta' \models \psi$, so $\Delta \models \varphi$.
- Let $\varphi = \varphi_1 \wedge \varphi_2$ and suppose $\Delta \models \varphi$. Then $\Delta \models \varphi_i$ for $i = 1, 2$ and hence, by induction, $\exists o_i \in \mathcal{A}^{\mathrm{d}}(T_i, \Delta) : o_i \leq v_i$. Thus, $o := p \cdot o_1 + (1 - p) \cdot o_2 \in \mathcal{A}^{\mathrm{d}}(T_\varphi, \Delta)$ by Lemma 6.39(5), and $o \leq v_\varphi$.

  Now, suppose $\exists o \in \mathcal{A}^{\mathrm{d}}(T_\varphi, \Delta) : o \leq v_\varphi$. Then, using Lemma 6.39(5), we know that $o = p \cdot o_1 + (1 - p) \cdot o_2$ for certain $o_i \in \mathcal{A}^{\mathrm{d}}(T_i, \Delta)$. Recall that T_1, T_2 are Ω-disjoint tests. One has $o_i \leq v_i$ for both $i = 1, 2$, for if $o_i(\omega) > v_i(\omega)$ for some $i = 1$ or 2 and $\omega \in \Omega$, then ω must occur in T_i and hence cannot occur in T_{3-i}. This implies $v_{3-i}(\omega) = 0$ and thus $o(\omega) > v_\varphi(\omega)$, in contradiction with the assumption. By induction, $\Delta \models \varphi_i$ for $i = 1, 2$, and hence $\Delta \models \varphi$.
- Let $\varphi = \varphi_1 {}_p\!\oplus \varphi_2$. Suppose $\Delta \models \varphi$. Then there are $\Delta_1, \Delta_2 \in \mathcal{D}_{sub}(\mathsf{sCSP})$ with $\Delta_1 \models \varphi_1$ and $\Delta_2 \models \varphi_2$ such that $\Delta \Longrightarrow p \cdot \Delta_1 + (1 - p) \cdot \Delta_2$. By induction, for

$i = 1, 2$ there are $o_i \in \mathcal{A}^{\mathrm{d}}(T_i, \Delta_i)$ with $o_i \leq v_i$. Hence, there are $o'_i \in \mathcal{A}^{\mathrm{d}}(T'_i, \Delta_i)$ with $o'_i \leq v'_i$. Thus, $o := p \cdot o'_1 + (1-p) \cdot o'_2 \in \mathcal{A}^{\mathrm{d}}(T_\varphi, \Delta)$ by Lemma 6.39(6), and $o \leq v_\varphi$.

Now, suppose $\exists o \in \mathcal{A}^{\mathrm{d}}(T_\varphi, \Delta) : o \leq v_\varphi$. Then, by Lemma 6.42, there are $q \in [0,1]$ and $\Delta_1, \Delta_2 \in \mathcal{D}_{sub}(\mathsf{sCSP})$ such that $\Delta \Longrightarrow q \cdot \Delta_1 + (1-q) \cdot \Delta_2$ and $o = q \cdot o'_1 + (1-q) \cdot o'_2$ for certain $o'_i \in \mathcal{A}^{\mathrm{d}}(T'_i, \Delta_i)$. Now $\forall i : o'_i(\omega_i) = v'_i(\omega_i) = \frac{1}{2}$, so, using that T_1, T_2 are Ω-disjoint tests,

$$\frac{1}{2}q = q \cdot o'_1(\omega_1) = o(\omega_1) \leq v_\varphi(\omega_1) = p \cdot v'_1(\omega_1) = \frac{1}{2}p$$

and likewise

$$\frac{1}{2}(1-q) = (1-q) \cdot o'_2(\omega_2) = o(\omega_2) \leq v_\varphi(\omega_2) = (1-p) \cdot v'_2(\omega_2) = \frac{1}{2}(1-p).$$

Together, these inequalities say that $q = p$. Exactly as in the previous case one obtains $o'_i \leq v'_i$ for both $i = 1, 2$. Given that $T'_i = T_i \,{}_{\frac{1}{2}}\!\oplus \omega_i$, using Lemma 6.39(5), it must be that $o'_i = \frac{1}{2}o_i + \frac{1}{2}\vec{\omega}_i$ for some $o_i \in \mathcal{A}^{\mathrm{d}}(T_i, \Delta_i)$ with $o_i \leq v_i$. By induction, $\Delta_i \models \varphi_i$ for $i = 1, 2$, and hence $\Delta \models \varphi$. □

Theorem 6.17 *If $\Theta \sqsubseteq^{\Omega}_{pmust} \Delta$ then $\Theta \sqsubseteq^{\mathcal{F}} \Delta$.*

Proof Suppose $\Theta \sqsubseteq^{\Omega}_{\mathrm{pmust}} \Delta$ and $\Delta \models \varphi$ for some $\varphi \in \mathcal{F}$. Let T_φ be a characteristic test of φ with target value v_φ. Then Proposition 6.14 yields $\exists o \in \mathcal{A}^{\mathrm{d}}(T_\varphi, \Delta) : o \leq v_\varphi$, and hence, given that $\Theta \sqsubseteq^{\Omega}_{\mathrm{pmust}} \Delta$, by the Smyth preorder we are guaranteed to have $\exists o' \in \mathcal{A}^{\mathrm{d}}(T_\varphi, \Theta) : o' \leq v_\varphi$. Thus $\Theta \models \varphi$ by Proposition 6.14 again. □

Corollary 6.13 *For any finitary processes P and Q, if $P \sqsubseteq_{pmust} Q$ then $P \sqsubseteq_{FS} Q$.*

Proof From Theorems 6.17 and 6.16, we know that if $P \sqsubseteq^{\Omega}_{\mathrm{pmust}} Q$ then $P \sqsubseteq_{FS} Q$. Theorem 4.7 tells us that Ω-testing is reducible to scalar testing. So the required result follows. □

Remark 6.4 Note that in our testing semantics we have allowed tests to be finitary. The proof of Proposition 6.14 actually tells us that if two processes are behaviourally different they can be distinguished by some characteristic tests that are always finite. Therefore, Corollary 6.13 still holds if tests are required to be finite.

6.8 Simulations and May Testing

In this section, we follow the same strategy as for failure simulations and testing (Sect. 6.6) except that we restrict our treatment to full distributions; this is possible because partial distributions are not necessary for this case; and it is desirable because the approach becomes simpler as a result.

Definition 6.28 (Simulation Preorder) Define $\sqsubseteq_S$ to be the largest relation in $\mathcal{D}(S) \times \mathcal{D}(S)$ such that if $\Delta \sqsubseteq_S \Theta$ then whenever $\Delta \stackrel{\alpha}{\Longrightarrow} (\sum_i p_i \Delta'_i)$, for finitely many p_i with $\sum_i p_i = 1$, there are Θ'_i with $\Theta \stackrel{\alpha}{\Longrightarrow} (\sum_i p_i \Theta'_i)$ and $\Delta'_i \sqsubseteq_S \Theta'_i$ for each i.
Note that, unlike for Definition 6.28, this summation cannot be empty.

Again it is trivial to see that $\sqsubseteq_S$ is reflexive and transitive; and again it is sometimes easier to work with an equivalent formulation based on a state-level "simulation" defined as follows.

Definition 6.29 (Simulation) Define $\lhd_s$ to be the largest relation in $S \times \mathcal{D}(S)$ such that if $s \lhd_s \Theta$ then whenever $s \stackrel{\alpha}{\longrightarrow} \Delta'$ there is some Θ' such that $\Theta \stackrel{\alpha}{\Longrightarrow} \Theta'$ and $\Delta' \, (\lhd_s)^\dagger \, \Theta'$.

Remark 6.5 Definition 6.29 entails the same simulation given in Definition 5.4 when applied to finite processes. It differs from the analogous Definition 6.21 in three ways: it is missing the clause for divergence, and for refusal; and it is (implicitly) limited to $\stackrel{\alpha}{\Longrightarrow}$-transitions that simulate by producing full distributions only.[2] Without that latter limitation, any simulation relation could be scaled down uniformly without losing its simulation properties, for example, allowing counter-intuitively $a.\mathbf{0}$ to be simulated by $a.\mathbf{0}\,_{\frac{1}{2}}\!\oplus \varepsilon$.

Lemma 6.43 *The above preorder and simulation are equivalent in the following sense: for distributions Δ, Θ we have $\Delta \sqsubseteq_S \Theta$ just when there is a distribution Θ^{match} with $\Theta \Rightarrow \Theta^{match}$ and $\Delta \, (\lhd_s)^\dagger \, \Theta^{match}$.*

Proof The proof is as for the failure case, except that in Theorem 6.12 we can assume total distributions, and so do not need the second part of its proof where divergence is treated. □

6.8.1 Soundness

In this section, we prove that simulations are sound for showing that processes are related via the may-testing preorder. We assume initially that we are using only one success action ω, so that $|\Omega| = 1$.

Because we prune our pLTS s before extracting values from them, we will be concerned mainly with ω-respecting structures, and for those we have the following.

Lemma 6.44 *Let Δ and Θ be two distributions. If Δ is stable and $\Delta \, (\lhd_s)^\dagger \, \Theta$, then $\mathcal{V}(\Delta) \leq_{\mathrm{Ho}} \mathcal{V}(\Theta)$.*

Proof We first show that if s is stable and $s \lhd_s \Theta$ then $\mathcal{V}(s) \leq_{\mathrm{Ho}} \mathcal{V}(\Theta)$. Since s is stable, we have only two cases:

[2] Even though for simplicity of presentation in Definition 6.2 the relation $\longrightarrow$ was defined by using subdistributions, it can be equivalently defined by using full distributions.

(i) $s \not\rightarrow$. Here $\mathcal{V}(s) = \{0\}$ and since $\mathcal{V}(\Theta)$ is not empty we clearly have that $\mathcal{V}(s) \leq_{\text{Ho}} \mathcal{V}(\Theta)$.

(ii) $s \xrightarrow{\omega} \Delta'$ for some Δ'. Here $\mathcal{V}(s) = \{1\}$ and $\Theta \Longrightarrow \Theta' \xrightarrow{\omega}$ with $\mathcal{V}(\Theta') = \{1\}$. By Lemma 6.32 specialised to full distributions, we have $1 \in \mathcal{V}(\Theta)$. Therefore, $\mathcal{V}(s) \leq_{\text{Ho}} \mathcal{V}(\Theta)$.

Now for the general case we suppose $\Delta\ (\lhd_s)^\dagger\ \Theta$. Use Proposition 6.2 to decompose Θ into $\sum_{s \in \lceil \Delta \rceil} \Delta(s) \cdot \Theta_s$ such that $s \lhd_s \Theta_s$ for each $s \in \lceil \Delta \rceil$, and recall each such state s is stable. From above, we have that $\mathcal{V}(s) \leq_{\text{Ho}} \mathcal{V}(\Theta_s)$ for those s, and so $\mathcal{V}(\Delta) = \sum_{\in \lceil \Delta \rceil} \Delta(s) \cdot \mathcal{V}(s) \leq_{\text{Ho}} \sum_{s \in \lceil \Delta \rceil} \Delta(s) \cdot \mathcal{V}(\Theta_s) = \mathcal{V}(\Theta)$. □

Lemma 6.45 *Let Δ and Θ be distributions in an ω-respecting finitary pLTS given by $\langle S, \{\tau, \omega\}, \rightarrow \rangle$. If $\Delta\ (\lhd_s)^\dagger\ \Theta$, then we have $\mathcal{V}(\Delta) \leq_{\text{Ho}} \mathcal{V}(\Theta)$.*

Proof Since $\Delta\ (\lhd_s)^\dagger\ \Theta$, we consider subdistributions Δ'' with $\Delta \Longrightarrow\!\!\!\succ \Delta''$; by distillation of divergence (Theorem 6.11) we have full distributions Δ', Δ'_1 and Δ'_2 and probability p such that $\overline{s} \Longrightarrow \Delta' = (\Delta'_1\, {}_p\!\oplus \Delta'_2)$ and $\Delta'' = p \cdot \Delta'_1$ and $\Delta'_2 \Longrightarrow \varepsilon$. There is, thus, a matching transition $\Theta \Longrightarrow \Theta'$ such that $\Delta'\ (\lhd_s)^\dagger\ \Theta'$. By Proposition 6.2, we can find distributions Θ'_1, Θ'_2 such that $\Theta' = \Theta'_1\, {}_p\!\oplus \Theta'_2$, $\Delta'_1\ (\lhd_s)^\dagger\ \Theta'_1$ and $\Delta'_2\ (\lhd_s)^\dagger\ \Theta'_2$.

Since $\lceil \Delta'_1 \rceil = \lceil \Delta'' \rceil$ we have that Δ'_1 is stable. It follows from Lemma 6.44 that $\mathcal{V}(\Delta'_1) \leq_{\text{Ho}} \mathcal{V}(\Theta'_1)$. Thus we finish off with

	$\mathcal{V}(\Delta'')$	
$=$	$\mathcal{V}(p \cdot \Delta'_1)$	$\Delta'' = p \cdot \Delta'_1$
$=$	$p \cdot \mathcal{V}(\Delta'_1)$	linearity of $\mathcal{V}$
$\leq_{\text{Ho}}$	$p \cdot \mathcal{V}(\Theta'_1)$	above argument based on distillation
$=$	$\mathcal{V}(p \cdot \Theta'_1)$	linearity of $\mathcal{V}$
$\leq_{\text{Ho}}$	$\mathcal{V}(\Theta')$	$\Theta' = \Theta'_1\, {}_p\!\oplus \Theta'_2$
$\leq_{\text{Ho}}$	$\mathcal{V}(\Theta)$.	Lemma 6.32 specialised to full distributions

Since Δ'' was arbitrary, we have our result. □

Lemma 6.46 *Let Δ and Θ be distributions in an ω-respecting finitary pLTS given by $\langle S, \{\tau, \omega\}, \rightarrow \rangle$. If $\Delta \sqsubseteq_S \Theta$, then it holds that $\mathcal{V}(\Delta) \leq_{\text{Ho}} \mathcal{V}(\Theta)$.*

Proof Suppose $\Delta \sqsubseteq_S \Theta$. By Lemma 6.43, there exists some distribution Θ^{match} such that $\Theta \Longrightarrow \Theta^{\text{match}}$ and $\Delta\ (\lhd_s)^\dagger\ \Theta^{\text{match}}$. By Lemmas 6.45 and 6.32 we obtain $\mathcal{V}(\Delta) \leq_{\text{Ho}} \mathcal{V}(\Theta') \subseteq \mathcal{V}(\Theta)$. □

Theorem 6.18 *For any finitary processes P and Q, if $P \sqsubseteq_S Q$ then $P \sqsubseteq_{pmay} Q$.*

Proof We reason as follows.

	$P \sqsubseteq_S Q$	
implies	$[P\vert_{\mathsf{Act}}T] \sqsubseteq_S [Q\vert_{\mathsf{Act}}T]$	the counterpart of Lemma 6.30 for simulation
implies	$\mathcal{V}([P\vert_{\mathsf{Act}}T]) \leq_{\mathrm{Ho}} \mathcal{V}([Q\vert_{\mathsf{Act}}T])$	$[\cdot]$ is ω-respecting; Lemma 6.46
iff	$\mathcal{A}^{\mathrm{d}}(T, P) \leq_{\mathrm{Ho}} \mathcal{A}^{\mathrm{d}}(T, Q)$	Definition 6.24
iff	$P \sqsubseteq_{\mathrm{pmay}} Q$.	Definition 6.11

□

6.8.2 Completeness

Let $\mathcal{L}$ be the subclass of $\mathcal{F}$ by skipping the **div** and **ref** (X) clauses. In other words, the formulae are exactly the same as those in the logic for characterising the simulation preorder in Sect. 5.6. The semantic interpretation is different now because the weak transition relation $\stackrel{\hat{\tau}}{\Longrightarrow}$ used there has been replaced in this chapter by a more general form $\Longrightarrow$ given in Definition 6.4. We continue to write $P \sqsubseteq^{\mathcal{L}} Q$ just when $[\![P]\!] \models \varphi$ implies $[\![Q]\!] \models \varphi$ for all $\varphi \in \mathcal{L}$.

We have the counterparts of Theorems 6.16 and 6.17, with similar proofs.

Theorem 6.19 *In a finitary pLTS $\Delta \sqsubseteq^{\mathcal{L}} \Theta$ if and only if $\Delta \sqsubseteq_S \Theta$.* □

Theorem 6.20 *If $\Delta \sqsubseteq^{\Omega}_{pmay} \Theta$ then $\Delta \sqsubseteq^{\mathcal{L}} \Theta$.* □

Corollary 6.14 *Suppose P and Q are finitary* rpCSP *processes. If $P \sqsubseteq_{pmay} Q$ then $P \sqsubseteq_S Q$.*

Proof From Theorems 6.19 and 6.20 we know that if $P \sqsubseteq^{\Omega}_{\mathrm{pmay}} Q$ then $P \sqsubseteq_S Q$. Theorem 4.7 says that Ω-testing is reducible to scalar testing. So the required result follows. □

As one would expect, the completeness result in Corollary 6.14 would fail for infinitary processes.

Example 6.26 Consider the state s_2 that we saw in Example 6.5. It turns out that

$$\tau.(\mathbf{0}\,{}_{\frac{1}{2}}\!\oplus a.\mathbf{0}) \sqsubseteq_{\mathrm{pmay}} s_2$$

However, we do not have

$$\tau.(\mathbf{0}\,{}_{\frac{1}{2}}\!\oplus a.\mathbf{0}) \lhd_s \overline{s_2}$$

because the transition

$$\tau.(\mathbf{0}\,{}_{\frac{1}{2}}\!\oplus a.\mathbf{0}) \xrightarrow{\tau} (\mathbf{0}\,{}_{\frac{1}{2}}\!\oplus a.\mathbf{0})$$

cannot be matched by a transition from s_2 as there is no *full distribution* Δ such that $s_2 \Longrightarrow \Delta$ and $(\mathbf{0}\,{}_{\frac{1}{2}}\!\oplus a.\mathbf{0})\ (\lhd_s)^{\dagger}\ \Delta$.

6.9 Real-Reward Testing

In Sect. 4.4, we introduced a notion of reward testing inspired by [12]. The idea is to associate with each success action a nonnegative reward, and performing a success action means accumulating some reward. The outcomes of this reward testing are nonnegative expected rewards.

In certain occasions, it is very natural to introduce negative rewards. For example, this is the case in the theory of Markov decision processes [6]. Intuitively, we could understand negative rewards as costs, while positive rewards are often viewed as benefits or profits. Consider, for instance, the (nonprobabilistic) processes Q_1 and Q_2 with initial states q_1 and q_2, respectively, in Fig. 6.5. Here a represents the action of making an investment. Assuming that the investment is made by bidding for some commodity, the τ-action represents an unsuccessful bid—if this happens one simply tries again. Now b represents the action of reaping the benefits of this investment. Whereas Q_1 models a process in which making the investment is always followed by an opportunity to reap the benefits, the process Q_2 allows, nondeterministically, for the possibility that the investment is unsuccessful, so that a does not always lead to a state where b is enabled. The test T with initial state t, which will be explained later, allows us to give a negative reward to action ω_1—its cost—and a positive reward to ω_2.

This leads to the question: *if both negative- and positive rewards are allowed, how would the original reward-testing semantics change?*[3] We refer to the more relaxed form of testing as *real-reward testing* and the original one as *nonnegative-reward testing*.

The power of real-reward testing is illustrated in Fig. 6.5. The two (nonprobabilistic) processes in the left- and central diagrams are equivalent under (probabilistic) may- as well as must testing; the τ-loops in the initial states cause both processes to fail any nontrivial must test. Yet, if a reward of -2 is associated with performing the action ω_1, and a reward of 4 with the subsequent performance of ω_2, it turns out that in the first process the net reward is either 0, if the process remains stuck in its initial state, or positive, whereas running the second process may yield a loss. See Example 6.27 for details of how these rewards are assigned, and how net rewards are associated with the application of tests such as T. This example shows that for processes that may exhibit divergence, real-reward testing is more discriminating than nonnegative-reward testing, or other forms of probabilistic testing. It also illustrates that the extra power is relevant in applications.

[3] One might suspect no change at all, for any assignment of rewards from the interval $[-1, +1]$ can be converted into a nonnegative assignment simply by adding 1 to all of them. But that would not preserve the testing order in the case of zero-outcomes that resulted from a process' s failing to reach any success state at all: those zeroes would remain zero.

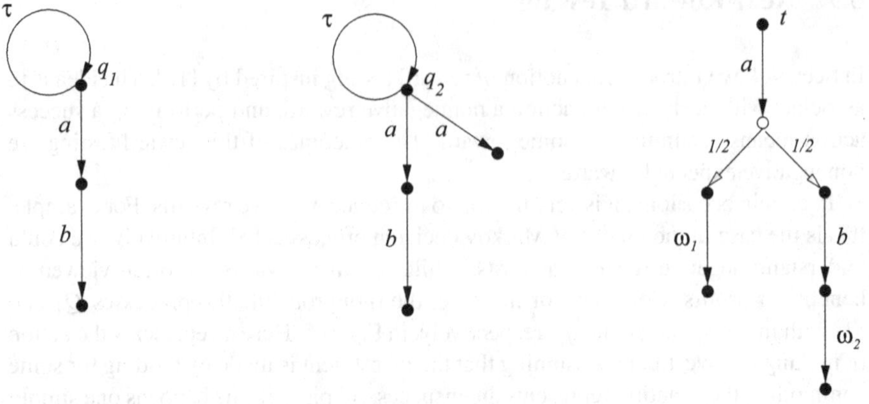

Fig. 6.5 Two processes with divergence and a test. Reprinted from [13], with kind permission from Elsevier

We will show that for real-reward testing may and must preorders are the inverse of each other, that is, for any processes P and Q,

$$P \sqsubseteq^{\Omega}_{\mathbf{rrmay}} Q \text{ iff } Q \sqsubseteq^{\Omega}_{\mathbf{rrmust}} P. \tag{6.27}$$

A more surprising result is that for finitary convergent processes real-reward must preorder coincides with nonnegative-reward must preorder, that is, for any finitary convergent processes P and Q,

$$P \sqsubseteq^{\Omega}_{\mathbf{rrmust}} Q \text{ iff } P \sqsubseteq^{\Omega}_{\mathbf{nrmust}} Q. \tag{6.28}$$

Here by convergence we mean that in the pLTS generated by a process there is no infinite sequence of internal transitions between distributions like

$$\Delta_0 \xrightarrow{\tau} \Delta_1 \xrightarrow{\tau} \cdots$$

Although, it is easy to see that in (6.28) the former is included in the latter, to prove that the latter is included in the former is far from being trivial. Our proof strategy is to make use of failure simulation preorder as a stepping stone, and adapt the soundness proof of failure simulation with respect to must testing (Theorem 6.13).

We now recall our testing framework. A *test* is simply a finite process in the language pCSP, except that it may in addition use special *success* actions for reporting outcomes; these are drawn from a set Ω of fresh actions not already in Act_τ. Here, we require tests to be finite processes because we will consider convergent processes; if P, T are finitary convergent then their parallel composition $P|_{\mathsf{Act}}T$ is not necessarily convergent unless T is very finite. As we have seen from Remark 6.4, restricting to finite tests does not weaken our testing semantics, as far as finitary processes are concerned. As in Sect. 4.2, to apply test T to process P, we form the process $T|_{\mathsf{Act}}P$ in which *all* visible actions of P must synchronise with T. The resulting composition is a process whose only possible actions are τ and the elements of Ω. Applying the test T to the process P gives rise to the set of testing outcomes $\mathcal{A}(T, P)$ defined in (4.2), exactly one of which results from each resolution of the choices in $T|_{\mathsf{Act}}P$. Each *testing outcome* is an Ω-tuple of real numbers in the interval [0,1], that is, a function $o : \Omega \to [0, 1]$, and its ω-component $o(\omega)$, for $\omega \in \Omega$, gives the probability that the resolution in question will reach an *ω-success state*, one in which the success action ω is possible.

In Sect. 4.4 two reward testing preorders are obtained by associating each success action $\omega \in \Omega$ a nonnegative reward. We refer to that approach of testing as *nonnegative-reward testing*. If we also allow negative rewards, which intuitively can be understood as costs, then we obtain an approach of testing called *real-reward testing*. Technically, we simply let reward tuples h range over the set $[-1, 1]^\Omega$. If $o \in [0, 1]^\Omega$, we use the dot-product $h \cdot o = \sum_{\omega\in\Omega} h(\omega) * o(\omega)$. It can apply to a set $O \subseteq [0, 1]^\Omega$ so that $h \cdot A = \{h \cdot o \mid o \in O\}$. Let $A \subseteq [-1, 1]$. We use the notation $\bigsqcup A$ for the supremum of set A, and $\bigsqcap A$ for the infimum.

Definition 6.30 (Real-Reward Testing Preorders)

(i) $P \sqsubseteq^{\Omega}_{\mathbf{rrmay}} Q$ if for every Ω-test T and real-reward tuple $h \in [-1, 1]^\Omega$, $\bigsqcup h \cdot \mathcal{A}(T, P) \leq \bigsqcup h \cdot \mathcal{A}(T, Q)$.
(ii) $P \sqsubseteq^{\Omega}_{\mathbf{rrmust}} Q$ if for every Ω-test T and real-reward tuple $h \in [-1, 1]^\Omega$, $\bigsqcap h \cdot \mathcal{A}(T, P) \leq \bigsqcap h \cdot \mathcal{A}(T, Q)$.

Note that for any test T and process P it is easy to see that

$$h \cdot \mathcal{A}(T, P) = \mathcal{A}^h(T, P).$$

Therefore, the nonnegative-reward testing preorders presented in Definition 4.6 can be equivalently formulated in the following way:

(i) $P \sqsubseteq^{\Omega}_{\mathbf{nrmay}} Q$ if for every Ω-test T and nonnegative-reward tuple $h \in [0, 1]^\Omega$, $\bigsqcup h \cdot \mathcal{A}(T, P) \leq \bigsqcup h \cdot \mathcal{A}(T, Q)$.
(ii) $P \sqsubseteq^{\Omega}_{\mathbf{nrmust}} Q$ if for every Ω-test T and nonnegative-reward tuple $h \in [0, 1]^\Omega$, $\bigsqcap h \cdot \mathcal{A}(T, P) \leq \bigsqcap h \cdot \mathcal{A}(T, Q)$.

Although the two nonnegative-reward testing preorders are in general incomparable, the two real-reward testing preorders are simply the inverse relations of each other.

Theorem 6.21 *For any processes P and Q, it holds that $P \sqsubseteq^{\Omega}_{\mathbf{rr}may} Q$ if and only if $Q \sqsubseteq^{\Omega}_{\mathbf{rr}must} P$.*

Proof We first notice that for any nonempty set $A \subseteq [0, 1]^\Omega$ and any reward tuple $h \in [-1, 1]^\Omega$,

$$\bigsqcup h \cdot A = -\left(\bigsqcap (-h) \cdot A\right), \tag{6.29}$$

where $-h$ is the negation of h, that is, $(-h)(\omega) = -(h(\omega))$ for any $\omega \in \Omega$. We consider the "if" direction; the "only if" direction is similar. Let T be any Ω-test and h be any real-reward tuple in $[-1,1]^\Omega$. Clearly, $-h$ is also a real-reward tuple. Suppose $Q \sqsubseteq_{\text{rrmust}} P$, then

$$\bigsqcap (-h) \cdot \mathcal{A}(T, Q) \leq \bigsqcap (-h) \cdot \mathcal{A}(T, P) \tag{6.30}$$

Therefore, we can infer that

$$\begin{aligned} \bigsqcup h \cdot \mathcal{A}(T, P) &= -\left(\bigsqcap (-h) \cdot \mathcal{A}(T, P)\right) && \text{by (6.29)} \\ &\leq -\left(\bigsqcap (-h) \cdot \mathcal{A}(T, Q)\right) && \text{by (6.30)} \\ &= \bigsqcup h \cdot \mathcal{A}(T, Q) && \text{by (6.29)} \end{aligned}$$ □

Our next task is to compare $\sqsubseteq^\Omega_{\text{rrmust}}$ with $\sqsubseteq^\Omega_{\text{nrmust}}$. The former is included in the latter, which directly follows from Definition 6.30. Surprisingly, it turns out that for finitary convergent processes the latter is also included in the former, thus the two preorders are in fact the same. The rest of this section is devoted to proving this result. However, we first show that this result does not extend to divergent processes.

Example 6.27 Consider the processes Q_1 and Q_2 depicted in Fig. 6.5. Using the characterisations of $\sqsubseteq^\Omega_{\text{pmay}}$ and $\sqsubseteq^\Omega_{\text{pmust}}$ in Sects. 6.6–6.8, it is easy to see that these processes cannot be distinguished by probabilistic may- and must testing, and hence not by nonnegative-reward testing either. However, let T be the test in the right diagram of Fig. 6.5 that first synchronises on the action a, and then with probability $\frac{1}{2}$ reaches a state in which a reward of -2 is allocated, and with the remaining probability $\frac{1}{2}$ synchronises with the action b and reaches a state that yields a reward of 4. Thus the test employs two success actions ω_1 and ω_2, and we use the reward tuple h with $h(\omega_1) = -2$ and $h(\omega_2) = 4$. Then the resolution of $\overline{q_1}$ that does not involve the τ-loop contributes the value $-2 \cdot \frac{1}{2} + 4 \cdot \frac{1}{2} = 1$ to the set $h \cdot \mathcal{A}(T, Q_1)$, whereas the resolution that only involves the τ-loop contributes the value 0. Due to interpolation, $h \cdot \mathcal{A}(T, Q_1)$ is in fact the entire interval $[0, 1]$. On the other hand, the resolution corresponding to the a-branch of q_2 contributes the value -1 and $h \cdot \mathcal{A}(T, Q_2) = [-1, 1]$. Thus $\bigsqcap h \cdot \mathcal{A}(T, Q_1) = 0 > -1 = \bigsqcap h \cdot \mathcal{A}(T, Q_2)$, and hence $\overline{q_1} \not\sqsubseteq^\Omega_{\text{rrmust}} \overline{q_2}$.

For convergent pLTSs, the results in Lemmas 6.31 and 6.33 as well as Theorem 6.13 can be strengthened.

Lemma 6.47 *Let Δ and Θ be distributions in an ω-respecting convergent pLTS $\langle S, \Omega_\tau, \rightarrow \rangle$. If distribution Δ is stable and $\Delta\ (\lhd^e_{FS})^\dagger\ \Theta$, then $\$\Delta \in \mathcal{V}(\Theta)$.*

Proof We first show that if s is stable and $s \lhd^e_{FS} \Theta$ then $\$\overline{s} \in \mathcal{V}(\Theta)$. Since s is stable, we have only two cases:

(i) $s \not\rightarrow$. Here $\$\overline{s} = \vec{0}$. Since $s \lhd^e_{FS} \Theta$ we have $\Theta \Longrightarrow \Theta'$ with $\Theta' \not\rightarrow$, whence in fact $\Theta \Longrightarrow\!\!\!\!\!\!\rightarrow \Theta'$ and $\$\Theta' = \vec{0}$. Thus it holds that $\$\overline{s} = \vec{0} \in \mathcal{V}(\Theta)$.

(ii) $s \xrightarrow{\omega} \Delta'$ for some Δ'. Here $\$\overline{s} = \vec{\omega}$, and since $s \lhd^e_{FS} \Theta$ we have $\Theta \Longrightarrow \Theta' \xrightarrow{\omega}$. As the pLTS we are considering is convergent and Θ is a distribution, we know that Θ' is also a distribution. Hence, we have $\$\Theta' = \vec{\omega}$. Because the pLTS is ω-respecting, in fact $\Theta \Longrightarrow\!\!\succ \Theta'$ and so again we have $\$\overline{s} = \vec{\omega} \in \mathcal{V}(\Theta)$.

Now for the general case we suppose $\Delta\ (\lhd^e_{FS})^\dagger\ \Theta$. It is not hard to show that we can decompose Θ into $\sum_{s\in\lceil\Delta\rceil} \Delta(s)\cdot\Theta_s$ such that $s \lhd^e_{FS} \Theta_s$ for each $s \in \lceil\Delta\rceil$, and recall that each such state s is stable. From above we have that $\$\overline{s} \in \mathcal{V}(\Theta_s)$ for those s, and so $\$\Delta = \sum_{\in\lceil\Delta\rceil} \Delta(s)\cdot\$\overline{s} \in \sum_{s\in\lceil\Delta\rceil} \Delta(s)\cdot\mathcal{V}(\Theta_s) = \mathcal{V}(\Theta)$. □

Lemma 6.48 *Let Δ and Θ be distributions in an ω-respecting convergent pLTS $\langle S, \Omega_\tau, \rightarrow\rangle$. If $\Theta \sqsubseteq_{FS} \Delta$, then it holds that $\mathcal{V}(\Theta) \supseteq \mathcal{V}(\Delta)$.*

Proof Let Δ and Θ be distributions in an ω-respecting convergent pLTS given by $\langle S, \Omega_\tau, \rightarrow\rangle$. We note that

(i) If $\Delta \Longrightarrow \Delta'$ then $\mathcal{V}(\Delta') \subseteq \mathcal{V}(\Delta)$.
(ii) If $\Delta\ (\lhd^e_{FS})^\dagger\ \Theta$, then we have $\mathcal{V}(\Delta) \subseteq \mathcal{V}(\Theta)$.

Here (i) follows from Lemma 6.32. For (ii), let us assume $\Delta\ (\lhd^e_{FS})^\dagger\ \Theta$. For any $\Delta \Longrightarrow\!\!\succ \Delta'$ we have the matching transition $\Theta \Longrightarrow \Theta'$ such that $\Delta'\ (\lhd^e_{FS})^\dagger\ \Theta'$. It follows from Lemmas 6.47 and (i) that $\$\Delta' \in \mathcal{V}(\Theta') \subseteq \mathcal{V}(\Theta)$. Consequently, we obtain $\mathcal{V}(\Delta) \subseteq \mathcal{V}(\Theta)$.

Now suppose $\Theta \sqsubseteq_{FS} \Delta$. By definition there exists some Θ' such that $\Theta \Longrightarrow \Theta'$ and $\Delta\ (\lhd^e_{FS})\dagger\ \Theta'$. By (i) and (ii) above we obtain $\mathcal{V}(\Delta) \subseteq \mathcal{V}(\Theta') \subseteq \mathcal{V}(\Theta)$. □

Theorem 6.22 *For any finitary convergent processes P and Q, if $P \sqsubseteq_{FS} Q$ then $P \sqsubseteq^{\Omega}_{\mathbf{rr}must} Q$.*

Proof We reason as follows.

	$P \sqsubseteq_{FS} Q$	
implies	$[P\vert_{\mathsf{Act}} T] \sqsubseteq_{FS} [Q\vert_{\mathsf{Act}} T]$	Lemma 6.30, for any Ω-test T
implies	$\mathcal{V}([P\vert_{\mathsf{Act}} T]) \supseteq \mathcal{V}([Q\vert_{\mathsf{Act}} T])$	$[\cdot]$ is ω-respecting; Lemma 6.48
iff	$\mathcal{A}^{\mathrm{d}}(T,P) \supseteq \mathcal{A}^{\mathrm{d}}(T,Q)$	Definitions 6.24 and 6.10
iff	$\mathcal{A}(T,P) \supseteq \mathcal{A}(T,Q)$	Corollary 6.3
implies	$h\cdot\mathcal{A}(T,P) \supseteq h\cdot\mathcal{A}(T,Q)$ for any $h \in [-1,1]^\Omega$	
implies	$\bigsqcap h\cdot\mathcal{A}(T,P) \le \bigsqcap h\cdot\mathcal{A}(T,Q)$ for any $h \in [-1,1]^\Omega$	
iff	$P \sqsubseteq^{\Omega}_{\mathbf{rrmust}} Q$.	

Note that in the second line above, both $[P\vert_{\mathsf{Act}} T]$ and $[Q\vert_{\mathsf{Act}} T]$ are convergent, since for any convergent process R and very finite process T, by induction on the structure of T, it can be shown that the composition $R\vert_{\mathsf{Act}} T$ is also convergent. □

We are now ready to prove the main result of the section which states that nonnegative-reward must testing is as discriminating as real-reward must testing.

Theorem 6.23 *For any finitary convergent processes P, Q, it holds that $P \sqsubseteq^{\Omega}_{\mathbf{rr}must} Q$ if and only if $P \sqsubseteq^{\Omega}_{\mathbf{nr}must} Q$.*

Proof The "only if" direction is obvious. Let us consider the "if" direction. Suppose P and Q are finitary processes. We reason as follows.

$$
\begin{array}{lll}
 & P \sqsubseteq^{\Omega}_{\mathrm{nrmust}} Q & \\
\text{iff} & P \sqsubseteq^{\Omega}_{\mathrm{pmust}} Q & \text{Theorem 4.5} \\
\text{iff} & P \sqsubseteq_{FS} Q & \text{Theorems 4.7 and 6.13, Corollary 6.13} \\
\text{implies} & P \sqsubseteq^{\Omega}_{\mathrm{rrmust}} Q\,. & \text{Theorem 6.22}
\end{array}
$$

□

In the presence of divergence, $\sqsubseteq^{\Omega}_{\mathrm{rrmust}}$ is strictly included in $\sqsubseteq^{\Omega}_{\mathrm{nrmust}}$. For example, let P and Q be the processes $\mathsf{rec}\ xx$ and $a.\mathbf{0}$, respectively. It holds that $P \sqsubseteq_{FS} Q$ because $P \stackrel{\varepsilon}{\Longrightarrow}$ and the empty subdistribution can failure simulate any processes. It follows that $P \sqsubseteq^{\Omega}_{\mathrm{nrmust}} Q$, by recalling the first two steps of reasoning in the proof of Theorem 6.23. However, if we apply the test $T = a.\omega$ and reward tuple h with $h(\omega) = -1$, then

$$
\begin{aligned}
\textstyle\bigsqcap h \cdot \mathcal{A}(T, P) &= \textstyle\bigsqcap h \cdot \{\varepsilon\} = \bigsqcap \{0\} = 0 \\
\textstyle\bigsqcap h \cdot \mathcal{A}(T, Q) &= \textstyle\bigsqcap h \cdot \{\vec{\omega}\} \;=\; \bigsqcap \{-1\} \;=\; -1
\end{aligned}
$$

As $\bigsqcap h \cdot \mathcal{A}(T, P) \not\leq \bigsqcap h \cdot \mathcal{A}(T, Q)$, we see that $P \not\sqsubseteq^{\Omega}_{\mathrm{rrmust}} Q$.

Below we give a characterisation of $\sqsubseteq^{\Omega}_{\mathrm{rrmust}}$ in terms of the set inclusion relation between testing outcome sets. As a similar characterisation for $\sqsubseteq^{\Omega}_{\mathrm{nrmust}}$ does not hold in general for finitary (nonconvergent) processes, this indicates the subtle difference between $\sqsubseteq^{\Omega}_{\mathrm{rrmust}}$ and $\sqsubseteq^{\Omega}_{\mathrm{nrmust}}$, and we see more clearly why our proof of Theorem 6.23 involves the failure simulation preorder.

Theorem 6.24 *Let P and Q be any finitary processes. Then $P \sqsubseteq^{\Omega}_{\mathrm{rrmust}} Q$ if and only if $\mathcal{A}(T, P) \supseteq \mathcal{A}(T, Q)$ for any Ω-test T.*

Proof ($\Leftarrow$) Let T be any Ω-test and $h \in [-1, 1]^{\Omega}$ be any real-reward tuple. Suppose $\mathcal{A}(T, P) \supseteq \mathcal{A}(T, Q)$. It is obvious that $h \cdot \mathcal{A}(T, P) \supseteq h \cdot \mathcal{A}(T, Q)$, from which it easily follows that

$$\textstyle\bigsqcap h \cdot \mathcal{A}(T, P) \;\leq\; \bigsqcap h \cdot \mathcal{A}(T, Q).$$

As this holds for an arbitrary real-reward tuple h, we see that $P \sqsubseteq^{\Omega}_{\mathrm{rrmust}} Q$.

($\Rightarrow$) Suppose for a contradiction there is some Ω-test T with $\mathcal{A}(T, P) \not\supseteq \mathcal{A}(T, Q)$. Then there exists some outcome $o \in \mathcal{A}(T, Q)$ lying outside $\mathcal{A}(T, P)$, i.e.

$$o \notin \mathcal{A}(T, P). \tag{6.31}$$

Since T is finite, it contains only finitely many elements of Ω, so that we may assume with loss of generality that Ω is finite. Since P and T are finitary, it is easy to see that the pruned composition $[P|_{\mathsf{Act}}T]$ is also finitary. By Theorem 6.1, the set $\{\Phi \mid [[P|_{\mathsf{Act}}T]] \Longrightarrow \Phi\}$ is convex and compact. With an analogous proof, using Corollary 6.4, it can be shown that so is the set $\{\Phi \mid [[P|_{\mathsf{Act}}T]] \Longrightarrow\!\!\!\!\!\succ \Phi\}$. It follows that the set

$$\{\$\Phi \mid [[P|_{\mathsf{Act}}T]] \Longrightarrow\!\!\!\succ \Phi\}$$

that is, $\mathcal{A}^{\mathrm{d}}(T, P)$, is also convex and compact. By Corollary 6.3, the set $\mathcal{A}(T, P)$ is, thus, convex and compact. Combining this with (6.31), and using the Separation theorem, Theorem 2.7, we infer the existence of some hyperplane whose normal is $h \in \mathbb{R}^{\Omega}$ such that $h \cdot o' > h \cdot o$ for all $o' \in \mathcal{A}(T, P)$. By scaling h, we obtain without loss of generality that $h \in [-1, 1]^{\Omega}$. It follows that

$$\bigsqcap h \cdot \mathcal{A}(T, P) \;>\; h \cdot o \;\geq\; \bigsqcap h \cdot \mathcal{A}(T, Q)$$

which is a contradiction to the assumption that $P \sqsubseteq^{\Omega}_{\mathbf{rrmust}} Q$. □

Note that in the above proof the normal of the separating hyperplane belongs to $[-1, 1]^{\Omega}$ rather than $[0, 1]^{\Omega}$. So we cannot repeat the above proof for $\sqsubseteq^{\Omega}_{\mathbf{nrmust}}$. In general, we do not have that $P \sqsubseteq^{\Omega}_{\mathbf{nrmust}} Q$ implies $\mathcal{A}(T, P) \supseteq \mathcal{A}(T, Q)$ for any Ω-test T and for arbitrary finitary processes P and Q, that is finitary processes that might not be convergent. However, when we restrict ourselves to finitary convergent processes, this property does indeed hold, as can be seen from the first five lines in the proof of Theorem 6.22. Note that in that proof there is an essential use of the failure simulation preorder, in particular the pleasing property stated in Lemma 6.48. Even for finitary convergent processes we cannot give a direct and simple proof of that property for $\sqsubseteq^{\Omega}_{\mathbf{nrmust}}$, analogous to that of Theorem 6.24.

Although for finitary convergent processes, real-reward must testing is no more powerful then nonnegative-reward must testing, a similar result does not hold for may testing. This follows immediately from our result that (the inverse of) real-reward may testing is as powerful as real-reward must testing, that is known not to hold for nonnegative-reward may and must testing. Thus, real-reward may testing is strictly more discriminating than nonnegative-reward may testing, even in the absence of divergence.

6.10 Summary

We have generalised the results in Chap. 5 of characterising the may preorder as a simulation relation and the must preorder as a failure-simulation relation, from finite processes to finitary processes. Although the general proof schema is inherited from Chap. 5, the details here are much more complicated. One important reason is the inapplicability of structural induction, an important proof principle used in proving some fundamental properties for finite processes, when we shift to finitary processes. So we have to make use of more advanced mathematical tools such as fixed points on complete lattices, compact sets in topological spaces, especially in complete metric spaces, etc. Technically, we develop weak transitions between probabilistic processes, elaborate their topological properties and capture divergence in terms of partial distributions. In order to obtain the characterisation results of testing preorders as simulation relations, we found it necessary to investigate fundamental structural properties of derivation sets (finite generability) and similarities (infinite approximations), which are of independent interest. The use of Markov decision processes and Zero-One laws was essential in obtaining our results.

We have studied a notion of real-reward testing that extends the nonnegative-reward testing (cf. Sect. 4.4) with negative rewards. It turned out that real-reward may preorder is the inverse of real-reward must preorder, and vice versa. More interestingly, for finitary convergent processes, real-reward must testing preorder coincides with nonnegative-reward testing preorder.

There is a great amount of work about probabilistic testing semantics and simulation semantics, as we have seen in Sects. 4.8 and 5.11. Here we mention the closely related work [9], where Segala defined two preorders called trace distribution precongruence ($\sqsubseteq_{TD}$) and failure distribution precongruence ($\sqsubseteq_{FD}$). He proved that the former coincides with an action-based version of $\sqsubseteq^{\Omega}_{\text{pmay}}$ and that for "probabilistically convergent" systems the latter coincides with an action-based version of $\sqsubseteq^{\Omega}_{\text{pmust}}$. The condition of probabilistic convergence amounts in our framework to the requirement that for $\Delta \in \mathcal{D}(S)$ and $\Delta \Longrightarrow \Delta'$ we have $|\Delta'| = 1$. In [4] it has been shown that $\sqsubseteq_{TD}$ coincides with a notion of simulation akin to $\sqsubseteq_S$.

In [14] by restricting the power of schedulers a testing preorder is proposed and shown to coincide with a probabilistic ready-trace preorder that is strictly coarser than our simulation preorder but is still a precongruence [15].

References

1. Deng, Y., van Glabbeek, R., Hennessy, M., Morgan, C.: Testing finitary probabilistic processes (extended abstract). Proceedings of the 20th International Conference on Concurrency Theory, Lecture Notes in Computer Science, vol. 5710, pp. 274–288. Springer (2009)
2. Jones, C.: Probabilistic non-determinism. Ph.D. thesis, University of Edinburgh (1990)
3. McIver, A.K., Morgan, C.C.: Abstraction, Refinement and Proof for Probabilistic Systems. Springer (2005)
4. Lynch, N., Segala, R., Vaandrager, F.W.: Observing branching structure through probabilistic contexts. SIAM J. Comput. **37**(4), 977–1013 (2007)
5. Milner, R.: Communication and Concurrency. Prentice Hall, New York (1989)
6. Puterman, M.L.: Markov Decision Processes. Wiley, New York (1994)
7. Rutten, J., Kwiatkowska, M., Norman, G., Parker, D.: Mathematical Techniques for Analyzing Concurrent and Probabilistic Systems. (Volume 23 of CRM Monograph Series). American Mathematical Society (2004)
8. Segala, R.: Modeling and verification of randomized distributed real-time systems. Tech. Rep. MIT/LCS/TR-676, PhD thesis, MIT, Dept. of EECS (1995)
9. Segala, R.: Testing probabilistic automata. Proceedings of the 7th International Conference on Concurrency Theory, Lecture Notes in Computer Science, vol. 1119, pp. 299–314. Springer (1996)
10. Deng, Y., van Glabbeek, R., Morgan, C.C., Zhang, C.: Scalar outcomes suffice for finitary probabilistic testing. Proceedings of the 16th European Symposium on Programming, Lecture Notes in Computer Science, vol. 4421, pp. 363–378. Springer (2007)
11. Hennessy, M., Milner, R.: Algebraic laws for nondeterminism and concurrency. J ACM **32**(1), 137–161 (1985)

12. Jonsson, B., Ho-Stuart, C., Yi, W.: Testing and refinement for nondeterministic and probabilistic processes. Proceedings of the 3rd International Symposium on Formal Techniques in Real-Time and Fault-Tolerant Systems, Lecture Notes in Computer Science, vol. 863, pp. 418–430. Springer (1994)
13. Deng, Y., van Glabbeek, R., Hennessy, M., Morgan, C.: Real-reward testing for probabilistic processes. Theor. Comput. Sci. **538**, 16–36 (2014)
14. Georgievska, S., Andova, S.: Probabilistic may/must testing: Retaining probabilities by restricted schedulers. Form. Asp. Comput. **24**, 727–748 (2012)
15. Georgievska, S., Andova, S.: Composing systems while preserving probabilities. Proceedings of the 7th European Performance Engineering Workshop, Lecture Notes in Computer Science, vol. 6342, pp. 268–283. Springer (2010)

Chapter 7
Weak Probabilistic Bisimulation

Abstract By taking the symmetric form of simulation preorder, we obtain a notion of weak probabilistic bisimulation. It provides a sound and complete proof methodology for an extensional behavioural equivalence, a probabilistic variant of the traditional reduction barbed congruence.

Keywords Weak probabilistic bisimulation · Reduction barbed congruence · Compositionality

7.1 Introduction

In Sect. 6.8, we have considered simulation preorder. By taking the symmetric form of Definition 6.28 we easily obtain a notion of weak probabilistic bisimulation.

Definition 7.1 (Weak Probabilistic Bisimulation) Let S be the set of states in a probabilistic labelled transitional system (pLTS). A relation $\mathcal{R} \subseteq \mathcal{D}(S) \times \mathcal{D}(S)$ is a *weak (probabilistic) bisimulation* if $\Delta \mathcal{R} \Theta$ implies, for each $\alpha \in \mathsf{Act}_\tau$ and all finite sets of probabilities $\{ p_i \mid i \in I\}$ satisfying $\sum_{i \in I} p_i = 1$,

(i) whenever $\Delta \stackrel{\alpha}{\Longrightarrow} \sum_{i \in I} p_i \cdot \Delta_i$, for any distributions Δ_i, there are some distributions Θ_i with $\Theta \stackrel{\alpha}{\Longrightarrow} \sum_{i \in I} p_i \cdot \Theta_i$, such that $\Delta_i \mathcal{R} \Theta_i$ for each $i \in I$;
(ii) symmetrically, whenever $\Theta \stackrel{\alpha}{\Longrightarrow} \sum_{i \in I} p_i \cdot \Theta_i$, for any distributions Θ_i, there are some distributions Δ_i with $\Delta \stackrel{\alpha}{\Longrightarrow} \sum_{i \in I} p_i \cdot \Delta_i$, such that $\Delta_i \mathcal{R} \Theta_i$ for each $i \in I$.

The largest weak probabilistic bisimulation, which is guaranteed to exist using standard arguments, is called *weak probabilistic bisimilarity* and denoted by $\approx$.

It is easy to see that $\approx$ is an equivalence relation. Moreover, it turns out that this provides a sound and complete proof methodology for a natural extensional behavioural equivalence between probabilistic systems, a generalisation of *reduction barbed congruence*, the well-known behavioural equivalence for a large variety of process description languages. Intuitively, reduction barbed congruence is defined to be the coarsest relation that

- is *compositional*; that is preserved by some natural operators for constructing systems

DOI 10.1007/978-3-662-45198-4_7

- *preserves barbs*; barbs are simple experiments which observers may perform on systems
- is *reduction-closed*; this is a natural condition on the reduction semantics of systems which ensures that nondeterministic choices are in some sense preserved.

The three criteria chosen above are very robust because they can be formalised in a similar way across different process description languages. In our setting, compositions of pLTSs and reduction semantics can be easily defined. For barbs, we use predicates like $P \Downarrow_a^{\geq p}$, which means that process P can expose the visible action a with probability at least p. The details are given in Sect. 7.4.

7.2 A Simple Bisimulation

Due to the use of weak arrows and the quantification over sets of probabilities, it is difficult to directly apply Definition 7.1 and exhibit witness bisimulations. We, therefore, give an alternative characterisation of $\approx$ in terms of a relation between *states* and distributions by adapting Definition 6.29.

Definition 7.2 (Simple Bisimulation) A relation $\mathcal{R} \subseteq S \times \mathcal{D}(S)$ is a *simple (weak probabilistic) bisimulation,* if $s \; \mathcal{R} \Theta$ implies, for each $\alpha \in \mathsf{Act}_\tau$,

(i) whenever $s \stackrel{\alpha}{\longrightarrow} \Delta'$, there is some $\Theta \stackrel{\alpha}{\Longrightarrow} \Theta'$, such that $\Delta' \; \mathcal{R}^\dagger \; \Theta'$;
(ii) there exists some $\Delta \in \mathcal{D}(S)$ such that $\overline{s} \stackrel{\tau}{\Longrightarrow} \Delta$ and $\Theta \; \mathcal{R}^\dagger \; \Delta$.

We use $\approx_s$ to denote the largest simple bisimulation. As in Remark 6.5, the bisimulation game is implicitly limited to weak transitions that simulate by producing full distributions only.

The precise relationship between the two forms of bisimulations is given by:

Theorem 7.1 *Let Δ and Θ be two distributions in a finitary pLTS.*

(i) *If $\Delta \approx \Theta$ then there is some Θ' with $\Theta \stackrel{\tau}{\Longrightarrow} \Theta'$ and $\Delta \; (\approx_s)^\dagger \; \Theta'$.*
(ii) *If $\Delta \; (\approx_s)^\dagger \; \Theta$ then $\Delta \approx \Theta$.*

The remainder of this subsection is devoted to the proof of this theorem; it involves first developing a number of subsidiary results.

Proposition 7.1 *Suppose $\Delta \; (\approx_s)^\dagger \; \Theta$ and $\Delta \stackrel{\alpha}{\longrightarrow} \Delta'$ in an arbitrary pLTS. Then there exists some Θ' such that $\Theta \stackrel{\alpha}{\Longrightarrow} \Theta'$ and $\Delta' \; (\approx_s)^\dagger \; \Theta'$.*

Proof Suppose $\Delta \; (\approx_s)^\dagger \; \Theta$ and $\Delta \stackrel{\alpha}{\longrightarrow} \Delta'$. By Lemma 6.1 there is a finite index set I such that (i) $\Delta = \sum_{i \in I} p_i \cdot \overline{s_i}$, (ii) $\Theta = \sum_{i \in I} p_i \cdot \Theta_i$ and (iii) $s_i \approx_s \Theta_i$ for each $i \in I$. By the condition $\Delta \stackrel{\alpha}{\longrightarrow} \Delta'$, (i) and Proposition 6.2, we can decompose Δ' into $\sum_{i \in I} p_i \cdot \Delta'_i$ for some Δ'_i such that $\overline{s_i} \stackrel{\alpha}{\longrightarrow} \Delta'_i$. By Lemma 6.1 again, for each $i \in I$, there is an index set J_i such that $\Delta'_i = \sum_{j \in J_i} q_{ij} \cdot \Delta'_{ij}$ and $s_i \stackrel{\alpha}{\longrightarrow} \Delta'_{ij}$ for each $j \in J_i$ and $\sum_{j \in J_i} q_{ij} = 1$. By (iii) there is some Θ'_{ij} such that $\Theta_i \stackrel{\alpha}{\Longrightarrow} \Theta'_{ij}$ and $\Delta'_{ij} \; (\approx_s)^\dagger \; \Theta'_{ij}$.

Let $\Theta' = \sum_{i\in I, j\in J_i} p_i q_{ij} \cdot \Theta'_{ij}$. By Theorem 6.5(i) the relation $\Longrightarrow$ is linear. Now, it is easy to see that $\stackrel{\alpha}{\Longrightarrow}$ is also linear. It follows that $\Theta = \sum_{i\in I} p_i \sum_{j\in J_i} q_{ij}\Theta_i \stackrel{\alpha}{\Longrightarrow} \Theta'$. By the linearity of $(\approx_s)^\dagger$, we notice that $\Delta' = (\sum_{i\in I} p_i \sum_{j\in J_i} q_{ij}\cdot\Delta'_{ij})\ (\approx_s)^\dagger\ \Theta'$. □

Theorem 7.2 *In a finitary pLTS, if* $s \approx_s \Theta$ *and* $\overline{s} \stackrel{\tau}{\Longrightarrow} \Delta'$ *then there is some* Θ' *with* $\Theta \stackrel{\tau}{\Longrightarrow} \Theta'$ *and* $\Delta'\ (\approx_s)^\dagger\ \Theta'$.

Proof The arguments are similar to those in the proof of Theorem 6.12. Suppose s is a state and Θ a distribution in a finitary pLTS such that $s \approx_s \Theta$ and $\overline{s} \stackrel{\tau}{\Longrightarrow} \Delta'$. Referring to Definition 6.4, there must be Δ_k, $\Delta_k^\rightarrow$ and $\Delta_k^\times$ for $k \geq 0$ such that $\overline{s} = \Delta_0$, $\Delta_k = \Delta_k^\rightarrow + \Delta_k^\times$, $\Delta_k^\rightarrow \stackrel{\tau}{\longrightarrow} \Delta_{k+1}$ and $\Delta' = \sum_{k=1}^{\infty} \Delta_k^\times$. Since $\Delta_0^\times + \Delta_0^\rightarrow = \overline{s} \approx_s^\dagger \Theta$, using Proposition 6.2 there exist some $\Theta_0^\times$ and $\Theta_0^\rightarrow$ such that $\Theta = \Theta_0^\times + \Theta_0^\rightarrow$, $\Delta_0^\times\ (\approx_s)^\dagger\ \Theta_0^\times$ and $\Delta_0^\rightarrow\ (\approx_s)^\dagger\ \Theta_0^\rightarrow$. Since $\Delta_0^\rightarrow \stackrel{\tau}{\longrightarrow} \Delta_1$ and $\Delta_0^\rightarrow\ (\approx_s)^\dagger\ \Theta_0^\rightarrow$, by Proposition 7.1 we have $\Theta_0^\rightarrow \Longrightarrow \Theta_1$ with $\Delta_1\ (\approx_s)^\dagger\ \Theta_1$.

Repeating the above procedure gives us inductively a series $\Theta_k, \Theta_k^\rightarrow, \Theta_k^\times$ of subdistributions, for $k \geq 0$, such that $\Theta_0 = \Theta$, $\Delta_k\ (\approx_s)^\dagger\ \Theta_k$, $\Theta_k = \Theta_k^\rightarrow + \Theta_k^\times$, $\Delta_k^\times\ (\approx_s)^\dagger\ \Theta_k^\times$, $\Delta_k^\rightarrow\ (\approx_s)^\dagger\ \Theta_k^\rightarrow$ and $\Theta_k^\rightarrow \stackrel{\tau}{\Longrightarrow} \Theta_k$. We define $\Theta' := \sum_i \Theta_i^\times$. By Additivity (Remark 6.2), we have $\Delta'\ (\approx_s)^\dagger\ \Theta'$. It remains to be shown that $\Theta \Longrightarrow \Theta'$.

For that final step, since the set $\{\Theta'' \mid \Theta \Longrightarrow \Theta''\}$ is closed by Lemma 6.17, we can establish $\Theta \Longrightarrow \Theta'$ by exhibiting a sequence Θ'_i with $\Theta \Longrightarrow \Theta'_i$ for each i and with the Θ'_i's being arbitrarily close to Θ'. Induction establishes for each i the weak transition $\Theta \Longrightarrow \Theta'_i := (\Theta_i^\rightarrow + \sum_{k\leq i} \Theta_k^\times)$. Since Δ' is a full distribution (cf. Definition 7.2), whose mass is 1, i.e. $|\Delta'| = 1$, we must have $\lim_{i\to\infty} |\Delta_i^\rightarrow| = 0$. It is easy to see that for any two subdistributions Γ_1, Γ_2 if $\Gamma_1\ (\approx_s)^\dagger\ \Gamma_2$ then they have the same mass. Therefore, it follows from the condition $\Delta_i^\rightarrow\ (\approx_s)^\dagger\ \Theta_i^\rightarrow$ that $\lim_{i\to\infty} |\Theta_i^\rightarrow| = 0$. Thus these Θ'_i's form the sequence we needed. □

Corollary 7.1 *In a finitary pLTS, suppose* $\Delta\ (\approx_s)^\dagger\ \Theta$ *and* $\Delta \stackrel{\alpha}{\Longrightarrow} \Delta'$. *Then there is some* Θ' *with* $\Theta \stackrel{\alpha}{\Longrightarrow} \Theta'$ *and* $\Delta'\ (\approx_s)^\dagger\ \Theta'$.

Proof Given the two previous results, this is fairly straightforward. Suppose that $\Delta \stackrel{\alpha}{\Longrightarrow} \Delta'$ and $\Delta\ (\approx_s)^\dagger\ \Theta$. If α is τ then the required Θ' follows by an application of Theorem 7.2, since the relation $\stackrel{\tau}{\Longrightarrow}$ is actually defined to be $\Longrightarrow$.

Otherwise, by definition we know $\Delta \Longrightarrow \Delta_1$, $\Delta_1 \stackrel{\alpha}{\longrightarrow} \Delta_2$ and $\Delta_2 \Longrightarrow \Delta'$. An application of Theorem 7.2 gives a Θ_1 such that $\Theta \Longrightarrow \Theta_1$ and $\Delta_1\ (\approx_s)^\dagger\ \Theta_1$. An application of Proposition 7.1 gives a Θ_2 such that $\Theta_1 \stackrel{\alpha}{\Longrightarrow} \Theta_2$ and $\Delta_2\ (\approx_s)^\dagger\ \Theta_2$. Finally, another application of Theorem 7.2 gives $\Theta_2 \Longrightarrow \Theta'$ such that $\Delta'\ (\approx_s)^\dagger\ \Theta'$.

The result now follows since the transitivity of $\Longrightarrow$, Theorem 6.6, gives the transition $\Theta \stackrel{\alpha}{\Longrightarrow} \Theta'$. □

Theorem 7.3 *In a finitary pLTS,* $\Delta\ (\approx_s)^\dagger\ \Theta$ *implies* $\Delta \approx \Theta$.

Proof Let $\mathcal{R}$ denote the relation $(\approx_s)^\dagger \cup ((\approx_s)^\dagger)^{-1}$. We show that $\mathcal{R}$ is a bisimulation relation, from which the result follows.

Suppose that $\Delta\ \mathcal{R}\ \Theta$. There are two possibilities:

(a) $\Delta\ (\approx_s)^\dagger\ \Theta$.
First suppose $\Delta \stackrel{\alpha}{\Longrightarrow} \sum_{i\in I} p_i \cdot \Delta'_i$. By Corollary 7.1 there is some distribution Θ' with $\Theta \stackrel{\alpha}{\Longrightarrow} \Theta'$ and $(\sum_{i\in I} p_i \cdot \Delta'_i)\ (\approx_s)^\dagger\ \Theta'$. But by Proposition 6.2 we know that the relation $(\approx_s)^\dagger$ is left-decomposable. This means that $\Theta' = \sum_{i\in I} p_i \cdot \Theta'_i$ for some distributions Θ'_i such that $\Delta'_i\ (\approx_s)^\dagger\ \Theta'_i$ for each $i \in I$. We, hence, have the required matching move from Θ.
For the converse, suppose $\Theta \stackrel{\alpha}{\Longrightarrow} \sum_{i\in I} p_i \cdot \Theta'_i$. We have to find a matching transition, $\Delta \stackrel{\alpha}{\Longrightarrow} \sum_{i\in I} p_i \cdot \Delta'_i$, such that $\Delta'_i\ \mathcal{R}\ \Theta'_i$. In fact it is sufficient to find a transition $\Delta \stackrel{\alpha}{\Longrightarrow} \Delta'$, such that $\sum_{i\in I} p_i \cdot \Theta'_i\ (\approx_s)^\dagger\ \Delta'$, since $((\approx_s)^\dagger)^{-1} \subseteq \mathcal{R}$ and the deconstruction of Δ' into the required sum $\sum_{i\in I} p_i \cdot \Delta'_i$ will again follow from the fact that $(\approx_s)^\dagger$ is left-decomposable. To this end let us abbreviate $\sum_{i\in I} p_i \cdot \Theta'_i$ to simply Θ'.
We know from $\Delta\ (\approx_s)^\dagger\ \Theta$, using the left-decomposability of $(\approx_s)^\dagger$, the convexity of $\approx_s$ and Remark 6.1, that $\Theta = \sum_{s\in\lceil\Delta\rceil} \Delta(s) \cdot \Theta_s$ for some Θ_s with $s \approx_s \Theta_s$. Then, by the definition of $\approx_s$, $\overline{s} \stackrel{\tau}{\Longrightarrow} \Delta_s$ for some Δ_s such that $\Theta_s\ (\approx_s)^\dagger\ \Delta_s$. Now using Theorem 6.5(ii) it is easy to show the left-decomposability of weak actions $\stackrel{\alpha}{\Longrightarrow}$. Then from $\Theta \stackrel{\alpha}{\Longrightarrow} \Theta'$, we can derive that $\Theta' = \sum_{s\in\lceil\Delta\rceil} \Delta(s) \cdot \Theta'_s$, such that $\Theta_s \stackrel{\alpha}{\Longrightarrow} \Theta'_s$, for each s in the support of Δ. Applying Corollary 7.1 to $\Theta_s\ (\approx_s)^\dagger\ \Delta_s$ we have, again for each s in the support of Δ, a matching move $\Delta_s \stackrel{\alpha}{\Longrightarrow} \Delta'_s$ such that $\Theta'_s\ (\approx_s)^\dagger\ \Delta'_s$. But, since $\overline{s} \stackrel{\tau}{\Longrightarrow} \Delta_s$, this gives $\overline{s} \stackrel{\alpha}{\Longrightarrow} \Delta'_s$ for each $s \in \lceil\Delta\rceil$; using the linearity of weak actions $\stackrel{\alpha}{\Longrightarrow}$, these moves from the states s in the support of Δ can be combined to obtain the action $\Delta \stackrel{\alpha}{\Longrightarrow} \sum_{s\in\lceil\Delta\rceil} \Delta(s) \cdot \Delta'_s$. The required Δ' is this sum, $\sum_{s\in\lceil\Delta\rceil} \Delta(s) \cdot \Delta'_s$, since linearity of $(\approx_s)^\dagger$ gives $\Delta'((\approx_s)^\dagger)^{-1}\Theta'$.

(b) The second possibility is that $\Delta\ ((\approx_s)^\dagger)^{-1}\ \Theta$, that is $\Theta\ (\approx_s)^\dagger\ \Delta$. But in this case the proof that the relevant moves from Θ and Δ can be properly matched is exactly the same as in case (a). □

We also have a partial converse to Theorem 7.3:

Proposition 7.2 *In a finitary pLTS, $\overline{s} \approx \Theta$ implies $s \approx_s \Theta$.*

Proof Let $\approx^s_{bis}$ be the restriction of $\approx$ to $S \times \mathcal{D}(S)$, in the sense that $s \approx^s_{bis} \Theta$ whenever $\overline{s} \approx \Theta$. We show that $\approx^s_{bis}$ is a simple bisimulation. Suppose $s \approx^s_{bis} \Theta$.

(i) First suppose $s \stackrel{\alpha}{\longrightarrow} \Delta'$. Then, since $\overline{s} \approx \Theta$ there must exist some $\Theta \stackrel{\alpha}{\Longrightarrow} \Theta'$ such that $\Delta' \approx \Theta'$. Now consider the degenerate action $\Delta' \stackrel{\tau}{\Longrightarrow} \sum_{t\in\lceil\Delta'\rceil} \Delta'(t) \cdot \overline{t}$. There must be a matching move from Θ', $\Theta' \stackrel{\tau}{\Longrightarrow} \Theta'' = \sum_{t\in\lceil\Delta'\rceil} \Delta'(t) \cdot \Theta'_t$ such that $\overline{t} \approx \Theta'_t$, that is $t \approx^s_{bis} \Theta'_t$ for each $t \in \lceil\Delta'\rceil$.
By linearity, this means $\Delta'\ (\approx^s_{bis})^\dagger\ \Theta''$ and by the transitivity of $\Longrightarrow$ we have the required matching move $\Theta \stackrel{\alpha}{\Longrightarrow} \Theta''$.

(ii) To establish the second requirement, consider the trivial move $\Theta \stackrel{\tau}{\Longrightarrow} \Theta$. Since $\overline{s} \approx \Theta$, there must exist a corresponding move $\overline{s} \stackrel{\tau}{\Longrightarrow} \Delta$ such that $\Delta \approx \Theta$. Since $\approx$ is a symmetric relation, we also have $\Theta \approx \Delta$. Now by an argument symmetric to that used in part (i) we can show that this implies the existence of some Δ' such that $\Delta \stackrel{\tau}{\Longrightarrow} \Delta'$, that is $\overline{s} \stackrel{\tau}{\Longrightarrow} \Delta'$ and $\Theta\ (\approx^s_{bis})^\dagger\ \Delta'$. □

But in general the relations $\approx$ and $(\approx_s)^\dagger$ do not coincide for arbitrary distributions. For example, consider the two processes $P = a.\mathbf{0}\ _{\frac{1}{2}}\!\oplus b.\mathbf{0}$ and $Q = P \sqcap P$. It is easy to see that $[\![P]\!] \approx [\![Q]\!]$ but not $[\![P]\!]\ (\approx_s)^\dagger\ [\![Q]\!]$; the latter follows because the point distribution $[\![Q]\!]$ cannot be decomposed as $\frac{1}{2} \cdot \Theta_a + \frac{1}{2} \cdot \Theta_b$ for some Θ_a and Θ_b so that $[\![a.\mathbf{0}]\!] \approx_s \Theta_a$ and $[\![b.\mathbf{0}]\!] \approx_s \Theta_b$.

The nearest to a general converse to Theorem 7.3 is the following:

Proposition 7.3 *Suppose $\Delta \approx \Theta$ in a finitary pLTS. Then there is some Θ' with $\Theta \stackrel{\tau}{\Longrightarrow} \Theta'$ and $\Delta\ (\approx_s)^\dagger\ \Theta'$.*

Proof Now suppose $\Delta \approx \Theta$. We can rewrite Δ as $\sum_{s \in \lceil \Delta \rceil} \Delta(s) \cdot \overline{s}$, and the reflexivity of $\stackrel{\tau}{\Longrightarrow}$ gives $\Delta \stackrel{\tau}{\Longrightarrow} \sum_{s \in \lceil \Delta \rceil} \Delta(s) \cdot \overline{s}$. Since $\approx$ is a bisimulation, this move can be matched by some $\Theta \stackrel{\tau}{\Longrightarrow} \Theta' = \sum_{s \in \lceil \Delta \rceil} \Delta(s) \cdot \Theta_s$ such that $\overline{s} \approx \Theta_s$. But we have just shown in the previous proposition that this means $s \approx_s \Theta_s$.

By Definition 6.2, $\Delta\ (\approx_s)^\dagger\ \Theta'$ and therefore $\Theta \stackrel{\alpha}{\Longrightarrow} \Theta'$ is the required move. □

The weak bisimilarity, $\approx$ from Definition 7.1, is our primary behavioural equivalence but we will often develop properties of it via the connection we have just established with $\approx_s$ from Definition 7.2; the latter is more amenable as it only required strong moves to be matched. However, we can also prove properties of $\approx_s$ by using this connection to weak bisimilarity; a simple example is the following:

Corollary 7.2 *Suppose $s \approx_s \Theta$ in a finitary pLTS, where $s \stackrel{\tau}{\not\longrightarrow}$. Then whenever $\Theta \stackrel{\tau}{\Longrightarrow} \Theta'$ it follows that $s \approx_s \Theta'$.*

Proof Suppose $s \approx_s \Theta$, which means $\overline{s}\ (\approx_s)^\dagger\ \Theta$ and therefore by Theorem 7.3 $\overline{s} \approx \Theta$. The move $\Theta \stackrel{\tau}{\Longrightarrow} \Theta'$ must be matched by a corresponding move from $\overline{s}$. However, since $s \stackrel{\tau}{\not\longrightarrow}$ the only possibility is the empty move, giving $\overline{s} \approx \Theta'$. Now by Proposition 7.2, we have the required $s \approx_s \Theta'$. □

Corollary 7.3 *In any finitary pLTS, the relation $\approx$ is linear.*

Proof Consider any collection of probabilities p_i with $\sum_{i \in I} p_i = 1$, where I is a finite index set. Suppose further that $\Delta_i \approx \Theta_i$ for each $i \in I$. We need to show that $\Delta \approx \Theta$, where $\Delta = \sum_{i \in I} p_i \cdot \Delta_i$ and $\Theta = \sum_{i \in I} p_i \cdot \Theta_i$.

By Proposition 7.3, there is some Θ'_i with $\Theta_i \stackrel{\tau}{\Longrightarrow} \Theta'_i$ and $\Delta_i\ (\approx_s)^\dagger\ \Theta'_i$. By Theorem 6.6 (i) and Definition 6.2, both $\stackrel{\tau}{\Longrightarrow}$ and $(\approx_s)^\dagger$ are linear. Therefore, we have $\Theta \stackrel{\tau}{\Longrightarrow} \Theta'$ and $\Delta\ (\approx_s)^\dagger\ \Theta'$, where $\Theta' = \sum_{i \in I} p_i \cdot \Theta'_i$. It follows from Theorem 7.3 that $\Delta \approx \Theta'$.

Now for any transition $\Delta \stackrel{\alpha}{\Longrightarrow} (\sum_{j \in J} q_j \cdot \Delta_j)$, where J is finite, there is a matching transition $\Theta' \stackrel{\alpha}{\Longrightarrow} (\sum_{j \in J} q_j \cdot \Theta_j)$ such that $\Delta_j \approx \Theta_j$ for each $j \in J$. Note that we also have the transition $\Theta \stackrel{\alpha}{\Longrightarrow} (\sum_{j \in J} q_j \cdot \Theta_j)$ according to the transitivity of $\stackrel{\tau}{\Longrightarrow}$. By symmetrical arguments, any transition $\Theta \stackrel{\alpha}{\Longrightarrow} (\sum_{j \in J} q_j \cdot \Theta_j)$ can be matched by some transition $\Delta \stackrel{\alpha}{\Longrightarrow} (\sum_{j \in J} q_j \cdot \Delta_j)$ such that $\Delta_j \approx \Theta_j$ for each $j \in J$. □

7.3 Compositionality

The main operator of interest in modelling concurrent systems is the parallel composition. We will use a parallel composition from the calculus of communicating systems (CCS) $|$. It is convenient for the proofs in Sect. 7.4, so we add it to our language rpCSP presented in Sect. 6.2. Its behaviour is specified by the following three rules.

$$\frac{s_1 \xrightarrow{\alpha} \Delta}{s_1 \mid s_2 \xrightarrow{\alpha} \Delta \mid s_2} \qquad \frac{s_2 \xrightarrow{\alpha} \Delta}{s_1 \mid s_2 \xrightarrow{\alpha} s_1 \mid \Delta}$$

$$\frac{s_1 \xrightarrow{a} \Delta_1,\ s_2 \xrightarrow{a} \Delta_2}{s_1 \mid s_2 \xrightarrow{\tau} \Delta_1 \mid \Delta_2}.$$

Intuitively in the parallel composition of two processes, each of them can proceed independently or synchronise on any visible action they share with each other. The rules use the obvious extension of the function $|$ on pairs of states to pairs of distributions. To be precise $\Delta|\Theta$ is the distribution defined by:

$$(\Delta|\Theta)(s) = \begin{cases} \Delta(s_1) \cdot \Theta(s_2) & \text{if } s = s_1 \mid s_2 \\ 0 & \text{otherwise} \end{cases}$$

This construction can also be explained as follows:

Lemma 7.1 *For any state t and distributions Δ, Θ,*

(i) $\Delta \mid \overline{t} = \sum_{s \in \lceil \Delta \rceil} \Delta(s) \cdot (\overline{s} \mid \overline{t})$
(ii) $\Delta \mid \Theta = \sum_{t \in \lceil \Theta \rceil} \Theta(t) \cdot (\Delta \mid \overline{t})$.

Proof Straightforward calculation. □

We show that both $\approx_s$ and $\approx$ are closed under the parallel composition. This requires some preliminary results, particularly on composing actions from the components of a parallel composition, as in Lemma 6.27.

Lemma 7.2 *In a pLTS,*

(i) $\Delta \xrightarrow{\alpha} \Delta'$ *implies* $\Delta \mid \Theta \xrightarrow{\alpha} \Delta' \mid \Theta$, *for* $\alpha \in \mathsf{Act}_\tau$
(ii) $\Delta_1 \xrightarrow{a} \Delta_1'$ *and* $\Delta_2 \xrightarrow{a} \Delta_2'$ *implies* $\Delta_1 \mid \Delta_2 \xrightarrow{\tau} \Delta_1' \mid \Delta_2'$, *for* $a \in \mathsf{Act}$.

Proof Each case follows by straightforward linearity arguments. As an example, we outline the proof of (i). $\Delta \xrightarrow{\alpha} \Delta'$ means that

$$\Delta = \sum_{i \in I} p_i \cdot \overline{s_i} \qquad s_i \xrightarrow{\alpha} \Delta_i \qquad \Delta' = \sum_{i \in I} p_i \cdot \Delta_i$$

For any state t, we have $s_i \mid t \xrightarrow{\alpha} \Delta_i \mid \overline{t}$. By linearity we can immediately infer that $\sum_{i \in I} p_i \cdot (\overline{s_i} \mid \overline{t}) \xrightarrow{\alpha} \sum_{i \in I} p_i \cdot (\Delta_i \mid \overline{t})$ and this may be rendered as

$$\Delta \mid \overline{t} \xrightarrow{\alpha} \Delta' \mid \overline{t}$$

By the second part of Lemma 7.1, $(\Delta|\Theta)$ may be written as $\sum_{t\in\lceil\Theta\rceil}\Theta(t)\cdot(\Delta|t)$ and therefore another application of linearity gives $\Delta|\Theta \xrightarrow{\alpha} \sum_{t\in\lceil\Theta\rceil}\Theta(t)\cdot(\Delta'|\bar{t})$ and by the same result this residual coincides with $(\Delta'|\Theta)$. □

Lemma 7.3 *In a pLTS,*

(i) $\Delta \Longrightarrow \Delta'$ *implies* $\Delta|\Theta \Longrightarrow \Delta'|\Theta$;
(ii) $\Delta \stackrel{\alpha}{\Longrightarrow} \Delta'$ *implies* $\Delta|\Theta \stackrel{\alpha}{\Longrightarrow} \Delta'|\Theta$, *for* $\alpha \in \mathsf{Act}_\tau$;
(iii) $\Delta_1 \stackrel{a}{\Longrightarrow} \Delta_1'$ *and* $\Delta_2 \stackrel{a}{\Longrightarrow} \Delta_2'$ *implies* $\Delta_1|\Delta_2 \stackrel{\tau}{\Longrightarrow} \Delta_1'|\Delta_2'$, *for* $a \in \mathsf{Act}$.

Proof Parts (ii) and (iii) follow from (i) and the corresponding result in the previous lemma.

For (i) suppose $\Delta \Longrightarrow \Delta'$. First note that a weak move from Δ to $\sum_{k=0}^{\infty}\Delta_k^{\times} = \Delta'$, as in Definition 6.4, can easily be transformed into a weak transition from $(\Delta \mid \bar{t})$ to $\sum_{k=0}^{\infty}(\Delta_k^{\times} \mid \bar{t})$. This means that for any state t we have a $(\Delta \mid \bar{t}) \Longrightarrow (\Delta' \mid \bar{t})$.

By the second part of Lemma 7.1, $(\Delta \mid \Theta)$ can be written as $\sum_{t\in\lceil\Theta\rceil}\Theta(t)\cdot(\Delta \mid \bar{t})$, and since $\Longrightarrow$ is linear, Theorem 6.5(i), this means $(\Delta \mid \Theta) \Longrightarrow \sum_{t\in\lceil\Theta\rceil}\Theta(t)\cdot(\Delta' \mid \bar{t})$ and again Lemma 7.1 renders this residual to be $(\Delta' \mid \Theta)$. □

Theorem 7.4 (Compositionality of $\approx_s$) *Let s, t be states and Θ a distribution in an arbitrary pLTS, if $s \approx_s \Theta$ then $s|t \approx_s \Theta|\bar{t}$.*

Proof We construct the following relation

$$\mathcal{R} = \{(s|t, \Theta|\bar{t}) \mid s \approx_s \Theta\}$$

and check that $\mathcal{R}$ is a simple bisimulation in the associated pLTS. This will imply that $\mathcal{R} \subseteq \approx_s$, from which the result follows. Note that by construction we have that

(a) $\Delta_1\ (\approx_s)^\dagger\ \Delta_2$, which implies $(\Delta_1|\Theta)\ \mathcal{R}^\dagger\ (\Delta_2|\Theta)$ for any distribution Θ

We use this property throughout the proof.

Let $(s|t,\ \Theta|\bar{t}) \in \mathcal{R}$. We first prove property (ii) in Definition 7.2, which turns out to be straightforward. Since $s \approx_s \Theta$, there is some Δ such that $\bar{s} \stackrel{\tau}{\Longrightarrow} \Delta$ and $\Theta\ (\approx_s)^\dagger\ \Delta$. An application of Lemma 7.3(ii) gives $\bar{s}|\bar{t} \stackrel{\tau}{\Longrightarrow} \Delta|\bar{t}$ and property (a) that $(\Theta|\bar{t})\ \mathcal{R}^\dagger\ (\Delta|\bar{t})$.

Let us concentrate on property (i) where we must prove that every transition from $s|t$ has a matching weak transition from $\Theta|\bar{t}$. Assume that $s|t \xrightarrow{\alpha} \Gamma$ for some action α and distribution Γ. There are three possibilities:

- Γ is $\Delta'|\bar{t}$ for some action α and distribution Δ' with $s \xrightarrow{\alpha} \Delta'$. Here we have $\Theta \stackrel{\alpha}{\Longrightarrow} \Theta'$ such that $\Delta'\ (\approx_s)^\dagger\ \Theta'$, since $s \approx_s \Theta$. Moreover by Lemma 7.3(ii), we can deduce $\Theta|\bar{t} \stackrel{\alpha}{\Longrightarrow} \Theta'|\bar{t}$. Again by (a), we have $(\Delta'|\bar{t},\ \Theta'|\bar{t}) \in \mathcal{R}^\dagger$, and therefore a matching transition.
- Suppose Γ is $\bar{s}|\Delta'$, where $t \xrightarrow{\alpha} \Delta'$. Here a symmetric version of Lemma 7.2(i) gives $\Theta|\bar{t} \xrightarrow{\alpha} \Theta|\Delta'$. This is the required matching transition since we can use (a) above to deduce $(\bar{s}|\Delta', \Theta|\Delta') \in \mathcal{R}^\dagger$.

- The final possibility for α is τ and Γ is $(\Delta_1|\Delta_2)$, where $s \xrightarrow{a} \Delta_1$ and $t \xrightarrow{a} \Delta_2$ for some $a \in A$. Here, since $s \approx_s \Theta$, we have a transition $\Theta \stackrel{a}{\Longrightarrow} \Theta'$ such that $\Delta_1\ (\approx_s)^\dagger\ \Theta'$. By combining these transitions using part (iii) of Lemma 7.3 we obtain $\Theta|\bar{t} \stackrel{\tau}{\Longrightarrow} \Theta'|\Delta_2$. Again this is the required matching transition since an application of (a) above gives $(\Delta_1|\Delta_2, \Theta'|\Delta_2) \in \mathcal{R}^\dagger$.

□

Corollary 7.4 *In an arbitrary pLTS, $\Delta\ (\approx_s)^\dagger\ \Theta$ implies $(\Delta|\Gamma)\ (\approx_s)^\dagger\ (\Theta|\Gamma)$*

Proof A simple consequence of the previous compositionality result, using a straightforward linearity argument. □

Theorem 7.5 (Compositionality of $\approx$) *Let Δ, Θ and Γ be any distributions in a finitary pLTS. If $\Delta \approx \Theta$ then $\Delta|\Gamma \approx \Theta|\Gamma$.*

Proof We show that the relation

$$\mathcal{R} = \{(\Delta|\Gamma, \Theta|\Gamma) \mid \Delta \approx \Theta\} \cup \approx$$

is a bisimulation, from which the result follows.

Suppose $(\Delta|\Gamma, \Theta|\Gamma) \in \mathcal{R}$. Since $\Delta \approx \Theta$, we know from Theorem 7.1 that some Θ' exists such that $\Theta \stackrel{\tau}{\Longrightarrow} \Theta'$ and $\Delta\ (\approx_s)^\dagger\ \Theta'$ and the previous corollary implies that $(\Delta|\Gamma)\ (\approx_s)^\dagger\ (\Theta'|\Gamma)$; by Theorem 7.1 this gives $(\Delta|\Gamma) \approx (\Theta'|\Gamma)$.

We now show that $\mathcal{R}$ is a weak bisimulation. Consider the transitions from both $(\Delta|\Gamma)$ and $(\Theta|\Gamma)$; by symmetry it is sufficient to show that the transitions of the former can be matched by the latter. Suppose that $(\Delta|\Gamma) \stackrel{\alpha}{\Longrightarrow} (\sum_i p_i \cdot \Delta'_i)$. Then, $(\Theta'|\Gamma) \stackrel{\alpha}{\Longrightarrow} (\sum_i p_i \cdot \Theta'_i)$ with $\Delta'_i \approx \Theta'_i$ for each i. But by part (i) of Lemma 7.3 $(\Theta|\Gamma) \stackrel{\tau}{\Longrightarrow} (\Theta'|\Gamma)$ and therefore by the transitivity of $\stackrel{\tau}{\Longrightarrow}$ we have the required matching transition $(\Theta|\Gamma) \stackrel{\alpha}{\Longrightarrow} (\sum_i p_i \cdot \Theta'_i)$. □

7.4 Reduction Barbed Congruence

We now introduce an extensional behavioural equivalence called reduction barbed congruence and show that weak bisimilarity is both sound and complete for it.

Definition 7.3 (Barbs) For $\Delta \in \mathcal{D}(S)$ and $a \in \mathsf{Act}$ let $\mathcal{V}_a(\Delta) = \sum\{\Delta(s) \mid s \xrightarrow{a}\}$. We write $\Delta \Downarrow_a^{\geq p}$ whenever $\Delta \Longrightarrow \Delta'$, where $\mathcal{V}_a(\Delta') \geq p$. We also we use the notation $\Delta \not\Downarrow_a^{>0}$ to mean that $\Delta \Downarrow_a^{\geq p}$ does not hold for any $p > 0$.

Then we say a relation $\mathcal{R}$ is *barb-preserving* if $\Delta \Downarrow_a^{\geq p}$ iff $\Theta \Downarrow_a^{\geq p}$ whenever $\Delta\mathcal{R}\Theta$. It is *reduction-closed* if $\Delta\mathcal{R}\Theta$ implies

(i) whenever $\Delta \Longrightarrow \Delta'$, there is a $\Theta \Longrightarrow \Theta'$ such that $\Delta'\mathcal{R}\Theta'$
(ii) whenever $\Theta \Longrightarrow \Theta'$, there is a $\Delta \Longrightarrow \Delta'$ such that $\Delta'\mathcal{R}\Theta'$.

Finally, we say that in a binary relation $\mathcal{R}$ is *compositional* if $\Delta_1\mathcal{R}\Delta_2$ implies $(\Delta_1|\Theta)\mathcal{R}(\Delta_2|\Theta)$ for any distribution Θ.

Definition 7.4 In a pLTS, let $\approx_{rbc}$ be the largest relation over the states that is barb-preserving, reduction-closed and compositional.

Theorem 7.6 (Soundness) *In a finitary pLTS, if $\Delta \approx \Theta$ then $\Delta \approx_{rbc} \Theta$.*

Proof Theorem 7.5 says that $\approx$ is compositional. It is also easy to see that $\approx$ is reduction-closed. So it is sufficient to prove that $\approx$ is barb-preserving.

Suppose $\Delta \approx \Theta$ and $\Delta \Downarrow_{\bar{a}}^{\geq p}$, for any action a and probability p, we need to show that $\Theta \Downarrow_{\bar{a}}^{\geq p}$. We see from $\Delta \Downarrow_{\bar{a}}^{\geq p}$ that $\Delta \Longrightarrow \Delta'$ for some Δ' with $\mathcal{V}_a(\Delta') \geq p$. Since the relation $\approx$ is reduction-closed, there exists Θ' such that $\Theta \Longrightarrow \Theta'$ and $\Delta' \approx \Theta'$. The degenerate weak transition $\Delta' \stackrel{\tau}{\Longrightarrow} \sum_{s \in \lceil \Delta' \rceil} \Delta'(s) \cdot \overline{s}$ must be matched by some transition

$$\Theta' \stackrel{\tau}{\Longrightarrow} \sum_{s \in \lceil \Delta' \rceil} \Delta'(s) \cdot \Theta'_s \tag{7.1}$$

such that $\overline{s} \approx \Theta'_s$. By Proposition 7.2 we know that $s \approx_s \Theta'_s$ for each $s \in \lceil \Delta' \rceil$. Now, if $s \stackrel{a}{\longrightarrow} \Gamma_s$ for some distribution Γ_s, then $\Theta'_s \stackrel{\tau}{\Longrightarrow} \Theta''_s \stackrel{a}{\longrightarrow} \stackrel{\tau}{\Longrightarrow} \Theta'''_s$ for some distributions Θ''_s and Θ'''_s with $\Gamma_s \; (\approx_s) \; \Theta'''_s$. It follows that $|\Theta''_s| \geq |\Theta'''_s| = |\Gamma_s| = 1$. Let S_a be the set of states $\{s \in \lceil \Delta' \rceil \mid s \stackrel{a}{\longrightarrow}\}$, and Θ'' be the distribution

$$\left(\sum_{s \in S_a} \Delta'(s) \cdot \Theta''_s \right) + \left(\sum_{s \in \lceil \Delta' \rceil \backslash S_a} \Delta'(s) \cdot \Theta'_s \right).$$

By the linearity and reflexivity of $\stackrel{\tau}{\Longrightarrow}$, Theorem 6.6, we have

$$\left(\sum_{s \in \lceil \Delta' \rceil} \Delta'(s) \cdot \Theta'_s \right) \stackrel{\tau}{\Longrightarrow} \Theta'' \tag{7.2}$$

By (7.1), (7.2) and the transitivity of $\stackrel{\tau}{\Longrightarrow}$, we obtain $\Theta' \stackrel{\tau}{\Longrightarrow} \Theta''$, thus $\Theta \Longrightarrow \Theta''$. It remains to show that $\mathcal{V}_a(\Theta'') \geq p$.

Note that for each $s \in S_a$ we have $\Theta''_s \stackrel{a}{\longrightarrow}$, which means that $\mathcal{V}_a(\Theta''_s) = 1$. It follows that

$$\begin{aligned}
\mathcal{V}_a(\Theta'') &= \textstyle\sum_{s \in S_a} \Delta'(s) \cdot \mathcal{V}_a(\Theta''_s) + \sum_{s \in \lceil \Delta' \rceil \backslash S_a} \Delta'(s) \cdot \mathcal{V}_a(\Theta'_s) \\
&\geq \textstyle\sum_{s \in S_a} \Delta'(s) \cdot \mathcal{V}_a \Theta''_s \\
&= \textstyle\sum_{s \in S_a} \Delta'(s) \\
&= \mathcal{V}_a(\Delta') \\
&\geq p.
\end{aligned}$$

□

In order to establish a converse to Theorem 7.6, *completeness*, we need to work in a pLTS that is expressive enough to provide appropriate contexts and barbs in order to distinguish processes that are not bisimilar. For this purpose we use the pLTS determined by the language rpCSP. For the remainder of this section we focus on this particular pLTS.

We will eventually establish the completeness by showing that $\approx_{rbc}$ is a bisimulation, but this requires that we first develop a few auxiliary properties of $\approx_{rbc}$ in this setting. The technique used normally involves examining the barbs of processes in certain contexts; the following lemma gives extra power to this technique. If an action name c does not appear in the processes under consideration, we say c is *fresh*.

Lemma 7.4 *In rpCSP suppose* $(\Delta|c.\mathbf{0})_p\oplus\Delta' \approx_{rbc} (\Theta|c.\mathbf{0})_p\oplus\Theta'$ *where* $p > 0$ *and* c *is a fresh action name. Then* $\Delta \approx_{rbc} \Theta$.

Proof Consider the relation

$$\mathcal{R} = \{(\Delta, \Theta) \mid (\Delta|c.\mathbf{0})_p\oplus\Delta' \approx_{rbc} (\Theta|c.\mathbf{0})_p\oplus\Theta' \text{ for some } \Delta', \Theta' \text{ and fresh } c\}$$

We show that $\mathcal{R} \subseteq \approx_{rbc}$, by showing that $\mathcal{R}$ satisfies the three defining properties of $\approx_{rbc}$.

(1) $\mathcal{R}$ is compositional. Suppose $\Delta\mathcal{R}\Theta$; we have to show $(\Delta|\Phi)\mathcal{R}(\Theta|\Phi)$, for any distribution Φ. Since $\Delta\mathcal{R}\Theta$ there are some Δ', Θ' and fresh c such that

$$\Lambda \approx_{rbc} \Gamma \text{ where } \Lambda = (\Delta|c.\mathbf{0})_p\oplus\Delta',\ \Gamma = (\Theta|c.\mathbf{0})_p\oplus\Theta'. \qquad (7.3)$$

Since $\approx_{rbc}$ is compositional, we have $(\Lambda|\Phi) \approx_{rbc} (\Gamma|\Phi)$. Therefore, it follows that $(\Delta|\Phi|c.\mathbf{0})_p\oplus(\Delta'|\Phi) \approx_{rbc} (\Theta|\Phi|c.\mathbf{0})_p\oplus(\Theta'|\Phi)$, which means, by definition, that $(\Delta|\Phi)\mathcal{R}(\Theta|\Phi)$.

(2) $\mathcal{R}$ is barb-preserving. Suppose $\Delta \Downarrow_a^{\geq q}$ for some action a and probability q, where $\Delta\mathcal{R}\Theta$. Again we may assume (7.3) above. Consider the testing process $a.c.b.\mathbf{0}$, where b is fresh. Since $\approx_{rbc}$ is compositional, we have

$$(\Lambda|a.c.b.\mathbf{0}) \approx_{rbc} (\Gamma|a.c.b.\mathbf{0})\ .$$

Note that $(\Lambda|a.c.b.\mathbf{0}) \Downarrow_b^{\geq pq}$, which implies $(\Gamma|a.c.b.\mathbf{0}) \Downarrow_b^{\geq pq}$. Since c is fresh for Θ', the latter has no potential to enable the action c, and thus $\Theta'|a.c.b.\mathbf{0}$ is not able to fire the action b. Therefore, the action is triggered by Θ and it must be the case that $(\Theta|c.\mathbf{0}|a.c.b.\mathbf{0}) \Downarrow_b^{\geq q}$, which implies $\Theta \Downarrow_a^{\geq q}$.

(3) $\mathcal{R}$ is reduction-closed. Suppose $\Delta\mathcal{R}\Theta$ and $\Delta \stackrel{\tau}{\Longrightarrow} \Delta''$ for some distribution Δ''. Let Γ and Λ be determined as in (7.3) above. Then $\Lambda \stackrel{\tau}{\Longrightarrow} (\Delta''|c.\mathbf{0})_p\oplus\Delta'$. Since $\Lambda \approx_{rbc} \Gamma$, there is some Γ' with $\Gamma \Longrightarrow \Gamma'$ and $(\Delta''|c.\mathbf{0})_p\oplus\Delta' \approx_{rbc} \Gamma'$. In other words, there are some Θ'', Θ''' such that $\Gamma' \equiv (\Theta''|c.\mathbf{0})_p\oplus\Theta'''$ with $\Theta \stackrel{\tau}{\Longrightarrow} \Theta''$ and $\Theta' \stackrel{\tau}{\Longrightarrow} \Theta'''$. Thus $(\Delta''|c.\mathbf{0})_p\oplus\Delta' \approx_{rbc} (\Theta''|c.\mathbf{0})_p\oplus\Theta'''$. Thus by definition $\Delta''\mathcal{R}\Theta''$.

□

Proposition 7.4 *In rpCSP, if* $(\sum_{i\in I} p_i \cdot \Delta_i) \approx_{rbc} \Theta$ *with* $\sum_{i\in I} p_i = 1$, *then there are some* Θ_i *such that* $\Theta \stackrel{\tau}{\Longrightarrow} \sum_{i\in I} p_i \cdot \Theta_i$ *and* $\Delta_i \approx_{rbc} \Theta_i$ *for each* $i \in I$.

Proof Without loss of generality, we assume that $p_i \neq 0$ for all $i \in I$. Suppose that $(\sum_{i\in I} p_i \cdot \Delta_i) \approx_{rbc} \Theta$. Consider the testing process $T = \bigsqcap_{i\in I}(a_i.\mathbf{0} \sqcap b.\mathbf{0})$, where all a_i and b are fresh and pairwise different actions. By the compositionality of $\approx_{rbc}$, we have $(\sum_{i\in I} p_i \cdot \Delta_i)|T \approx_{rbc} \Theta|T$. Now $(\sum_{i\in I} p_i \cdot \Delta_i)|T \stackrel{\tau}{\Longrightarrow} \sum_{i\in I} p_i \cdot (\Delta_i|a_i.\mathbf{0})$. Since $\approx_{rbc}$ is reduction-closed, there must exist some Γ such that $\Theta|T \Longrightarrow \Gamma$ and $\sum_{i\in I} p_i \cdot (\Delta_i|a_i.\mathbf{0}) \approx_{rbc} \Gamma$.

The barbs of $\sum_{i\in I} p_i \cdot (\Delta_i|a_i.\mathbf{0})$ constrain severely the possible structure of Γ. For example, since $\Gamma \not\Downarrow_b^{>0}$, we have $\Gamma \equiv \sum_{k\in K} q_k \cdot (\Theta_k|a_{k_i}.\mathbf{0})$ for some index set K, where $\Theta \stackrel{\tau}{\Longrightarrow} \sum_k q_k \cdot \Theta_k$ and $k_i \in I$. For any indices k_1 and k_2, if $a_{k_1} = a_{k_2}$, we can combine the two summands $q_{k_1} \cdot \Theta_{k_1}$ and $q_{k_2} \cdot \Theta_{k_2}$ into one $(q_{k_1} + q_{k_2}) \cdot \Theta_{k_{12}}$ where $\Theta_{k_{12}} = (\frac{q_{k_1}}{q_{k_1}+q_{k_2}} \cdot \Theta_{k_1} + \frac{q_{k_2}}{q_{k_1}+q_{k_2}} \cdot \Theta_{k_2})$. In this way, we see that Γ can be written as $\sum_{i\in I} q_i \cdot (\Theta_i|a_i.\mathbf{0})$. Since $\Gamma \Downarrow_{a_i}^{\geq p_i}$, $q_i \geq p_i$ and $\sum_{i\in I} p_i = 1$, we have $p_i = q_i$ for each $i \in I$.

Therefore, the required matching move is $\Theta \stackrel{\tau}{\Longrightarrow} \sum_{i\in I} p_i \cdot \Theta_i$. This follows because $\sum_{i\in I} p_i \cdot (\Delta_i|a_i.\mathbf{0}) \approx_{rbc} \sum_{i\in I} p_i \cdot (\Theta_i|a_i.\mathbf{0})$, from which Lemma 7.4 implies the required $\Delta_i \approx_{rbc} \Theta_i$ for each $i \in I$.

Proposition 7.5 *Suppose that* $\Delta \approx_{rbc} \Theta$ *in* rpCSP. *If* $\Delta \stackrel{\alpha}{\Longrightarrow} \Delta'$ *with* $\alpha \in \mathsf{Act}_\tau$ *then* $\Theta \stackrel{\alpha}{\Longrightarrow} \Theta'$ *such that* $\Delta' \approx_{rbc} \Theta'$.

Proof We can distinguish two cases.

(1) α is τ. This case is trivial because $\approx_{rbc}$ is reduction-closed.
(2) α is a, for some $a \in \mathsf{Act}$. Let T be the process $b.\mathbf{0}\Box a.c.\mathbf{0}$ where b and c are fresh actions. Then $\Delta|T \Longrightarrow \Delta'|c.\mathbf{0}$ by Lemma 7.3(iii). Since $\Delta \approx_{rbc} \Theta$, it follows that $\Delta \mid T \approx_{rbc} \Theta \mid T$. Since $\approx_{rbc}$ is reduction-closed, there is some Γ such that $\Theta|T \Longrightarrow \Gamma$ and $\Delta' \mid c.\mathbf{0} \approx_{rbc} \Gamma$.
Since $\approx_{rbc}$ is barb-preserving, we have $\Gamma \not\Downarrow_b^{>0}$ and $\Gamma \Downarrow_c^{\geq 1}$. By the construction of the test T it must be the case that Γ has the form $\Theta'|c.\mathbf{0}$ for some Θ' with $\Theta \stackrel{a}{\Longrightarrow} \Theta'$. By Lemma 7.4 and $\Delta'|c.\mathbf{0} \approx_{rbc} \Theta'|c.\mathbf{0}$, it follows that $\Delta' \approx_{rbc} \Theta'$.

□

Theorem 7.7 (Completeness) *In* rpCSP, $\Delta \approx_{rbc} \Theta$ *implies* $\Delta \approx \Theta$.

Proof We show that $\approx_{rbc}$ is a bisimulation. Because of symmetry it is sufficient to show that if $\Delta \stackrel{\alpha}{\Longrightarrow} \sum_{i\in I} p_i \cdot \Delta_i$ with $\sum_{i\in I} p_i = 1$, where $\alpha \in \mathsf{Act}_\tau$ and I is a finite index set, there is a matching move $\Theta \stackrel{\alpha}{\Longrightarrow} \sum_{i\in I} p_i \cdot \Theta_i$ for some Θ_i such that $\Delta_i \approx_{rbc} \Theta_i$.

In fact, because of Proposition 7.4 it is sufficient to match a simple move $\Delta \stackrel{\alpha}{\Longrightarrow} \Delta'$ with a simple move $\Theta \stackrel{\alpha}{\Longrightarrow} \Theta'$ such that $\Delta' \approx_{rbc} \Theta'$. But this can easily be established using Propositions 7.5. □

7.5 Bibliographic Notes

We have considered a notion of weak bisimilarity, which is induced from simulation preorder. It turns out that this provides both a sound and a complete proof methodology for an extensional behavioural equivalence that is a probabilistic version of of the well-known reduction barbed congruence. This result is extracted from [1] where the proofs are carried out for a more general model called *Markov automata* [2], which describe systems in terms of events that may be nondeterministic, may occur probabilistically, or may be subject to time delays. As a distinguishing feature, the weak bisimilarity is based on distributions. It is strictly coarser than the state-based bisimilarity investigated in Sect. 3.5, even in the absence of invisible actions [3]. A decision algorithm for the weak bisimilarity is presented in [4]. For a state-based weak bisimilarity, a decision algorithm is given in [5]. Note that by taking the symmetric form of Definition 6.19 we can obtain a notion of weak failure bisimulation. Its characteristics are yet to be explored.

The idea of using barbs as simple experiments [6] and then deriving reduction barbed congruence originated in [7] but has now been widely used for different process description languages; for example see [8, 9] for its application to higher-order process languages, [10] for mobile ambients and [11] for asynchronous languages. Incidentally, there are also minor variations on the formulation of reduction barbed congruence, often called *contextual equivalence* or *barbed congruence*, in the literature. See [11, 12] for a discussion of the differences.

References

1. Deng, Y., Hennessy, M.: On the semantics of Markov automata. Inf. Comput. **222**, 139–168 (2013)
2. Eisentraut, C., Hermanns, H., Zhang, L.: On probabilistic automata in continuous time. Proceedings of the 25th Annual IEEE Symposium on Logic in Computer Science, pp. 342–351. IEEE Computer Society (2010)
3. Hennessy, M.: Exploring probabilistic bisimulations, part I. Form. Asp. Comput. **24**(4–6), 749–768 (2012)
4. Eisentraut, C., Hermanns, H., Krämer, J., Turrini, A., Zhang, L.: Deciding bisimilarities on Distributions. Proceedings of the 10th International Conference on Quantitative Evaluation of Systems, Lecture Notes in Computer Science, vol. 8054, pp. 72–88. Springer (2013)
5. Cattani, S., Segala, R.: Decision Algorithms for Probabilistic Bisimulation. Proceedings of the 13th International Conference on Concurrency Theory, Lecture Notes in Computer Science, vol. 2421, pp. 371–385. Springer (2002)
6. Rathke, J., Sobocinski, P.: Making the unobservable, unobservable. Electron. Notes Comput. Sci. **229**(3), 131–144 (2009)
7. Honda, K., Yoshida, N.: On reduction-based process semantics. Theoret. Comput. Sci. **151**(2), 437–486 (1995)
8. Jeffrey, A., Rathke, J.: Contextual equivalence for higher-order pi-calculus revisited. Log. Methods Comput. Sci. **1**(1:4) (2005)
9. Sangiorgi, D., Kobayashi, N., Sumii, E.: Environmental bisimulations for higher-order languages. Proceedings of the 22nd IEEE Symposium on Logic in Computer Science, pp. 293–302. IEEE Computer Society (2007)

10. Rathke, J., Sobocinski, P.: Deriving structural labelled transitions for mobile ambients. Proceedings of the 19th International Conference on Concurrency Theory, Lecture Notes in Computer Science, vol. 5201, pp. 462–476. Springer (2008)
11. Fournet, C., Gonthier, G.: A hierarchy of equivalences for asynchronous calculi. J. Log. Algebraic Program. **63**(1), 131–173 (2005)
12. Sangiorgi, D., Walker, D.: The π-Calculus: a Theory of Mobile Processes. Cambridge University Press, Cambridge (2001)

Index

DOI 10.1007/978-3-662-45198-4

Printed in the United States
By Bookmasters